ANTENNAS
Rigorous Methods of Analysis and Synthesis

Boris Levin
Holon Institute of Technology, Israel

CRC Press
Taylor & Francis Group
Boca Raton London New York

CRC Press is an imprint of the
Taylor & Francis Group, an **informa** business

A SCIENCE PUBLISHERS BOOK

CRC Press
Taylor & Francis Group
6000 Broken Sound Parkway NW, Suite 300
Boca Raton, FL 33487-2742

© 2021 by Taylor & Francis Group, LLC
CRC Press is an imprint of Taylor & Francis Group, an Informa business

No claim to original U.S. Government works

Version Date: 20200910

International Standard Book Number-13: 978-0-367-48923-6 (Hardback)

This book contains information obtained from authentic and highly regarded sources. Reasonable efforts have been made to publish reliable data and information, but the author and publisher cannot assume responsibility for the validity of all materials or the consequences of their use. The authors and publishers have attempted to trace the copyright holders of all material reproduced in this publication and apologize to copyright holders if permission to publish in this form has not been obtained. If any copyright material has not been acknowledged please write and let us know so we may rectify in any future reprint.

Except as permitted under U.S. Copyright Law, no part of this book may be reprinted, reproduced, transmitted, or utilized in any form by any electronic, mechanical, or other means, now known or hereafter invented, including photocopying, microfilming, and recording, or in any information storage or retrieval system, without written permission from the publishers.

For permission to photocopy or use material electronically from this work, please access www.copyright.com (http://www.copyright.com/) or contact the Copyright Clearance Center, Inc. (CCC), 222 Rosewood Drive, Danvers, MA 01923, 978-750-8400. CCC is a not-for-profit organization that provides licenses and registration for a variety of users. For organizations that have been granted a photocopy license by the CCC, a separate system of payment has been arranged.

Trademark Notice: Product or corporate names may be trademarks or registered trademarks, and are used only for identification and explanation without intent to infringe.

Library of Congress Cataloging-in-Publication Data
Names: Levin, Boris, 1937- author. Title: Antennas : rigorous methods of analysis and synthesis / Boris Levin, (Retired), Holon Institute of Technology, Israel. Description: Boca Raton : CRC Press, Taylor & Francis Group, CRC Press is an imprint of the Taylor & Francis Group, an Informa Business, [2021]	"A Science Publishers book."	Includes bibliographical references and index.	Summary: "The book contains new variant of solving the equation of Leontovich, which is most rigorous equation for the current in a thin linear antenna. This solution based in particular on the method of constants variation permits to calculate the sum of any members of series, as well as the total sum of the series members, i.e. solve a problem that has agitated specialists for more than half a century. The book gives a consistent exposition of the theory of integral equations and discusses the features of their application for analyzing different types of antennas. The book considers new ways of analyzing antennas: method of calculating an antenna gain based on its main radiation patterns, procedure of calculating the directional characteristics of radiators with known distribution of current amplitude, method of electrostatic analogy which permits to compare the electromagnetic fields of high-frequency currents with electrostatic charges fields located on linear conductors and to improve the directional characteristics of log-periodic and director-type antennas"-- Provided by publisher. Identifiers: LCCN 2020041783	ISBN 9780367489236 (hardcover) Subjects: LCSH: Antennas (Electronics)--Mathematical models.	Antenna radiation patterns--Mathematical models.	Electromagnetic waves--Mathematical models. Classification: LCC TK6565.A6 L455 2021	DDC 621.382/40151545--dc23 LC record available at https://lccn.loc.gov/2020041783

Visit the Taylor & Francis Web site at
http://www.taylorandfrancis.com

and the CRC Press Web site at
http://www.routledge.com

Preface

Modern rigorous methods of antenna calculations are based on solving integral equations for currents along their axes. The calculation of thin linear radiators: dipoles and monopoles—forms the foundation of an antenna analysis, since these radiators are widely used in practice both as individual antennas and as part of complex antennas, which are divided into simpler elements during the analysis. The method of integral equations for a current in the antenna provides accurate results and is applicable to the calculation of characteristics of a wide class of radiators. The application of the method of variation of constants, described in the present book, to for solving the equation of Leontovich, makes it possible to look at the solution method from a new point of view. A consistent transition from an integral equation for a metal antenna to an equation for a slot antenna and further to equations for metal and slot antennas with a surface impedance permits to acquaint a reader with the deep works of Feld and Miller, whose content is relatively little known and, unfortunately, is rarely used which sometimes leads to misunderstandings and mistakes. The proposed book gives a systematic exposition of the theory of integral equations and discusses the features of their application for analyzing different types of antennas: metal and slot, rectilinear and curvilinear, with reactive and resistive surface impedance, with distributed and concentrated loads. For the physical interpretation of obtained results, the method of the equivalent long line is used.

The book considers various ways of analyzing antennas, including the method of impedance line and the method of a long line with loads, the theory of electrically connected lines, the method of a complex potential for a conical (and also parabolic and pyramidal) line, for a homogeneous and a non-homogeneous medium. The method for calculating an antenna gain based on its main radiation patterns is described. The procedure of calculating the directional characteristics of linear and self-complementary radiators with different laws for current distribution, based on the calculation of the amplitude and phase of the electromagnetic field at different angles, is proposed.

Results of solving the problem related to creating a broadband radiator are summed up. The problem was solved by employment of concentrated loads (both capacitive and resistive) incorporated into the radiator wire for creation an in-phase current with a required law of its distribution along the antenna. The decision was based on understanding the advantages of the in-phase current distribution and on Hallen's hypothesis about the utility of capacitive loads, with values changing along the radiator axis in accordance with a linear or exponential law. The chosen approach confirmed Hallen's hypothesis and demonstrated the effectiveness of the proposed approximate methods for calculating capacitive loads: the methods of impedance long line and long line with loads. The loads calculated by these methods were used as initial values for the synthesis of flat and three-dimensional antennas by means of the methods of the mathematical programming. The procedure for computer synthesis of antennas with loads is described.

The history of creating in-phase radiators providing in the wide frequency range a high level of an antenna matching with a cable and a pressed to the ground directional pattern permitting to increase the communication range shows that the inclusion of concentrated capacitive loads along the antenna axis allows to improve significantly its electrical characteristics because the phase of its current weakly changed along the entire antenna length.

Antennas are widely used for a variety of purposes and find application in diverse areas of our lives. These goals and areas of use necessitate the development of antenna designs that meet their purpose and location. One of the most important applications of antennas is radio communication and broadcasting. The specific objects of antenna installation are ships for various purposes. The analysis of the features of antennas used in ships for providing radio communications and broadcasting is dealt with in a special chapter demonstrating the influence of specific tasks and conditions on their characteristics.

The author described in the book new interesting results. In particular, a new solution of Leontovich's equation for a current in a thin linear radiator permits to find an arbitrary number of the series members and the total sum of this series in the antenna input, as well as the total sum of the series members into the any antenna point, i.e., permits to solve a problem that has tormented specialists for more than half a century. A method for analyzing curvilinear metal antennas made of thin conductors based on their division into separate rectilinear and curvilinear radiators located along different coordinates is proposed. It is shown that the coincidence of the points of new radiators with the projections of the points of the original antenna provides a rigorous calculation of the characteristics of the original antenna due to the proximity of its current value with the new radiators' current values.

The new method of analysis, is called the electrostatic analogy method by the author, which makes it possible to compare the electromagnetic fields of high-frequency currents of linear radiators with electrostatic fields of charges located on linear conductors and to improve the directional characteristics of director-type radiators and log-periodic antennas.

The procedure has been developed for calculating the radiation resistance of metal and impedance radiators (with reactive and resistive impedances) by using the method of Poynting vector.

A new approach to the analysis of the electrical characteristics of a microstrip antenna is proposed. It is based on the comparison of this antenna with the antenna located on a horizontal pyramid and consisting of electric and slot radiators. The method of analyzing an antenna located on the horizontal pyramid was proposed by the author in the course of working on complementary and self-complementary antennas. A new look at the structure of the microstrip antenna clearly shows that the so-called transmitting long line radiates all along its length, and does not simply transfer energy along a closed tube from the line entrance to the direction of an aperture on the opposite end of the line. A method for extending the frequency range of the microstrip antenna is proposed. Reflector arrays consisting of the microstrip antennas are described.

This new book describes various specific tasks and gives methods for their solution. Among them are the issues of reducing the dimensions of the log-periodic antennas, the task of rotating the radiation pattern of a cellular base station, the development of a multi-tiered antenna with a constant height, whose radiation pattern is transferred to the ground in a wide frequency range.

The author has also tried to define more exactly, and correct and explain many of the previously obtained results and to include detailed links to previously published matter in various sources, for which there is not enough space available in this new book.

Contents

Preface   iii

1. Straight Metal Radiator   1

   1.1 Electromagnetic field   1
   1.2 Currents of radiators   4
   1.3 Wire antennas of complicate shapes   6
   1.4 Sinusoidal character of the currents   8
   1.5 Equivalent long line   12
   1.6 Equality of two powers and Poynting method   20
   1.7 Oscillating power theorem   34
   1.8 Method of induced emf   38
   1.9 Generalized method of induced emf   42
   1.10 Multi-wire antennas and cables   48

2. Integral Equation of Leontovich   63

   2.1 Integral equations for linear metal radiators   63
   2.2 Derivation of Leontovich's equation   66
   2.3 Method of variation of constants   70
   2.4 Integral equation for two metal radiators   79
   2.5 Radiators with distributed loads   83
   2.6 Radiators with resistive impedance   95
   2.7 Radiators with concentrated loads   105
   2.8 Curvilinear radiators   107
   2.9 Slot radiators   111
   2.10 Impedance magnetic antennas   114
   2.11 Distribution of the current over the antenna wire surface   121

3. Inverse Problems of Antenna Theory   124

   3.1 Wide-range linear radiator and impedance line   124
   3.2 Method of a metallic long line with loads. Synthesis of a current distribution   133
   3.3 Wide-range V-radiator   137

3.4	Method of mathematical programming	141
3.5	Application of results to concrete tasks	147
3.6	Calculating directional characteristics of linear and self-complementary radiators	157
3.7	Method of electrostatic analogy	166
3.8	Application of the method of electrostatic analogy to log-periodic antennas	174

4. New Methods of Analysis — 189

4.1	Reduction of three-dimensional problems to a plane task	189
4.2	Distribution of currents over the cross section of a long line	204
4.3	Microstrip antennas	207
4.4	Reflector arrays of microstrip antennas	215
4.5	Calculating directivity on the basis of main directional patterns	224
4.6	Diversity reception	229
4.7	Adaptive array	232
4.8	Struggle with environmental influences	237
4.9	Turn of the directional pattern of a cellular base station as concrete task	245

5. Problems of Design and Placement of Antennas — 250

5.1	Ship antennas of medium-frequency waves	250
5.2	Ship antenna of high-frequency waves	258
5.3	Antennas of meter and decimeter waves	272
5.4	Influence of ship designs on antenna characteristics	275
5.5	Multi-tiered antenna with a directional pattern pressed to the ground	295

Instead of a Conclusion — 307

References — 308

Index — 313

Chapter 1
Straight Metal Radiator

1.1 Electromagnetic field

The simplest radiators are symmetric and asymmetric electric radiators of finite length (Fig. 1.1). A symmetrical radiator (dipole) is a straight-line conductor, whose currents at points symmetrical about its center are equal in magnitude and have the same direction. To create such a current distribution, an electromotive force (emf) source is included in the middle of the radiator, which divides it into two arms with the same length L (Fig. 1.1a). An asymmetrical radiator is a straight conductor, excited by an emf source displaced from its middle (Fig. 1.1b). A special case of an asymmetric radiator is a monopole (Fig. 1.1c). This is a conductor with an emf at the base, mounted on a metal surface. The metal surface radiation replaces the radiation of the lower arm of the dipole. All these radiators are widely used in antenna technology. In addition, more complex antennas can be analyzed by breaking them into simple elements in the form of symmetrical or asymmetrical radiators. The significance of the theory of simple radiators is connected with this.

The modern theory of antennas is based on Maxwell's equations—the basic equations of electrodynamics. The first two of them in the differential form appear as

$$curl\vec{H} = \vec{J} + \frac{\partial \vec{D}}{\partial t}, \; curl\vec{E} = -\frac{\partial \vec{B}}{\partial t}, \qquad (1.1)$$

where $\vec{H}$ is a vector representing magnetic field strength, $\vec{J}$ is a vector representing volume density of conduction current, $\vec{D}$ is a vector representing electric displacement, t is time, $\vec{E}$ is a vector representing electric field strength, $\vec{B}$ is a vector representing magnetic induction. Hereinafter the International System of Units is used. The equations (1.1) are to be complemented with an equation of continuity

$$\operatorname{div}\vec{J} = -\frac{\partial \rho}{\partial t}, \qquad (1.2)$$

where ρ is a volume density of an electrical charge.

Typically, two more equations are included into a system of Maxwell's equations:

$$\operatorname{div}\vec{D} = \rho, \text{ and } \operatorname{div}\vec{B} = 0, \qquad (1.3)$$

but they follow from the equations (1.1) and (1.2) [1].

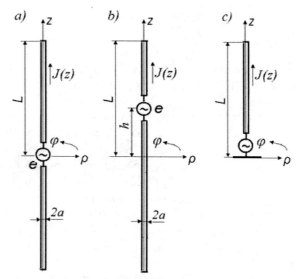

Fig. 1.1. Symmetrical (*a*) and asymmetrical (*b*) metal radiators and monopole (*c*) from cylindrical wires.

The equations (1.1) interconnect the electromagnetic fields and currents in free space. It would be wrong to consider the left- or the right-hand side of an equation as the cause and, accordingly, the other side as the consequence. Both currents and electric and magnetic components of fields exist jointly only. None of the quantities are the prime cause of appearance of others. From (1.1) and (1.2) it follows for a harmonic field time-varying as $exp(j\omega t)$ and an isotropic medium ($D = \varepsilon E$, $B = \mu H$) that

$$\operatorname{curl}\vec{H} = j + j\omega\varepsilon\vec{E}, \ \operatorname{curl}\vec{E} = -j\omega\mu\vec{H}, \ \operatorname{div}\vec{J} + j\omega\rho = 0. \qquad (1.4)$$

The antenna field is created by the power received from the radio transmitter. This means that the radiation of an antenna is a result of the action of another medium. In order to take it into account, the set of equations should have extraneous (impressed) currents and fields as the original sources of excitation in accordance with the equivalence theorem. They are introduced as summands in quantities of $\vec{J}$, $\vec{E}$ and $\vec{H}$. Their nature and placement depend on the model of the area near a generator, which is commonly called the 'excitation zone'. The total electromagnetic field of an antenna is equal to a sum of the field produced by the excitation zone and the field produced by the currents in the wires, which arise on switching on of the sources. As a rule, at a great distance from the antenna the first field is substantially less than the second one and can be neglected.

Maxwell's equations for electromagnetic fields, which are complemented with boundary conditions for any antenna, allow writing the equation for the current in the conductor of the antenna. By solving it and finding the current distribution along the conductor, one can determine the electrical characteristics of the radiator.

An investigation of an electromagnetic field can be simplified by introducing auxiliary functions, which are called potentials. The auxiliary vector field (vector potential) is introduced as follows. Comparing the mathematical identity $div curl \vec{A} = 0$, where $\vec{A}$ is an arbitrary vector, with the second equation of (1.3), we are convinced that the vector $\vec{B}$ can be presented as a curl of some vector $\vec{A}$:

$$\vec{B} = curl \vec{A}. \tag{1.5}$$

Yet, equality (1.5) determines the vector $\vec{A}$ ambiguously. To define it unambiguously, one should also specify the value of $div \vec{A}$.

Substituting (1.5) into the second equation of system (1.4) and using mathematical identity of

$$curl grad U = 0,$$

where U is an arbitrary scalar function (a scalar potential of field), we obtain

$$\vec{E} = -j\omega\vec{A} - grad U. \tag{1.6}$$

Substituting (1.5) and (1.6) into the first equation of system (1.4) and taking into account the mathematical identity of

$$curl curl \vec{A} = grad div \vec{A} - \Delta \vec{A},$$

we find:

$$\Delta \vec{A} + \omega^2 \mu \varepsilon \vec{A} - grad(div A + j\omega\mu\varepsilon U) = -\mu j. \tag{1.7}$$

Let us define $div \vec{A}$ in such a way as to simplify the last expression as far as possible. For this purpose, we set

$$div A = -j\omega\mu\varepsilon U. \tag{1.8}$$

This equality is known as the calibration condition, or Lorentz's condition. In accordance with (1.7) and (1.8)

$$\vec{A} + k^2\vec{A} = -\mu j, \qquad (1.9)$$

where $k = \omega\sqrt{\mu\varepsilon}$ is a wave propagation constant in the medium surrounding the antenna. Equation (1.9) is called the vector wave equation. Its solution allows us to find the vector potential $\vec{A}$, and then the electric and magnetic fields of the antenna. Indeed, according to (1.5), (1.6) and (1.8)

$$\vec{E} = -j\frac{\omega}{k^2}\left(graddiv\vec{A} + k^2\vec{A}\right), \vec{H} = \frac{1}{\mu}curl\vec{A}. \qquad (1.10)$$

If the electromagnetic field sources are distributed continuously in a certain region V, and the medium surrounding the region V is a homogeneous isotropic dielectric, then the solution of equation (1.9) for a harmonic field has the form

$$A = \frac{\mu}{4\pi}\iiint_{(v)} \vec{J}\frac{e^{-jkR}}{R}dV, \qquad (1.11)$$

where R is the distance from the observation point (in the medium) to the point of integration (in the area V). The expression (1.11) can be written as

$$A = \mu\iiint_{(v)} \vec{J}GdV, \qquad (1.12)$$

where $G = \dfrac{exp(-ikR)}{4\pi R}$ is the Green function.

In accordance with (1.8), (1.12) and the continuity equation, we can obtain a similar expression for the scalar potential:

$$U = j\frac{1}{\omega\varepsilon}\iiint_{(v)} Gdiv\vec{J}dV = \frac{1}{\varepsilon}\iiint_{(v)} \rho GdV. \qquad (1.13)$$

It should be noted that the region V, where the sources of the electromagnetic field are located, can be multiply connected (if, for example, radiation of several antennas or metal bodies are located close to the antenna).

1.2 Currents of radiators

We will begin to consider the variants of radiators from the particular case when the source of the electromagnetic field is electric currents, having axial symmetry and located parallel to the z axis in a certain region V:

$$j = j_z\vec{e}_z, j_z = j_z(z) = const(\varphi). \qquad (1.14)$$

Here, the cylindrical system of coordinates (ρ, φ, z) is used, with unit vectors (orts) $\vec{e}_\rho$, $\vec{e}_\varphi$, $\vec{e}_z$ along the axes. As seen from (1.12), the vector potential in this case has only the component A_z:

$$\vec{A} = A_z(\rho, z)\vec{e}_z, \tag{1.15}$$

i.e.,

$$div\vec{A} = \frac{dA_z}{dz}, \quad grad\,div\vec{A} = \frac{\partial^2 A_z}{\partial\rho\,\partial z}\vec{e}_\rho + \frac{\partial^2 A_z}{\partial z^2}\vec{e}_z, \quad curl\vec{A} = \frac{\partial A_z}{\partial\rho}\vec{e}_\varphi,$$

and in accordance with (1.10)

$$E_z(\rho,z) = -j\frac{\omega}{k^2}\left(k^2 A_z + \frac{\partial^2 A_z}{\partial z^2}\right), E_\rho(\rho,z) = -\frac{j\omega}{k^2}\frac{\partial^2 A_z}{\partial\rho\,\partial z}, H_\varphi(\rho,z) = -\frac{1}{\mu}\frac{\partial A_z}{\partial\rho}, E_\varphi = H_z = H_\rho = 0. \tag{1.16}$$

Obviously, if the distribution of current $J(z)$ along the radiator is known, one can calculate the electromagnetic field of the current with the help of presented formulas. If the antenna is excited at some point (e.g., $z = 0$) by a generator with concentrated emf e, the antenna input impedance at the driving point is

$$Z_A = \frac{e}{J(0)}, \tag{1.17}$$

and in order to determine this impedance, it is enough to know the current magnitude at the corresponding point. When calculating the power absorbed in the load of a receiving antenna, the current magnitude is also needed. So, the current distribution along the antenna constitutes a very important characteristic.

As a simple and clear model of a vertical linear radiator, one can use a straight perfectly conducting filament, coinciding with the z-axis (Fig. 1.2a), along which the conduction current flows. Current density $\vec{J}(s, z)$ is related to this current by a relation

$$J(z) = \iint_{(s)} j(s,z)\,ds, \tag{1.18}$$

where s is the filament cross-section. From (1.12) and (1.16)

$$A_z(\rho,z) = \mu\int_{-L}^{L} J(z')\,G\,dz', E_z(\rho,z) = \frac{1}{j\omega\varepsilon}\int_{-L}^{L} J(z')\left(k^2 G + \frac{\partial^2 G}{\partial z^2}\right)dz'. \tag{1.19}$$

Here $G_1 = exp(-jkR_1)/(4\pi R_1)$, distance R_1 from observation point M to integration point P is equal to $\sqrt{(z-z')^2 + \rho^2}$.

In the considered model the radiator radius is zero. The model of a radiator shaped as a straight circular thin-wall cylinder with radius a (Fig. 1.2b) has

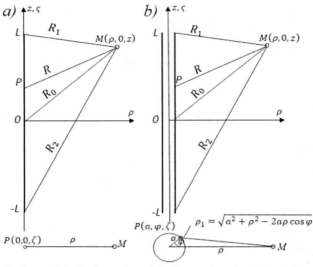

Fig. 1.2. The dipole models in the shape of a straight filament (*a*) and a thin-wall circular cylinder (*b*).

finite dimensions. Both ends of the cylinders are left open, without covers, in order that the current as before has only a longitudinal component. The surface density of current along the cylinder is $J_s(z) = \dfrac{J(z)}{2\pi a}$. Since a volume element in the cylindrical system of coordinates is equal to $dV = \rho d\rho d\varphi dz$, and $\rho = a$ on the cylinder surface, so, in accordance with (1.12) and (1.16),

$$A_z(\rho,z) = \frac{\mu}{4\pi}\int_{-L}^{L} J(z') \int_0^{2\pi} G d\varphi' dz', \quad E_z(\rho,z) = \frac{1}{2\pi j\omega\varepsilon}\int_{-L}^{L} J(z')\int_0^{2\pi}\left(k^2 G + \frac{\partial^2 G}{\partial z^2}\right)d\varphi' dz',$$
(1.20)

where $G_2 = exp(-jkR_2)/(4\pi R_2)$, and the distance R_2 from observation point M to integration point P is $\sqrt{(z-z')^2 + \rho^2 + a^2 - 2a\rho\cos\varphi'}$. In particular, if the observation point is located on the radiator surface, $R = \sqrt{(z-z')^2 + 4a^2 \sin^2 \varphi/2}$.

Sometimes the dipole model shaped as a filament with finite radius a, i.e., expressions (1.19) is used for A_z and E_z, but distance from the observation point to the integration point is equal to $R = \sqrt{(z-z')^2 + \rho^2 + a^2}$.

All three models of a cylindrical radiator were considered and used. These are the simplest radiators of finite length.

1.3 Wire antennas of complicate shapes

A more complex radiator is a wire antenna consisting of straight wire segments located arbitrarily and joined partially with each other (Fig. 1.3*a*).

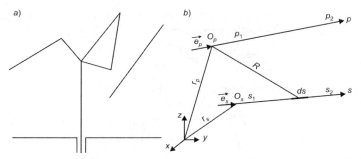

Fig. 1.3. Antenna consisting of straight wire segments (*a*), a filament of two segments (*b*).

It is considered that, as in the first example, the current here flows along thin perfectly conductive filaments. A filament of two segments is shown in Fig. 1.3*b*. The distance from point O_p of the segment *p* to element *ds* of the segment *s* is

$$R = |\vec{r}_p + p\vec{e}_p - \vec{r}_s - s\vec{e}_s|_{p=0}, \quad (1.21)$$

where $\vec{r}_p$ and $\vec{r}_s$ are radii vectors from the coordinate origin O to points O_p and O_s of the corresponding segments, *p* and *s* are coordinates, measured along the segments, and $\vec{e}_p$ and $\vec{e}_s$ are the unit vectors, with directions coinciding with the wire axes.

Let us write for the current along the segment *s*: $\vec{J}_s = j_s(s)\vec{e}_s$. According to (1.20)

$$\vec{A}_s(\vec{J}_s) = A_s(j_s)\vec{e}_s = \mu\vec{e}_s \int_{S_1}^{S_2} J(s')Gds', \quad (1.22)$$

where S_1 and S_2 are the coordinates of the beginning and end of the segment *s* on the *s*-axis. In order to find the vector potential of the total field, one has to sum up the vector potentials of fields of all segments:

$$\vec{A} = \sum_{n=1}^{N} \vec{A}_{sn}(\vec{J}_{sn}) = \sum_{n=1}^{N} A_{sn}(j_{sn})\vec{e}_{sn}, \quad (1.23)$$

where *n* is the segment's number, *N* is the number of segments, and

$$A_{sn}(j_{sn}) = \mu \int_{Sn_1}^{Sn_2} J(s_n)Gds_n.$$

In accordance with (1.20) and (1.21), the field of the segment *s* at point O_p is

$$E_s(O_p) = \frac{1}{j\omega\varepsilon}\int_{S_1}^{S_2} J(s)[k^2 G\vec{e}_s + graddiv(G\vec{e}_s)]ds.$$

The differentiation of the last term is performed with respect to the coordinates of the observation point. Since in the rectangular coordinate system the distance between the observation point with coordinates x_p, y_p, z_p and the integration point with coordinates x_s, y_s, z_s is equal to

$$R = \sqrt{(x_p - x_s)^2 + (y_p - y_s)^2 + (z_p - z_s)^2},$$

then as a result of symmetry $grad_p G = -grad_s G$ (the differentiation is performed with respect to coordinates of the point s instead of coordinates of the point p). Let us take into account that in accordance with the gradient definition $\vec{e}_s grad_s G = \dfrac{\partial G}{\partial s}$. Using the mathematical identity of expressions $div(G\vec{e}_s)$ and $\vec{e}_s gradG$, we find

$$div_p(G\vec{e}_s) = -\dfrac{\partial G}{\partial s},$$

i.e.,

$$\vec{E}_s(O_p) = \dfrac{1}{j\omega\varepsilon}\int_{S_1}^{S_2} J(s)\left[k^2 G\vec{e}_s - grad_p\dfrac{\partial G}{\partial s}\right]ds.$$

The projection of the field of wire s onto direction p is calculated as a product of $\vec{E}_s(O_p)$ and $\vec{e}_p$:

$$E_{ps} = \vec{E}_s(O_p)\vec{e}_p = \dfrac{1}{j\omega\varepsilon}\int_{S_1}^{S_2} J(s)\left[k^2 G\vec{e}_s\vec{e}_p - \dfrac{\partial^2 G}{\partial p \partial s}\right]ds. \quad (1.24)$$

The projection of the total field is the sum of the field's projections of all segments

$$E_p = \sum_{n=1}^{N} E_{psn} = \dfrac{1}{j\omega\varepsilon}\sum_{n=1}^{N}\int_{S_{n1}}^{S_{n2}} J(s_n)\left[k^2 G\vec{e}_{s_n}\vec{e}_p - \dfrac{\partial^2 G}{\partial p \partial s_n}\right]ds_n. \quad (1.25)$$

The expression (1.25) allows us to calculate the directivity pattern of a complex wire structure.

1.4 Sinusoidal character of the currents

As shown in the previous Sections, knowledge of a current distribution along the radiator allows us to determine all its electrical characteristics. An assumption that this distribution has a sinusoidal form plays an important role in the theory of antennas. The character of expressions (1.16), (1.19) and (1.20) for a longitudinal component of an electric field created by the antenna current confirms that. Partly this assumption was based on measurement results. But mainly it is based on a simple understanding that the sinusoidal current distribution existing in the two-wire long line open at the end does not change significantly, if the wires move and diverge from each other. Later on, in the derivation and solution of integral equations for currents in radiators, it was rigorously shown that the sinusoidal distribution is the first approximation to the true current distribution along the antenna, and from this point of view, its use is quite reasonable and has a reliable justification.

Suppose that the linear current along a symmetrical cylindrical radiator (see Fig. 1.1a) obeys the law

$$J(z) = I_n \frac{\sin k(L-|z|)}{\sin kL}. \qquad (1.26)$$

If we use a perfectly conducting filament as a model of a symmetrical radiator, then in accordance with (1.19)

$$E_z = \frac{1}{4\pi j\omega\varepsilon} \int_0^L J(z') \left(k^2 + \frac{\partial^2}{\partial z^2} \right) \left(\frac{e^{-jkR}}{R} + \frac{e^{-jkR_+}}{R_+} \right) dz'. \qquad (1.27)$$

This expression takes into account the symmetry of the currents in the radiator arms. Accordingly, the substitution of variable ($-\varsigma$ for ς) is performed in the lower arm, and the new R is designated as $R_+ = \sqrt{(z+\varsigma)^2 + \rho^2}$. Since

$$\frac{\partial R}{\partial z'} = -\frac{\partial R}{\partial z}, \frac{\partial R_+}{\partial \varsigma} = \frac{\partial R_+}{\partial z}, \frac{\partial^2 R}{\partial \varsigma^2} = \frac{\partial^2 R}{\partial z^2}, \frac{\partial^2 R_+}{\partial \varsigma^2} = \frac{\partial^2 R_+}{\partial z^2},$$

Then

$$E_z = \frac{1}{4\pi j\omega\varepsilon} \int_0^L J(\varsigma) \frac{\partial^2}{\partial \varsigma^2} \left[\frac{exp(-jkR)}{R} + \frac{exp(-jkR_+)}{R_+} \right] d\varsigma$$

$$+ \frac{k^2}{4\pi j\omega\varepsilon} \int_0^L J(\varsigma) \left[\frac{exp(-jkR)}{R} + \frac{exp(-jkR_+)}{R_+} \right] d\varsigma.$$

Twice integrating the first term of the expression by parts, we get

$$E_z = \frac{1}{4\pi j\omega\varepsilon} \left\{ \int_0^L \left[\frac{d^2 J(\varsigma)}{d\varsigma^2} + k^2 J(\varsigma) \right] \left[\frac{exp(-jkR)}{R} + \frac{exp(-jkR_+)}{R_+} \right] d\varsigma \right.$$

$$\left. + J(\varsigma) \frac{\partial}{\partial \varsigma} \left[\frac{exp(-jkR)}{R} + \frac{exp(-jkR_+)}{R_+} \right] \Big|_0^L - \frac{dJ(\varsigma)}{d\varsigma} \left[\frac{exp(-jkR)}{R} + \frac{exp(-jkR_+)}{R_+} \right] \Big|_0^L \right\}.$$
$$(1.28)$$

If the current along the radiator is distributed in accordance with (1.26), the first multiplier in the integrand and hence the first term of the expression is zero. It can be easily verified that the second summand is zero too, since the first multiplier becomes zero at $\varsigma = L$, making the second multiplier vanish at $\varsigma = 0$. If we calculate the derivative of the distribution function of a current

$$\frac{dJ(\varsigma)}{d\varsigma} = -kJ(0) \frac{\cos k(L-|\varsigma|)}{\sin kL} \, signz \qquad (1.29)$$

(where *signz* means sign of z), and taking into account that $\dfrac{k}{4\pi\omega\varepsilon_0} = 30$, we obtain

$$E_z = -j\frac{30J(0)}{\varepsilon_r \sin kL}\left[\frac{exp(-jkR_1)}{R_1} + \frac{exp(-jkR_2)}{R_2} - 2\cos kL\frac{exp(-jkR_0)}{R_0}\right]. \quad (1.30)$$

Here $R_1 = \sqrt{(z-L)^2+\rho^2}$, $R_2 = \sqrt{(z+L)^2+\rho^2}$, $R_0 = \sqrt{z^2+\rho^2}$ are the distances between the observation point M and the upper and lower ends and the middle of the radiator, respectively (see Fig. 1.2*a*).

As can be seen from (1.16), if the electric current, which excites the electromagnetic field, is directed parallel to the z axis and has axial symmetry, then together E_z, the normal component of the electric field E_ρ and the tangential component of the magnetic field H_φ do not equal zero. These components can be calculated similarly to E_z, i.e., in accordance with (1.16). However, it is easier to use the first Maxwell equation. Let's to write this equation in a cylindrical coordinate system:

$$-\frac{\partial H_\varphi}{\partial z} = j\omega\varepsilon E_\rho, \quad \frac{1}{\rho}\frac{\partial}{\partial \rho}(\rho H_\varphi) = j\omega\varepsilon E_z + j_z. \quad (1.31)$$

Since $\dfrac{\partial}{\partial \rho}[exp(-jkR_i)] = \dfrac{k\rho}{jR_i}exp(-ikR_i)$, where $i = 0,1,2$, (1.30) can be rewritten as

$$E_z = \frac{30J(0)}{k\rho\varepsilon_r \sin kL}\frac{\partial}{\partial \rho}\left[exp(-jkR_1) + exp(-jkR_2) - 2\cos kL exp(-jkR_0)\right] \quad (1.32)$$

$$H_\varphi = j\frac{J(0)}{4\pi\rho\sin kL}\left[exp(-jkR_1) + exp(-jkR_2) - 2\cos kL exp(-jkR_0)\right]. \quad (1.33)$$

Substituting (1.33) into the first expression of (1.31), we get

$$E_\rho = j\frac{30J(0)}{\varepsilon_r\rho\sin kL}\left[\frac{(z-L)\,exp(-jkR_1)}{R_1} + \frac{(z+L)\,exp(-jkR_2)}{R_2} - 2z\cos kL\frac{exp(-jkR_0)}{R_0}\right]. \quad (1.34)$$

If we use the radiator model in the form of a straight circular cylinder, then when calculating the field, one should proceed from the expression (1.20). Repeating the arguments and carrying out calculations similar to the case with a perfectly conducting filament, we find instead of (1.30)

$$E_z = -j\frac{30J(0)}{2\pi\varepsilon_r \sin kL} \int_0^{2\pi} \left[\frac{exp(-jkR_1)}{R_1} + \frac{exp(-jkR_2)}{R_2} - 2\cos kL \frac{exp(-jkR_0)}{R_0}\right] d\varphi, \quad (1.35)$$

where $R_1 = \sqrt{(z-L)^2 + \rho^2 + a^2 - 2a\rho \cos \varphi}$, $R_2 = \sqrt{(z+L)^2 + \rho^2 + a^2 - 2a\rho \cos \varphi}$, $R_0 = \sqrt{z^2 + \rho^2 + a^2 - 2a\rho \cos \varphi}$ are distances from the observation point to points on the radiator surface in the upper, lower and middle cross-sections of the radiator, respectively (see Fig. 1.2b). As for the other components of the electromagnetic field (H_φ and E_ρ), their calculation is complicated by the fact that $\frac{\partial}{\partial \rho}[exp(-jkR_i)] = \frac{k(\rho - a \cos \varphi)}{jR_i} exp(-jkR_i)$. Therefore, the field E_z cannot be represented in a form similar to (1.32).

Expressions (1.30) and (1.35) for the fields E_z, created by the currents flowing in radiator models in the form of a straight ideally conducting filament and a straight circular cylinder are valid for arbitrary observation points, including those lying on the radiator surface. It is easy to verify, however, that the tangential component of the electric field does not vanish when we put $\rho = 0$ into the first expression and $\rho = a$ into the second one. On the contrary, at the indicated values of ρ, the value of E_z sometimes becomes maximum.

In Fig. 1.4 the typical curves presented in the book [2] for the real and imaginary components of the field are given as functions of the z coordinate. The curves are constructed in accordance with the expression (1.30), i.e., for a perfectly conductive filament. When the length of the radiator arm is equal to $L = \lambda/4$, both components of the field (real and imaginary) are given by continuous curves. It can be seen from the figure that the real component of the field is almost constant along the entire radiator, while the imaginary component increases indefinitely at radiator ends. This increase is associated with the decrease of the distances $R_{1(2)} = L \pm z$ to 0, located in the denominators

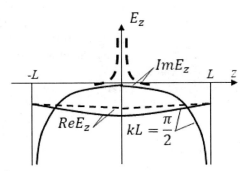

Fig. 1.4. Distribution of the real ($Re\ E_z$) and imaginary ($Im\ E_z$) components of the electric field strength tangent to the radiator surface.

of first and second terms of the square bracket in expression (1.30). The third term with R_0 in the denominator in this case does not give the maximum, because $\cos kz = 0$. But, with a small change in the radiator length both curves change (see dotted line). The unlimited growth of the curve is caused by the fact that in this model the radiator radius is zero.

From this it follows that in a metal cylinder or a filament with a sinusoidal current, the field is not identical to the field of the cylinder excited in the middle by a concentrated emf in the form of a narrow slot. To ensure a sinusoidal current distribution, it is necessary to create a continuous emf along the radiator, coincident in shape and opposite in sign to the curves shown in Fig. 1.4.

However, the approximate sinusoidal current distribution along the radiator creates an electromagnetic field in the surrounding space, which is close to the actual field, including the near zone. Indeed, as can be seen from (1.33) and (1.34), the components of the field H_φ and E_ρ on the antenna surface are inversely proportional to the radius a of the antenna and increase infinitely as the radius decreases to zero, while the field E_z, defined in accordance with (1.30) remains finite almost everywhere with the exception of the ends of the radiator and the point of excitation, i.e., electric field lines approach the axis of the radiator almost at a right angle, as it should be near a metal surface.

So much attention was paid to the sinusoidal distribution of the current along the radiator because it was used for a long time for antennas calculation, based on comparing the currents of the antenna with the currents of the equivalent long line and equating the power of the source to the power passing through the surrounding closed surface.

1.5 Equivalent long line

As already mentioned, the current distribution along the antenna in the main determines its characteristics. Therefore, a significant role in the development of the calculating antennas method was played by comparing the properties of radiators with the properties of two-wire long lines. The simplest metal radiators are shown in Fig. 1.1. The circuit of a long line equivalent to a symmetrical metal radiator (Fig. 1.1a), is presented in Fig. 1.5a. An infinitesimal element with length dz of such a line consists of inductance $d\Lambda = \Lambda_1 dz$ and capacitance $dC = C_1 dz$. Here, C_1 and Λ_1 are the capacitance and inductance per unit length of the line.

The impedance long line (Fig. 1.5b) is a similar analogue of an impedance radiator. An example of such a radiator is a metal rod with a shell made of a magnetodielectric. This is a radiator with distributed surface impedance $Z(z)$. Unlike a metal radiator, the tangential component E_z of the electric field on

Straight Metal Radiator   13

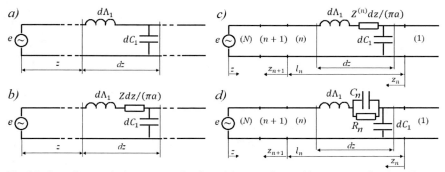

Fig. 1.5. Long lines equivalent to a metal radiator (*a*), to a radiator with constant surface impedance $Z(z)$ (*b*), to a radiator with piecewise constant surface impedance $Z^{(n)}$ (*c*), to a radiator with concentrated loads Z_n (*d*).

the surface of the impedance radiator is not zero, which leads to an additional voltage drop

$$dU = \frac{Z}{2\pi a} J(z) dz$$

on each of its elements. Thus, an infinitesimal element of the line, which is equivalent to a symmetric impedance radiator, contains an additional impedance $\frac{Z}{\pi a} dz$. Factor 2 accounts for the two wires in the radiator.

If the material and dimensions of the shell do not change along its length, the impedance is constant, and in case they do change—it is variable. If the radiator consists of several sections of length l_n (Fig. 1.5c), on each of which the magnitude of $Z^{(n)}$ is constant, the surface impedance is considered piecewise constant. Surface impedance allows us to obtain an in-phase current distribution in the radiator with a propagation constant, significantly different from the propagation constant in the air. By including concentrated loads along the radiator length (Fig. 1.5d) similar results can be achieved. Also, concentrated loads allow us to obtain such in-phase current distribution, which ensures better electrical characteristics of the radiator.

The current distribution in the wires of a metal dipole with arms of the same length L is incrementally different from the current distribution in identical wires of a two-wire metal line open at the end. The significant difference between them results only due to the fact that the long line radiates weakly, if the distance between parallel wires is small. The circuit of the long line, consisting of two metal wires, corresponds to a symmetrical radiator (dipole) with a generator located at its center, and to a grounded asymmetrical radiator (monopole) with a generator at its base. The input impedance of such a two-wire long line open at the end is, approximately, equal to the reactance of each radiator:

$$Z_l = -jW_l \cot kL,$$

where W_l is the wave impedance of the line, L is its length.

If the generator is displaced from the radiator center by a distance h, then the radiator reactance is equal to the sum of two impedances: these are the impedances of the long lines located on opposite sides of the generator. In the case of dipoles lines open at the ends have lengths $L-h$ and $L+h$, and in the case of the monopole the line open at the end has length $L-h$ and, and another line closed at the end has length h.

The shape of the current curve in the arm of an impedance radiator coincides with the shape of the current curve in an impedance long line of the same wires. The current curve in the antenna arm with a varying surface impedance along its length is the same as the current curve in a long line with the same impedance. Finally, the current in the radiator with concentrated loads along its length, consisting, for example, of capacitors and resistors connected in parallel, is similar to the current in a long line with similar concentrated loads.

The impedance long line, as already mentioned, is the same analogue of an impedance radiator as a customary line is an analogue of a metallic one. Unlike a metal antenna, the boundary condition on the surface of the wire with distributed surface impedance has the form

$$\frac{E_z(a,z)+K(z)}{H_\varphi(a,z)} = Z(z), \quad -L \le z \le L. \qquad (1.36)$$

Here $E_z(a, z)$ and $H_\varphi(a, z)$ are the tangential component of the electric field and the azimuthal component of the magnetic field, respectively, $K(z)$ is the external emf, and $Z(z)$ is the surface impedance, which in the general case depends on the coordinate z. The boundary conditions of this type are valid, if the structure of the field in one of the media (for example, in the magnetodielectric shell of the antenna) is known and does not depend on the structure of the field in another medium—in the surrounding space [3].

If the boundary conditions (1.36) are performed on the surface of long lines and radiators and the surface impedance is large enough to significantly change the current distribution along the antennas and lines already in the first approximation, they called impedance ones. Surface impedance leads to an additional voltage drop on each element of the radiator and the line. The choice of impedance provides an additional degree of freedom in the process of optimizing the characteristics of antennas.

The telegraph equations for the impedance long line are

$$-\frac{dU(z)}{dz} = J(z)\left(j\omega\Lambda_1 + \frac{Z}{\pi a}\right), -\frac{dJ(z)}{dz} = j\omega C_1 U(z), \quad (1.37)$$

whence

$$\frac{d^2 J}{dz^2} + k_1^2 J(z) = 0, \quad \frac{d^2 U}{dz^2} + k_1^2 U(z) = 0, \quad (1.38)$$

where $k_1^2 = k^2 - \frac{jz\omega C_1}{\pi a}$. Here k_1 is the wave propagation constant in the radiator and in its equivalent two-wire long line, C_1 is the capacitance per unit length between the radiator arms and between the wires of the long line. This capacitance is half as much as the capacitance C_0 between each radiator arm and the surface of zero potential. Because the total length $2L$ is the maximum dimension of the radiator, and this dimension is significantly larger than other dimensions (in particular, $2a$), in the first approximation the surface of zero potential can be placed at a distance of $2L$ from the radiator and it can be assumed that the capacitance C_0 is equal to the self-capacitance of the infinitely long wire with a radius a, i.e.,

$$C_0 = \frac{2\pi\varepsilon}{\ln(2L/a)}, C_1 = C_0/2 = \frac{\pi\varepsilon}{\ln(2L/a)}, \quad (1.39)$$

where from

$$k_1^2 = k^2 - j\frac{\omega\varepsilon Z}{a\ln(2L/a)}. \quad (1.40)$$

Solving equations (1.37) in the ordinary way, we find the current $J(z)$ and the wave impedance W_I of an open-end impedance line:

$$J(z) = J(0)\sin k_1(L-z)/\sin k_1 L; W_I = \frac{\omega\Lambda_1 - j(z/\pi a)}{k_1} = \frac{k_1}{\omega C_1} = 120\frac{k_1}{k}\ln\frac{2L}{a}. \quad (1.41)$$

Here $J(0)$ is the generator current. Thus, a dipole with constant surface impedance can be considered in a first approximation as an equivalent long line differing from a customary long line by the presence of the load impedance $Z/(\pi a)$ per unit length. The current in the dipole has sinusoidal character, its propagation constant differs from a propagation constant k in a metal antenna. The wave impedance of the impedance dipole is greater than W_I of a metal antenna by a factor k_1/k. The input impedance of a two-wire long line open at the end with a constant surface impedance is equal to

$$Z_I = -j\frac{k_1}{k}W_I \cot k_1 L. \quad (1.42)$$

When the power point is displaced in a symmetrical and grounded asymmetric impedance radiator, i.e., in an impedance dipole and monopole, the input impedance of equivalent impedance lines changes in the same way as the input impedance of a customary metal line.

If the surface impedance is changing along the antenna (more precisely, if this magnitude is piecewise constant), the equivalent impedance long line is a non-uniform line, in other words, it is a stepped line. It consists of N uniform segments of length l_m with wave impedance W_m, current J_m and voltage u_m. Infinitesimal lengths dz of each segment (see Fig. 1.5c) contain resembling elements of different size. We shall designate the surface impedance of segment m as $Z^{(m)}$. Comparison of each radiator segment and a corresponding segment of an equivalent line allows us to obtain expressions for the propagation constant k_m and the wave impedance W_m, similar to (1.40) and (1.41).

From the theory of long lines, it is known that on each segment of the stepped line

$$u_m = U_m \cos(k_m z_m + \varphi_m), J_m = jI_m \sin(k_m z_m + \varphi_m), \qquad (1.43)$$

and $I_m = \dfrac{U_m}{W_m}$. Because voltage and current along the stepped long line, equivalent to the radiator, are continuous:

$$u_m\big|_{z_m=0} = u_{m-1}\big|_{z_{m-1}=l_{m-1}}, \quad J_m\big|_{z_m=0} = J_m\big|_{z_{m-1}=l_{m-1}},$$

then

$$I_m = I_{m+1} \frac{\sin(k_{m-1} l_{m-1} + \varphi_{m-1})}{\sin \varphi_m}, \quad U_m = U_{m-1} \frac{\cos(k_{m-1} l_{m-1} + \varphi_{m-1})}{\cos \varphi_m}. \qquad (1.44)$$

Dividing the first of expressions (1.44) onto the second expression and taking into account (1.41) and (1.43), we find:

$$\tan \varphi_m = \frac{k_m}{k_{m-1}} \tan(k_{m-1} l_{m-1} + \varphi_{m-1}). \qquad (1.45)$$

Equality (1.45) enable us to express the amplitude and phase of the current in any segment through the parameters of segments and one of the currents:

$$\varphi_m = \tan^{-1}\left\{ \frac{k_m}{k_{m-1}} \tan\left[k_{m-1} l_{m-1} + \tan^{-1}\left\{ \frac{k_{m-1}}{k_{m-2}} \tan\left\langle k_{m-2} l_{m-2} + \cdots + \tan^{-1}\left(\frac{k_2}{k_1}\right) \cdots \right\rangle \right\} \right] \right\},$$

$$I_m = I_N \prod_{p=m+1}^{N} \frac{\sin \varphi_p}{\sin(k_{p-1} l_{p-1} + \varphi_{p-1})}. \qquad (1.46)$$

The second expression (1.46) is true for the segment N, if to adopt that $\prod_{p=N+1}^{N} = 1$.

Since the current of the generator is

$$J(0) = J_N \sin(k_N l_N + \varphi_N),$$

then

$$I_m = A_m J(0), \quad (1.47)$$

where

$$A_m = \prod_{p=m+1}^{N} \sin \varphi / \prod_{p=m}^{N} \sin(k_p l_p + \varphi_p).$$

Expressions (1.43) together with (1.46) and (1.47) present the approximate laws of voltage and current distribution along the radiator with piecewise constant surface impedance. The input voltage of the line and the input current are equal to

$$e = U_N \cos(k_N l_N + \varphi_N) \ \& \ J(0) = jI_N \sin(k_N l_N + \varphi_N), \text{ respectively}$$

i.e., the input impedance of the stepped long line is

$$Z_l = e/J(0) = -jW_N \cot(k_N l_N + \varphi_N).$$

The required value of the surface impedance and a predetermined law for its change along the radiator can be realized by means of concentrated loads, located along an antenna wire (Fig. 1.5d). Let these loads be placed uniformly at distance b from each other. If the distance b is small ($kb \ll 1$), the current distribution along the antenna undergoes practically no change, if concentrated loads are replaced by continuous surface impedance $Z(z)$, distributed across the length of each segment. Assume that the surface impedance of antenna segment m is constant and equal to $Z^{(m)}$. In that case

$$Z_m = bZ^{(m)}/(2\pi a_m). \quad (1.48)$$

Here a_m is the radius of the segment m. As will be shown later on, the best matching results in a wide frequency range are provided by means of a capacitor C_n and a resistor R_n. In this case, Z_m is the impedance formed by the resistor and capacitor, connecting in parallel, where a_m is the segment radius. The equivalent long line permits to analyze an antenna structure.

In the general case the number of wires can be significantly greater than two. They form for example the structure of electrically coupled lines located above the ground. The theory of coupled lines designed by A.A. Pistolkors [4] formed the foundation of the analysis of antennas consisting of parallel wires. This theory permits to analyze multi-conductor structures of antennas and cables.

18 ANTENNAS: Rigorous Methods of Analysis and Synthesis

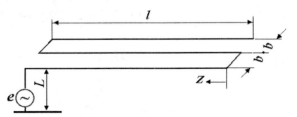

Fig. 1.6. Horizontal load of the antenna in the shape of a structure of three series-connected wires located parallel to the ground.

As an example, in Fig. 1.6 the horizontal load of the antenna is shown. It is made in the shape of a meander and allows us to significantly increase the effective height of the vertical antenna compared to the customary horizontal load consisting of three parallel-connected wires of the same length.

If a structure consisting of N parallel metal wires is located above the ground, expressions for a current and a potential of wire n have the form

$$i_n = I_n \cos kz + j\left[\frac{2U_n}{W_{nn}} - \sum_{s=1}^{N}\frac{U_s}{W_{ns}}\right]\sin kz, \quad u_n = U_n \cos kz + j\sum_{s=1}^{N}\rho_{ns}I_s \sin kz. \tag{1.49}$$

Here I_n and U_n are the current and potential at the beginning of wire n (at point $z = 0$). W_{ns} and ρ_{ns} are mutual electrostatic and electrodynamic characteristic (wave) impedances of wires n and s respectively:

$$\rho_{ns} = \frac{\alpha_{ns}}{c}, \quad W_{ns} = \begin{cases} 1/(c\beta_{ns}), & n = s, \\ -1/(c\beta_{ns}), & n \neq s, \end{cases}$$

where α_{ns} is the potential coefficient (taking into account a mirror image with respect to the perfectly conducting ground surface), β_{ns} is the coefficient of electrostatic induction, c is the velocity of light. The coefficients β_{ns} and α_{ns} are connected by an expression:

$$\beta_{ns} = \Delta_{ns}/\Delta_N,$$

where $\Delta_N = |\alpha_{ns}|$ is the $N \times N$ determinant, and Δ_{ns} is the cofactor (adjunct) of the determinant Δ_N. The coefficients are defined by the method of mean potentials in accordance with the actual position of antenna wires. The simplest variant of this method is the Howe's method.

Finally, if the wires of an asymmetrical line have unequal lengths, or if concentrated loads are connected into them, one must divide the line to segments. The expressions for the current and potential of wire n at segment m take the form:

$$i_n^{(m)} = I_n^{(m)} \cos kz_m + j\left[\frac{2U_n^{(m)}}{W_{nn}^{(m)}} - \sum_{s=1}^{N}\frac{U_s^{(m)}}{W_{ns}^{(m)}}\right]\sin kz_m,$$

$$u_n^{(m)} = U_n^{(m)} \cos kz_m + j \sum_{s=1}^{N} \rho_{ns}^{(m)} I_s^{(m)} \sin kz_m, \qquad (1.50)$$

where $I_n^{(m)}$ and $U_n^{(m)}$ are the current and potential of wire n at the beginning of segment m (at point $z_m = 0$), respectively, M is the number of wires in segment m, and $W_{ns}^{(m)}$ and $\rho_{ns}^{(m)}$ are the electrostatic and electrodynamic characteristic impedances between wires n and s at segment m. These equations generalize the expressions (1.49).

In particular, for an asymmetrical line of two wires, we can write

$$\frac{1}{W_{11}} = \frac{\rho_{22}}{\rho_{11}\rho_{22} - \rho_{12}^2}, \frac{1}{W_{22}} = \frac{\rho_{11}}{\rho_{11}\rho_{22} - \rho_{12}^2}, \frac{1}{W_{12}} = \frac{\rho_{12}}{\rho_{11}\rho_{22} - \rho_{12}^2},$$

i.e.,

$$\left(\frac{1}{W_{12}} - \frac{1}{W_{11}}\right) : \left(\frac{1}{W_{12}} - \frac{1}{W_{22}}\right) = \frac{\rho_{22} - \rho_{12}}{\rho_{11} - \rho_{12}} = g.$$

It is easily show that the mutual potential coefficient of two parallel wires of equal lengths, with size and position as given in Fig. 1.7, is,

$$\alpha_{ns} = \frac{\alpha(L,l,b)}{(2\pi\varepsilon)}, \qquad (1.51)$$

where

$$\alpha(L,l,b) = \frac{1}{2L}\left[(L+l)\sinh^{-1}\frac{L+l}{b} + (L-l)\sinh^{-1}\frac{L-l}{b} - 2L\sinh^{-1}\frac{L}{b} - \sqrt{(L+l)^2 + b^2} - \sqrt{(L-l)^2 + b^2} + 2\sqrt{l^2 + b^2}\right],$$

i.e.,

$$\rho_{ns} = \alpha_{ns}/c = \alpha(L, l, b)/(2\pi\varepsilon_0\varepsilon_r c) = 60\alpha(L, l, c)/\varepsilon_r.$$

Calculation of the currents, potentials and input impedances in different structures consisting of parallel metal wires, located above the ground, requires solving the corresponding set of equations, using the relevant boundary conditions. These conditions consist of the continuity of current

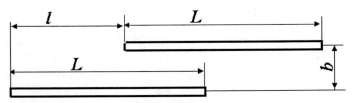

Fig. 1.7. The mutual location of wires.

and potential along each wire in the absence of currents at free ends, with abrupt changes in potential at the points of connecting loads and generator e. Calculating the current magnitude $J(0)$ at the exciting point permits finding the input impedance of the asymmetrical line,

$$Z_l = e/J(0). \qquad (1.52)$$

The input impedance is equal in the first approximation to the reactive impedance of the antenna, whose equivalent is the given asymmetrical line. In order to find the antenna impedance more accurately, we may consider an antenna as a linear radiator, the current along which is equal to the total current along the line.

While calculating the antenna input impedance, one needs, as a rule, to find field E_ς at the antenna surface. And it should be kept in mind that in the general case the magnitude of the field is determined by the expression (1.28), which goes into (1.30) and (1.32), if the current is continuous along the entire antenna and the function $dJ/d\varsigma$, which is equal to the derivative of the current with respect to the antenna length, does not have abrupt jumps on the boundaries of the segments. The first position is not always observed in the folded and multi-folded antennas, and the voltage jumps exist in piecewise uniform (stepped) antennas. If the number of such boundaries in each arm of the radiator (in a stepped line) is M, then taking into account the mirror image, their number is twice as large. In this case

$$E_\varsigma = \frac{1}{4\pi j\omega\varepsilon} \left\{ \frac{2\exp(-jkR_0)}{R_0} \frac{dJ(0)}{d\varsigma} - \left[\frac{\exp(-jkR_{11})}{R_{11}} + \frac{\exp(-jkR_{12})}{R_{12}} \right] \frac{dJ(L)}{d\varsigma} \right.$$
$$\left. + \sum_{m=1}^{M} \left[\frac{\exp(-jkR_{m1})}{R_{m1}} \frac{\exp(-jkR_{m2})}{R_{m2}} \right] \left[\frac{dJ(l_m + 0)}{d\varsigma} - \frac{dJ(l_m - 0)}{d\varsigma} \right] \right\},$$

where $R_0 = \sqrt{a^2 + \varsigma^2}$, $R_{m1} = \sqrt{a^2 + (l_m - \varsigma)^2}$, $R_{m2} = \sqrt{a^2 + (l_m + \varsigma)^2}$, $\dfrac{dJ(l_m + 0)}{d\varsigma}$ and $\dfrac{dJ(l_m - 0)}{d\varsigma}$ are values of derivatives to the right and left of the point $z = l_m$, a is a radius at the point ς.

1.6 Equality of two powers and Poynting method

One of the main characteristics of any radiator is its input impedance. A matching level of transmitter with the antenna and a losses level of during this transfer, i.e., a quantity of power transferred by a signal source, depend in particular on this magnitude. The input impedance of the radiator can be determined proceeding from the equality of power created by the source of emf and power passing through a closed surface surrounding the antenna. In the capacity of the closed surface a cylindrical surface is usually used in the method of induced emf and the spherical surface is usually used in the method of Poynting vector. In particular, the cylinder with height $2H$ and radius b, along the axis of which the symmetric radiator is located, is shown in Fig. 1.8.

The article [5], published in 1884, is devoted to the calculation of the active power of the radiated signal. In this article it was shown that the power flux density coming out of a volume bounded by a closed surface is determined by the Poynting vector

$$\vec{P} = [\vec{E}, \vec{H}], \quad (1.53)$$

more precisely, by a projection of this vector on the normal to the corresponding surface. For the side surface and the upper cover of the cylinder, these projections have the form

$$P_\rho = -E_z H_\varphi, \ P_z = -E_\rho H_\varphi.$$

Let's combine the cylindrical closed surface with the radiator surface, by putting $H = L$, $b = a$. Then, with a small radius of the radiator, the power flows passing through the cover and the bottom of the cylinder will also be small. If the radiator is a thin filament with a radius tending to zero, then the integrals over its upper and lower bases (the cover and the bottom of the cylinder) will tend to zero. If the radiator has the shape of a circular cylinder with a finite radius, then these integrals also tend to zero, since the longitudinal

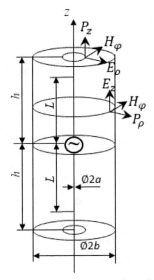

Fig. 1.8. Closed surface around a radiator.

current inside the cylinder decays rapidly, i.e., the tangential component of the magnetic field H_φ in both bases is close to zero when $\rho < a$.

Therefore, the power passing through a closed surface is determined only by an integral along the lateral surface of the cylinder:

$$P = -\int_{-L}^{L} \int_{0}^{2\pi} E_z H_\varphi a \, d\varphi \, dz, \qquad (1.54)$$

and the value P_ρ does not depend on the coordinate φ, since the components of the electromagnetic field do not depend on φ.

In this case, as was said, the power is understood as the so-called active power, equal to the average value over the oscillation period. Therefore, the power is equal to the product of two real quantities (without an asterisk). The spherical closed surface, in the center of which an antenna is located, is used in the Poynting vector method, which allows to calculate an antenna radiation resistance. We will consider that the sphere radius R is so large that the sphere is located in the far zone of the antenna, and a field passing through it has the character of a plane wave. In a plane wave, the vectors $\vec{E}$ and $\vec{H}$ are mutually perpendicular, and their ratio E/H is equal to Z_0 ($Z_0 = 120\pi$ is the wave resistance of a free space). This relation can be easily verified by comparing expressions (1.32) and (1.33) for E_z and H_φ. This means that the active power is proportional to the square of an electric field strength, i.e., $P = E_\theta^2 / Z_0$, where E_θ is the component of the electric field strength, tangent to the surface of the sphere.

In the general case, the electric and magnetic fields of an antenna are defined by the expression (1.10) and depend on a vector-potential $\vec{A}$, which is a function of currents in the region V. Let a symmetric linear radiator (dipole) located along z axis is a source of field, and a perfectly conductive filament is used as the radiator model. The current of this radiator is distributed in accordance with a sinusoidal fashion, i.e., in accordance with (1.26) follows the law

$$J(z) = I_0 \frac{\sin k(L-|z|)}{\sin kL} = J(0) \sin k(L-|z|).$$

One can consider that its field consists of the sum of fields of elementary radiators (Hertz dipoles), directed at the same angle. The elementary radiator n is a filament of a length l_n with unaltered current J_{n0}. The field of such a radiator in the far zone in the horizontal direction (in the perpendicular direction to its axis) is

$$E_\theta = j \frac{60 k L \, J(0) \exp(-jkR_0) \sin\theta}{\varepsilon_r R_0}.$$

Here $k = \omega\sqrt{\mu\varepsilon}$ is the wave propagation constant in the environment, ω is the circular frequency of the signal, μ and ε are permeability and permittivity of the medium respectively ($\varepsilon = \varepsilon_r\varepsilon_0$, where ε_r is relative permittivity of the medium, and ε_0 is absolute permittivity of air).

Let us compare the fields of the symmetric linear radiator and the Hertz dipole in the indicated direction. We assume that the fields are the same if the area under the current curve of the linear radiator coincides with the sum of the areas under the current curves of the Hertz dipoles equivalent to the sections of the linear radiator. The first of these areas is known to be equal to

$$S_1 = 2\frac{I_0}{\sin kL}\int_0^L \sin k(L-z)dz = \frac{2(1-\cos kL)}{k\sin kL} = \frac{2}{k}\frac{2\cos^2\frac{kL}{2}}{2\sin\frac{kL}{2}\cos\frac{kL}{2}} = \frac{2}{k}\tan\frac{kL}{2}.$$

For the second magnitude, one can write

$$S_2 = \sum_{n=1}^N J_{n0}l_n.$$

Here N is the number of sections, n is the section number. The magnitude S_1 in accordance with the method used to obtain it, is customary called by the equivalent effective length of the linear radiator l_e.

To determine the magnitude of the field at an arbitrary angle θ to the antenna, it is necessary to take into account the difference of distances from the central and other points of the antenna to the observation point (Fig. 1.9a). This difference is called the path difference, it is equal to $r = z\cos\theta$. For the field of symmetric radiator located along z axis with the center at the origin, we obtain

$$E_\theta = j\frac{30k\exp(-jkR)\sin\theta}{\varepsilon_r R}\int_{-L}^L J(z)\exp(jkz\cos\theta)dz. \qquad (1.55)$$

In the case of a sinusoidal current distribution along the radiator, we find

$$E_\theta = j\frac{60I_0}{\varepsilon_r \sin kL}\frac{\exp(-jkR)}{R}\frac{\cos(kL\cos\theta) - \cos kL}{\sin\theta}.$$

Here, the last multiplier determines the directional pattern of the radiator in a vertical plane.

Using the expression for the antenna field in an arbitrary direction, one can determine the active power radiated by the antenna and passing through the surface of the sphere:

$$P = \frac{R_0^2}{120\pi}\int_0^{2\pi}d\varphi\int_0^\pi E_\theta^2\sin\theta d\theta.$$

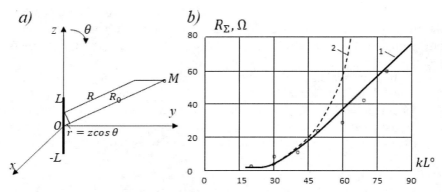

Fig. 1.9. Accounting the difference of the pathlength of the antenna rays (*a*) and results of calculating radiation resistances (*b*).

Dividing this power to the square of the current at the output of the generator, we find the radiation resistance of the antenna:

$$R_\Sigma = \frac{P}{J^2(0)}.$$

This method of determining of the radiation resistance is widely known as the Poynting method. The calculations commonly are performing by numerical methods. Analytical methods are encountering insurmountable difficulties in integrating squares of electric fields at arbitrary angles θ. For a long time, it was believed that the integrals could not be calculated. As shown below, this opinion turned out to be erroneous.

In accordance with the above expressions, we write

$$R_\Sigma = \frac{R^2}{120\pi J^2(0)} \int_0^{2\pi} d\varphi \int_0^\pi E_\theta^2 \sin\theta d\theta = \frac{R^2}{30 J^2(0)} \int_0^{\pi/2} E_\theta^2 \sin\theta d\theta.$$

As follows from above, it is proportional to the square of the effective antenna length, calculated taking into account the different magnitudes of the signals at different angles θ. If in this expression to replace the integral $\int_0^{2\pi} d\varphi \int_0^\pi E_\theta^2 \sin\theta d\theta$ by an approximate expression $4\pi E_\theta^2$, i.e., if to adopt that the field of the dipole of the length $2L$ is the same under all angles and is equal to the field in the horizontal direction, then

$$R_\Xi = 30(kh_e^2)$$

This value is clearly overestimated: for example, if $kL = \pi/2$, we get $R_\Xi = 120$. When the beam deviates from perpendicular to the antenna axis, the effective length rapidly decreases. A strict calculation in this case gives a value of the order of 80.

Using the rigorous expression for E_0, introducing the notation $\alpha = kL$ and assuming that $\varepsilon_r = 1$, we obtain:

$$R_\Sigma = 120 \int_0^{\pi/2} \frac{[\cos(\alpha\cos\theta) - \cos\alpha]^2}{\sin\theta} d\theta.$$

We introduce the new variable $x = \alpha \cos\theta$. If $\theta = 0$, then $\cos\theta = 1$, $x = \alpha$. With increasing θ, the magnitude x is decreased. Because $dx = -\alpha \sin\theta d\theta$, the replacement of a variable leads to the expression

$$R_\Sigma = 120 \int_0^\alpha \frac{(\cos x - \cos\alpha)^2}{\alpha \sin^2\theta} dx = 120 \propto \int_0^\alpha \frac{(\cos x - \cos\alpha)^2}{\alpha^2 - x^2} dx.$$

To calculate the resulting expression, we apply integration by parts. Since $\cos\theta \leq 1$, then $x \leq \propto$, and in this case $\int \frac{dx}{\alpha^2 - x^2} = \frac{1}{2\propto} \ln \frac{\propto + x}{\propto - x}$. The value of the derivative $\frac{d}{dx}(\cos x - \cos\alpha)^2$ is equal to $-\sin 2x + 2\cos\alpha \sin x$. As a result, we obtain

$$R_\Sigma = 60(\cos x - \cos\alpha)^2 \ln\frac{\propto + x}{\propto - x} \{_0^\alpha + 60 \int_0^\alpha \ln\frac{\propto + x}{\propto - x}(\sin 2x - 2\cos\alpha \sin x) dx.$$

It is not difficult to be convinced that the first term of this expression is equal to zero. Indeed, at the upper limit (at $x = \propto$), the coefficient at the logarithm has the form $60(\cos\propto - \cos\alpha)^2$. At the lower limit (at $x = 0$) the logarithm is $\ln\frac{\propto}{\propto} = \ln 1 = 0$.

To the second term of the expression one may also apply the integration by parts. The derivative of the first multiplier of the integrand is

$$\frac{d}{dx}(\ln\frac{\propto + x}{\propto - x}) = \frac{1}{\propto + x} + \frac{1}{\propto - x}.$$

The integral from the second multiplier has the form $-\frac{\cos 2x}{2} + 2\cos\alpha \cos x$. Accordingly, the second term of the expression for R_Σ takes the form

$$30(-\cos 2x + 4\cos\alpha \cos x)\ln\frac{\propto + x}{\propto - x}\{_0^\alpha - 30\int_0^\alpha (-\cos 2x + 4\cos\alpha \cos x)\left(\frac{1}{\propto + x} + \frac{1}{\propto - x}\right) dx.$$

Here the free member is

$$30(-\cos 2\propto + 4\cos^2\propto)\ln\frac{2\propto}{\propto \cdot 0} = (2 + \cos 2\propto)\ln\frac{2}{0},$$

and the integral can be reduced to a set of integral sines and cosines. This method is demonstrated by the example of the integrand summand $\dfrac{\cos 2x}{\alpha + x}$:

$$\frac{\cos 2x}{\alpha + x} = \frac{e^{j2(\alpha+x)}e^{-j2\alpha} + e^{-j2(\alpha+x)}e^{j2\alpha}}{2(\alpha + x)} =$$

$$e^{-j2\alpha}[\cos 2(\alpha + x) + j\sin 2(\alpha + x)] + e^{j2\alpha}[\cos 2(\alpha + x) - j\sin 2(\alpha + x)] =$$

$$2[\cos 2\alpha \cos 2(\alpha + x) + \sin 2\alpha \sin 2(\alpha + x)],$$

$$\frac{\sin 2x}{\alpha + x} = 2[\cos 2\alpha\, Si2(\alpha + x) - \sin 2\alpha\, Ci2(\alpha + x)].$$

The integral is reduced to the sum

$$60\{\cos 2\alpha[Ci2(\alpha + x) - Ci2(\alpha - x)] - 2\cos\alpha \cos x[Ci(\alpha + x) - Ci(\alpha - x)] +$$
$$\sin 2\alpha[Si2(\alpha + x) - Si2(\alpha - x)] - \sin 2\alpha[Si(\alpha + x) - Si(\alpha - x)]\}\Big\{^{\alpha},$$

whence, substituting the upper and lower limits of integration and taking into account the free member, we obtain

$$R_\Sigma = 60\{(1 + 0.5\cos 2\alpha)\ln\frac{2}{0} + \cos 2\alpha[Ci4\alpha - Ci(2\alpha \cdot 0)] - \qquad (1.56)$$
$$2\cos^2\alpha[Ci2\alpha - Ci(\alpha \cdot 0)] + \sin 2\alpha[Si4\alpha - Si2\alpha]\}.$$

When using the expression (1.56), one must first determine the functions included in it. To do this, it is need to use the tables of integral sines and cosines, for example, the book [63]. To obtain the required accuracy, it is necessary to apply the interpolation method. The values of these functions for several options of the radiator arm length are given in the Table 1.1. In the calculations, it is necessary to take into account that the integral cosine $Ci\, x$ has a singularity at the point $x = 0$. If $x = b\delta$ and the value δ tends to zero, then $Ci\, x = 0.5(lnb + ln\gamma)$, where $ln\gamma = C = 0.577$ is the Euler constant. If one of the terms is $-ln(b \cdot 0)$, then the quantity $Ci(b \cdot 0)$ allows us to compensate for this term.

In accordance with the foregoing we start with a symmetric radiator whose arm length at the frequency of the first serial resonance is $L = \dfrac{\lambda}{4}$, i.e., $\alpha = \pi/2$. In accordance with the general expression, we find

$$R_\Sigma = 60\left\{0.5\ln\frac{2}{0} + \cos 2\alpha[Ci\, 2\pi - Ci(\pi \cdot 0)]\right\}.$$

Table 1.1. The values of the functions included in expression for R_Σ.

α	$Si\,\alpha$	$Ci\,\alpha$	$ln\,\alpha$
$\pi/8$	0.3894	−0.3931	−0.9347
$\pi/6$	0.5157	−0.1377	−0.647
$\pi/4$	0.7586	0.1854	−0.242
$\pi/3$	0.983	0.357	0.04593
$\pi/2$	1.369	0.4673	0.4516
$2\pi/3$	1.643	0.389	0.739
π	1.8506	0.0746	1.1447
$4\pi/3$	1.72	−0.166	1.432
$3\pi/2$	1.6095	−0.1973	1.5502
2π	1.4189	−0.0228	1.8379
$8\pi/3$	1.6175	0.1071	2.1256

We take into account that

$ln2 = 0.693$, $Ci\,(\pi \cdot 0) = 0.5\,(ln\pi + C)$, $ln\pi = 1.1447$.

Substituting magnitudes into the expression for R_Σ, we obtain

$R_\Sigma = 60[0.3466 + 0.0228 + 0.5 \cdot 1.1447 + 0.288] = 73.8\Omega$.

This result is close to magnitudes obtained by different authors when solving the integral equation for the current in a metal radiator. In particular, the solution of the Leontovich's equation in the first approximation gives 71.2Ω for the radiation resistance of a dipole at the point of the first serial resonance. The proximity of the results confirms the validity of the proposed calculation method, which is based on the Poynting vector method.

The described procedure gives the following specific values for the radiation resistance of the dipole with an arm length equal to $\lambda/4.8$, $\lambda/6$, $\lambda/8$, $\lambda/12$ and $\lambda/16$: 43.4, 24.6, 12.6, 5.3 и 2.9Ω. The calculation results depending on the electric length of the dipole are shown in Fig. 1.9b in the form of a smooth curve 1. For comparison, in the same figure, the values of active components of the input impedance are given by means of the dots. These magnitudes were obtained using the new method of a solving the Leontovich's equation described in the second chapter. They correspond to the total input current of the antenna, equal to the total sum of the series for the current, and from this point of view are similar to the result of the solution obtained using the Poynting vector.

Since during the calculations all the upper limits were chosen equal to $\lambda/2$, this expression allows us to find the radiation resistance of the antenna with an electrical arm length equal to $kL = k\lambda/2 = \alpha$.

In order to determine the radiation resistance of an antenna with a different arm length, one must multiply the radiation resistance found by the above formula by $tan^2\left(\dfrac{kL}{2}\right)/tan^2\alpha$. The possibility of such a simple calculation method is explained primarily by the fact that for the same structure of the radiator a simultaneous change of all its dimensions leads to the same change in the magnitude of the field in all directions. The proposed simple method is valid if only because the radiator field in the horizontal direction is proportional to $\tan\dfrac{kL}{2}$, and the radiation resistance is proportional to the square of this field. If we take into account that the fields are calculated in the far zone, then the condition for applying the method to radiators with the same structure becomes redundant.

The numerical results obtained in accordance with above general formula coincide in the first approximation with the results of calculations by the simple formula. For relatively short radiators (with arm length $L < 0.3\lambda$) this coincidence is accurate. When the arm length is increased the accuracy is decreased. Repeat again. The validity of this simple method is based on the fact that as the radiator field in the horizontal direction as the radiator field in an arbitrary direction is proportional to $\tan\dfrac{kL}{2}$, and the radiation resistance is proportional to the square of this field.

The results of calculations by the above general formula confirm the applicability of the described simple method. The direct relationship of the simple method with the formula for E_θ also evidences this fact: substituting $\theta = \pi/2$ into this formula, we obtain an expression proportional to the previously calculated effective antenna length l_e.

The proposed methods can be generalized to the case of linear impedance radiators. The current of a symmetric impedance radiator (dipole), located along z axis with the center at the origin, is defined by the expression (1.41). In order to compare with each other the fields of the metal and impedance radiators in the far zone, we use the expression (1.28) for the field E_z. If the current along the metal radiator is distributed according to a sinusoidal law, then the first multiplier under the sign of the integral, and therefore the first term of this expression, are equal to zero. The second term is also equal to zero, and the expression for E_z is transformed to (1.30). The calculation of a field of an impedance antenna leads to a similar result. In this case the value k in the first multiplier of the integrand is replaced by the value k_1, but this multiplier as before is zero. The expression for the vertical field component of such an antenna differs only by another wave propagation constant along the antenna and by an additional factor k_1/k.

The difference between the other components of the field is of the same nature. In particular

$$E_\theta = j \frac{60 I_0}{\sin k_1 L} \frac{k_1}{k} \frac{\exp(-jkR)}{\varepsilon_r R} \frac{\cos(k_1 L \cos\theta) - \cos k_1 L}{\sin\theta}.$$

Substituting this value into the expression for R_Σ, we take into account that the current of the impedance radiator is

$$J_1(z) = j I_0 \frac{ke}{60 k_1 \cos k_1 L} \sin k_1(L - |z|),$$

i.e., for the same emf of the source the current amplitude of the impedance radiator is k_1/k times less than in the metal one. Accordingly, we write for its radiation resistance

$$R_{\Sigma 1} = 120 \left(\frac{k}{k_1}\right)^2 \int_0^{\frac{\pi}{2}} \frac{[\cos(k_1 L \cos\theta) - \cos k_1 L]^2}{\sin\theta} d\theta.$$

Introducing notations $\alpha = kL$, $\beta = k_1 L$ and the new variable $x = \beta \cos\theta$, we obtain, assuming that $\varepsilon_r = 1$ and taking into account that $dx = -\beta \sin\theta d\theta$:

$$R_{\Sigma 1} = 120 \left(\frac{\alpha}{\beta}\right)^2 \int_0^\beta \frac{(\cos x - \cos\beta)^2}{\beta \sin^2\theta} dx = 120 \frac{\alpha^2}{\beta} \int_0^\beta \frac{(\cos x - \cos\beta)^2}{\beta^2 - x^2} dx. \quad (1.57)$$

This expression differs from the similar expression for a metal radiator by replacing the magnitude α by β and by a factor $\left(\frac{\alpha}{\beta}\right)^2$. Accordingly, for the radiation resistance of the impedance radiator we write

$$R_{\Sigma 1} = 60 \left(\frac{\alpha}{\beta}\right)^2 \{(1 + 0.5 \cos 2\beta) \ln \frac{2}{0} + \cos 2\beta [Ci\ 4\beta - Ci(2\beta \cdot 0)] - \\ 2\cos^2\beta [Ci\ 2\beta - Ci\ (\beta \cdot 0)] + \sin 2\beta [Si\ 4\beta - Si\ 2\beta]\}. \quad (1.58)$$

The analogy of the expressions for the radiation resistances of metal and impedance antennas allows us to obtain specific results. The values R_Σ were previously given for metal radiators with a geometric arm length L, equal to $\frac{\lambda}{4}, \frac{\lambda}{4.8}, \frac{\lambda}{6}, \frac{\lambda}{8}, \frac{\lambda}{12}$ и $\frac{\lambda}{16}$, i.e., with an electrical arm length kL. With their help, it is possible to determine $R_{\Sigma 1}$ of the impedance radiators with electric arm length $k_1 L$, i.e., with geometric arm length $L_1 = kL/k_1$:

$$R_{\Sigma 1} = R_\Sigma (k/k_1)^2.$$

If, for example, $k_1/k = 2$, then for $L_1 = \dfrac{\lambda}{8}, \dfrac{\lambda}{9.6}, \dfrac{\lambda}{12}, \dfrac{\lambda}{16}, \dfrac{\lambda}{24}, \dfrac{\lambda}{32}$ the resistances are equal accordingly to 18.5, 10.9, 6.2, 3.2, 1.3 and 0.7.

Calculations show that for the same geometric length of the radiator arm the radiation resistance of the impedance radiator is greater than that of a metal one. For the same electric length, this resistance is significantly less (for the slowing factor 2 it is smaller approximately four times, i.e., inversely proportional to the square of the slowing)—in full accordance with the theory of impedance antennas.

For relatively short radiators (with arm length $L < 0.3\lambda$), the radiation resistance is proportional to the square of the geometric length of the radiator arm. And in accordance with described above simple calculating method one can write for the radiation resistance of an impedance radiator

$$R_{\Sigma 1} = 80(k/k_1)^2 \tan^2\left(\dfrac{k_1 L}{2}\right),$$

If $L \ll \lambda$, the effective length of the metal and impedance radiators is the same. In this case it is equal to L for a dipole and to $L/2$ for a monopole.

Next, we consider an antenna with a surface impedance varying along its length, or rather, its specific version with a stepwise change of the impedance along the radiator length. If the radiator is symmetric with respect to the central point, at which the emf is located, then for the current in the section n, counting from the free end of each radiator arm, we can write

$$J_n(z) = I_n \sin(k_n z_n + \varphi_n),$$

where I_n is the amplitude of the sinusoidal current in this section, k_n is the propagation constant, z_n is the coordinate measured from the section boundary that is farthest from the radiator center, φ_n is the current phase at this point. The continuity conditions of the current and charge at the boundaries of the sections allow us to express the amplitude and phase of the current in each section of length l_n in terms of the corresponding values of one of the currents and the parameters of the sections. If the number of sections is N, and the section N is adjacent to the generator, then the current of the section N is

$$J(z_N) = I_N \sin(k_N z_N + \varphi_N),$$

and

$$I_n = I_N \prod_{p=N}^{n+1} \dfrac{\sin \varphi_p}{\sin(k_{p-1} l_{p-1} + \varphi_{p-1})},$$

$$\varphi_n = \tan^{-1}\left\{\dfrac{k_n}{k_{n-1}} \tan\left[k_{n-1} l_{n-1} + \tan^{-1}\langle\dfrac{k_{n-1}}{k_{n-2}} \tan\left[k_{n-2} l_{n-2} + \cdots + \tan^{-1}\left(\dfrac{k_2}{k1}\right)\cdots\right]\rangle\right]\right\}.$$

Each section consists of two segments located on the upper and lower arms of the antenna symmetrically relative to its center. Comparing the segments fields with the arms fields of the radiator with a constant impedance, it is easy to be convinced of their analogy, from which it follows that the field of the section n of the step antenna is

$$E_{\theta n} = j60 \frac{I_n}{\sin\varphi_n} \frac{k_n}{k} \frac{\exp(-jkR_n)}{\varepsilon_r R_n \sin\theta}[\cos\langle(k_n l_n + \varphi_n)\cos\theta\rangle - \cos k_n l_n].$$

Here R_n is the distance from the spherical surface to the end of the section n closest to the antenna center. The total field of the antenna is equal to the sum of the fields of all sections:

$$E_\theta = \sum_{n=1}^{N} E_{\theta n},$$

The input current of the antenna is

$$J(0) = I_N \sin(k_N z_N + \varphi_N),$$

Substituting this value into the expression for R_Σ, we take into account that the wave impedance of the section at the input of the impedance radiator is k_N/k times greater than the wave impedance of the metal radiator, i.e., for the same emf the amplitude of input current in the impedance radiator is k_N/k times smaller than in the metal one. Accordingly, for radiation resistance we write

$$R_{\Sigma 2} = \frac{P}{J^2(0)} = 120\left(\frac{k}{k_N}\right)^2 \int_0^{\pi/2} (E_\theta/60)^2 d\theta.$$

Introducing the notation $\alpha = kL$, $\beta_n = k_n L$ and the new variables $x_n = \beta_n \cos\theta$, we find, assuming that $\varepsilon_r = 1$ and taking into account that $dx_n = -\beta n \sin\theta d\theta$:

$$R_{\Sigma 2} = 120(\alpha/\beta_N)^2 \sum_{n=1}^{N} \int_0^{\beta_n} \frac{(\cos x_n - \cos\beta_n)^2}{\beta_n \sin^2\theta} dx_n. \qquad (1.59)$$

Each member of the sum in this expression can be replaced by the corresponding expression for the effective length of this section.

Suppose, for example, that each arm of a radiator consists of two sections with the length $\lambda/8$. The first section adjacent to the free end of the antenna is made of metal, i.e., the wave propagation constant along it is equal to k, and the wave propagation constant along the second section, which is adjacent to the generator, is equal to $2k$. The results presented earlier allow in a first approximation to determine the radiation resistance of a given antenna. It is equal

$$R_{\Sigma 2(1)} = 12.6/4 + 18.5 = 21.6\Omega.$$

32 ANTENNAS: Rigorous Methods of Analysis and Synthesis

If to change places of the metal and impedance sections, the radiation resistance will change:

$$R_{\Sigma 2(2)} = 18.5 + 12.6 = 31.1\,\Omega.$$

A special case of a surface impedance is a resistive coating. Distributed resistive load leads to a slow change in the amplitude of the sinusoidal current, to its decrease with distance from the generator:

$$J(z) = J(0)\,e^{-\delta z}\sin k(L - |z|).$$

The decrement δ depends on the magnitude of losses in the coating material (see Section 2.6). In this case, for the field of a symmetric radiator at an arbitrary angle θ to the antenna, we can write, taking into account the difference of distances from the central and other points of the antenna to the observation point and generalizing the expression obtained for a metal antenna without losses:

$$E_\theta = j\frac{60 I_0 \sin\theta}{\exp(j\delta L)\sin kL}\,\frac{\exp(-jkR)}{\varepsilon_r R}\,\frac{\cos[kL(\cos\theta + j\delta L)] - \cos kL}{1-(\cos\theta + j\delta/k)^2} =$$

$$j60 J(0)\,\alpha^2\,\sin\theta\,\frac{(\cos x - \cos\alpha)}{X}.$$

Here the notation $\alpha = kL$, $X = \alpha^2 - x^2$ is used, it is taken into account that $\varepsilon_r = 1$, and a new variable $x = \alpha(\cos\theta + j\delta/k)$ is introduced. Since $x = -\alpha \sin\theta\,d\theta$, we obtain for radiation resistance

$$R_{\Sigma 3} = 120\,\alpha^3 \int_{j\delta L}^{\alpha + j\delta L}\frac{(\cos x - \cos\alpha)^2 \sin^2\theta}{(\alpha^2 - x^2)^2}\,dx,$$

where

$$\frac{\sin^2\theta}{(\alpha^2 - x^2)^2} = \frac{1 - \cos^2\theta}{X^2} = \frac{1}{X^2}\left[1 - \left(\frac{x}{\alpha} - j\frac{\delta}{k}\right)^2\right] = \left(1 + \frac{\delta^2}{k^2}\right)\frac{1}{X^2} + 2j\frac{\delta}{\alpha k}\frac{x}{X^2} - \frac{1}{\alpha^2}\frac{x^2}{X^2}.$$

Applying the integration by parts, we calculate the derivative $\dfrac{d}{dx}(\cos x - \cos\alpha)^2$ and integrate the second multiplier using known integrals

$$\int\frac{dx}{X^2} = \frac{x}{2\alpha^2 X} + \frac{1}{2\alpha^3}Y,\quad \int\frac{x\,dx}{X^2} = +\frac{1}{2X},\quad \int\frac{x^2 dx}{X^2} = +\frac{x}{2X} - \frac{Y}{2\alpha}$$

(here $Y = \dfrac{1}{2}\ln\dfrac{\alpha + x}{\alpha - x}$). Further we find the free member and the new integral, and thereafter we need to perform the integration by parts twice more. The last integral can be reduced to a set of integral sines and cosines:

$$60\Bigg|\Bigg[-\frac{\delta^2\alpha}{k^2}\{[\cos 2\alpha\, Si\, 2(\alpha+x) - \sin 2\alpha Ci\, 2(\alpha+x)] -$$
$$2\cos\alpha[\cos\alpha\, Si(\alpha+x) - \sin\alpha Ci(\alpha+x)]\}$$
$$+\frac{\delta^2\alpha}{k^2}\{[\cos 2\alpha\, Si\, 2(\alpha-x) - \sin 2\alpha Ci\, 2(\alpha-x)] -$$
$$2\cos\alpha[\cos\alpha\, Si(\alpha-x) - \sin\alpha Ci(\alpha-x)]\} -$$
$$\left(j\frac{\delta\alpha}{k}+1+\frac{\delta^2}{2k^2}\right)\{(\alpha+x)[\ln(\alpha+x)-1] + (\alpha+x)[\ln(\alpha-x)-1]\}\Bigg]\Bigg|_{j\delta L}^{\alpha+j\delta L}.$$

The sum of free members is

$$60(\cos x - \cos\alpha)^2\Bigg[\left(1+\frac{\delta^2}{k^2}\right)\frac{x\alpha}{\alpha^2-x^2}+2j\frac{\delta\alpha^2}{k(\alpha^2-x^2)} - \alpha\frac{x}{\alpha^2-x^2}+\left(2+\frac{\delta^2}{k^2}\right)Y\Bigg]+$$

$$120(\sin 2x - 2\cos\alpha\sin x)\Bigg[-\frac{\delta^2\alpha}{2k^2}\ln(\alpha^2-x^2)+2j\frac{\delta\alpha}{k(\alpha^2-x^2)}Y+$$

$$\left(1+\frac{\delta^2}{2k^2}\right)\{(\alpha+x)[\ln(\alpha+x)-1]-(\alpha-x)[\ln(\alpha-x)-1]\}\Bigg]dx\Big\{_{j\delta L}^{\alpha+j\delta L}.$$

An analysis of metal and impedance antennas with reactive surface impedance has shown that the rigorous method of calculating these antennas can be replaced by simple and easy-to-interpret method, leading to similar results. It is based on calculating the effective length of the radiator, calculated in accordance with the magnitude of the antenna field in the direction of the perpendicular to its axis. A similar approach can be applied to the antenna with the resistive coating. The expression for the field in an arbitrary direction has the form of a fraction whose numerator consists of two terms: $cos(kL\cos\theta + j\delta L)$ and $-\cos kL$. If in the first summand to put $\theta = \frac{\pi}{2}$, then the numerator is equal to

$$\cos(j\delta L) - \cos kL = \cos(j\delta L) - 1 + 1 - \cos kL = \cos(j\delta L) - 1 + 2\sin^2\frac{kL}{2}.$$

The denominator taking into account new multipliers has the form

$$\exp(j\delta L)\sin kL[1-(j\delta/k)^2] = 2\sin\frac{kL}{2}\cos\frac{kL}{2}(1+\delta^2/k^2)\exp(j\delta L).$$

For the effective length, we obtain

$$h_e = \frac{2}{k(1+\delta^2/k^2)}\left[\tan\frac{kL}{2} + \frac{\cos(j\delta L)-1}{\sin kL\exp(j\delta L)}\right]. \tag{1.60}$$

Let, for example, the antenna be made of a transparent film CEC005P with a surface resistance

$$R_{sq1} = 4.5 \ \Omega/\text{square},$$

i.e., the exponential decrement δ of its current is close to 7.1/m. If the frequency is equal to 5 GHz, then $\lambda = 0.06$ м, $k = 104.7$, $L = \delta/k = 0.0678$, $\cos j\delta L = \cosh 0.1065 = 1.0057$, $\exp(j\delta L) = j1.11$. Substitution of magnitudes shows that the effective length at an unaltered magnitude of k practically does not change.

The main disadvantage of antennas with a resistive coating is that the wave propagation constant in such an antenna is many times greater than in a metal one (in this example, k is 104.7). This is due to the small geometric length of the antenna arm. It is impossible to increase this length, since at a distance $L = 2/\lambda$ from the generator, the current drops to zero. It is known from the theory of impedance antennas that the radiation resistance of antennas with the same electrical length is inversely proportional to the square of the slowing. Therefore, when k increases, the magnitude of R_Σ sharply drops.

Calculation of the electrical characteristics of the antenna, based on the use of the Poynting vector method, is one of the three main methods for analyzing the properties of the antenna along with solving integral equations for the current along the antenna wire and the method of induced emf. Each of these methods is applicable to different types of radiators: metal, slotted, impedance with reactive and resistive (distributed and concentrated) loads (both constant and stepwise along the length of the antenna). The methods naturally complement each other and give similar results, confirming the rigour of the theoretical approach. Unfortunately, for various reasons, smaller attention was paid to using the Poynting vector method to analyze the characteristics of different antenna variants. The obtained results confirm the usefulness and effectiveness of its application.

The next step in the theory of linear radiators was made in the twentieth century. It is known as the induced emf method.

1.7 Oscillating power theorem

Before discussing the method of induced emf it is necessary to consider the "oscillating power" theorem.

The theorem and its proof for the first time were published in the book [6]. The book arose on the basis of lectures delivered by the author to undergraduate and graduate students and was devoted to electromagnetic waves of ultra-high frequencies. The reaction of many specialists to the theorem were extremely negative. In their view, the development of this theorem was caused by misunderstanding of the sense of reactive power, although this statement clearly conflicts with the well-known postulate, which these experts constantly repeated in articles and lectures. The postulate contends that the reactive power has no physical meaning.

Over the years since the first publication, the famous theorem has answered many questions. Let us start with the so-called symbolic method, i.e., with writing equations of the electromagnetic field in a complex form. Widely used electromagnetic fields, time-varying in accordance with sinusoidal law, are called harmonic or monochromatic fields. Both in the theory of alternating currents and in the field theory it is expedient in mathematical researches of harmonic processes, which are described by linear equations, to introduce complex magnitudes. The transition to these designations is performed in the following way. Complex magnitudes, designated as $E(\omega)$ and $H(\omega)$, correspond to magnitudes of electric $\vec{E}(t)$ and magnetic $\vec{H}(t)$ fields at a given point.

The relation between the physical quantities and their complex values is given by the following expressions:

$$\vec{E}(t) = Re[E(\omega)exp(j\omega t)], \vec{H}(t) = Re[H(\omega)exp(j\omega t)], \quad (1.61)$$

where ReA is a real part of a complex magnitude, located in the square brackets, ω is the circular frequency of the investigated process. Complex magnitudes $E(\omega)$ and $H(\omega)$, related with the instantaneous values by the relations of the type (1.53), correspond to two scalar physical magnitudes $E(t) = E \cos\omega t$ and $H(t) = H \cos\omega t$. If $E(\omega)$ and $H(\omega)$ are complex magnitudes:

$$E(\omega) = Eexp(j\alpha), H(\omega) = Hexp(j\beta), \quad (1.62)$$

where E and H are the amplitudes, and α and β are the arguments of the complex magnitudes, then

$$E(t) = E \cos(\omega t + \alpha), H(t) = H \cos(\omega t + \beta). \quad (1.63)$$

Thus, the amplitudes of the complex magnitudes are the amplitudes of the corresponding instantaneous values of the physical quantities, and the arguments of the complex magnitudes determine the phases of the instantaneous values of these quantities. Similarly, complex magnitudes are introduced for all physical magnitudes, in the Maxwell equations. Formal coupling of complex equations with the initial equations is simple: in order to obtain complex equations, one must replace the differentiation operator $\partial/\partial t$ by the operator of multiplication $j\omega$.

As is well known, energy quantities are determined by products (or squares) of instantaneous values of fields and currents. Calculation of an average for the oscillation period T product magnitude

$$a(t)b(t) = \frac{1}{2} AB[\cos(\alpha - \beta) + \cos(2\omega t + \alpha + \beta)] \quad (1.64)$$

gives

$$\overline{a(t)b(t)} = \frac{1}{T}\int_0^T a(t)b(t)dt = \frac{1}{2}AB\cos(\alpha - \beta) = \frac{1}{2}\mathrm{Re}\,a(\omega)b^*(\omega). \quad (1.65)$$

Similar expressions are true also for vector magnitudes. These expressions permit the computation of the average value (constant part) of the energy quantity in accordance with the known complex amplitudes. A similar method can be used for definition of the average value of the variable energy fraction (of an oscillating energy). Indeed, according to (1.64)

$$a(t)b(t) = \overline{a(t)b(t)} + \tilde{\Theta},$$

where it is natural to assume that the time-dependent second term

$$\tilde{\Theta} = \frac{1}{2}AB\cos(2\omega t + \alpha + \beta) \quad (1.66)$$

is the oscillating fraction of the product $a(t)b(t)$. This part oscillates in time with a frequency 2ω, and its average value is zero. One can rewrite the expression (1.66) as

$$\tilde{\Theta} = \frac{1}{2}\mathrm{Re}[a(\omega)b(\omega)\exp(2j\omega t)]. \quad (1.67)$$

It is seen that half the product of complex amplitudes is the complex amplitude of the oscillating fraction of the product $a(t)b(t)$.

As is well known, the energy conservation law for the electromagnetic field is given by expression

$$\frac{dW}{dt} + P + \Sigma = 0. \quad (1.68)$$

Here W is electromagnetic energy contained in a volume V, P is an outgoing power (which flows out of the volume through its bounding surface), and Σ is the radiation power. Passing from the differential formulation to the integral formulation and using the appropriate complex magnitudes, one can write the oscillating power theorem in the form

$$-\tilde{\Sigma} = \tilde{P} + 2j\omega\tilde{W}. \quad (1.69)$$

In deriving this expression, each term in (1.68) is considered as the sum of the active magnitude (average for the period of oscillation) and the oscillating (variable) fraction. In particular, for an instantaneous value of the power flux one can write according to (1.65)

$$p(t) = \overline{P} + \tilde{P}. \quad (1.70)$$

where

$$\overline{P} = \frac{1}{2}\text{Re}(EH^*), \quad \tilde{P} = \frac{1}{2}\text{Re}[EH\exp(2j\omega t)].$$

From here the physical meaning of magnitudes EH^* and EH is clear. The first magnitude is the complex amplitude of the active part of the power flow, equal to its average value. The second magnitude is the complex amplitude of the oscillating part of the power flow. In accordance with the law of conservation of energy, if the source of radiation is located inside of a closed surface, then the active (average for the period of oscillation) power, supplied by the source, is equal to the active power, passing through a closed surface. It is natural to assume that this equality of powers is true for any time, i.e., the oscillating part of the power, supplied by the source, is equal to the oscillating part of the power, passing through a closed surface. Equality (1.70) shows that the instantaneous power consists of the active power and the oscillating power. Reactive power $\frac{1}{2}\text{Im}(EH^*)$ does not have a physical meaning.

As will be shown in the next Section, the oscillating power theorem has significantly changed the understanding of the induced emf method. Another example is the losses of asymmetric vertical antenna in earth and grounding. For a long time, the calculation of losses in grounding was executed according to the procedure of Brown [7]. This method proceeds from the idea of high ground conductivity, owing to which a magnetic field at the ground surface is virtually identical to a magnetic field of an antenna, located above a perfectly conducting ground, and its strength is equal to the density of a surface current in the ground. The calculation of the resistance loss is reduced to an integral, and the result of its calculation unlimitedly grows, and, consequently, the resistance also unlimitedly grows.

New method of calculating an additional antenna resistance caused by a non-ideal conductivity of the ground (as authors called the resistance of losses in the ground), was proposed and published in 1954 [8]. The magnitude of the loss resistance turned out to be finite. But the calculation method was an intricate procedure, using the direct and inverse Fourier-Bessel transformation. A similar result was obtained, using the oscillating power theorem [9]. It was shown that the cause of the infinite growth of resistance loss in the Brown procedure was the use of the concept of complex power. Use of oscillating power not only allowed simply obtaining a clear result defined by using an intricate procedure, but enabled the discovery of the mistake. A detailed comparative description of the different calculation methods and the obtained results is given in [10].

1.8 Method of induced emf

The induced emf method was proposed in 1922 by Rojansky and Brillouin simultaneously. Klazkin was the first who used it for calculating radiator characteristics. Later on, Pistolkors, Tatarinov, Carter, Brown et al. have contributed to its development. A reference list is in the book [11], which is dedicated to checking and generalizing the results available in the literature, consisting of 96 titles. The method of induced emf allows determining both the active and reactive components of the antenna input impedance. Since the active component can be calculated with a similar accuracy by a simpler method of Poynting (see Section 1.5), the method of induced emf for practical purposes, as emphasized in [12], is actually only one of the methods for determining an input reactance of an antenna.

The oscillating power theorem has significantly changed the understanding of the induced emf method, which for a long time has been the only way to calculate an antenna input reactance. The method of induced emf is formulated as follows. A cylindrical radiator of height $2L$ and radius a is placed inside a closed surface. A power, created by the emf source (a generator), is equal to a complex power, passing through this surface. Assume that the closed surface is a circular cylinder of a height $2H$ and radius b, along the axis of which the symmetrical radiator is located (see Fig. 1.8). A density of power flux, which deserts a volume, bounded by a closed surface, is determined by the Poynting vector, or precisely by its projections onto the normal to the surface of sections which are the side surface and the end faces of the cylinder. These projections have the following form

$$P_\rho = -0.5 E_z H_\varphi^*, \; P_z = 0.5 E_\rho H_\varphi^*. \tag{1.71}$$

Let the cylinder surface coincide with a surface of the radiator, i.e., $H = L$, $b = a$. Then, if the radiator radius is small, power fluxes, passing through the upper and lower covers of the cylinder, will also be small. Therefore, power fluxes passing through a closed surface is determined by integrating only over the side surface of the cylinder

$$P_1 = \int_{-L}^{L} \int_0^{2\pi} P_\rho a d\varphi dz.$$

Here P_ρ is determined from (1.70) and does not depend on the coordinate φ, because the field components do not depend on it. Taking into account that $H_\varphi^* = J^*(z)/(2\pi a)$, we obtain

$$P_1 = -0.5 \int_{-L}^{L} E_z J^*(z) dz. \tag{1.72}$$

If a current $J(z)$ is excited by a single generator, located in the middle of the radiator, then the power, created by it, is

Straight Metal Radiator 39

$$P_2 = 0.5|J(0)|^2 Z_A, \qquad (1.73)$$

where Z_A is the antenna input impedance. Equating the complex power, created by the source of emf, to the complex power, passing through the closed surface, we obtain

$$Z_{AI} = -\frac{1}{|J(0)|^2}\int_{-L}^{L} E_z J^*(z) dz. \qquad (1.74)$$

The expression (1.74) reveals the essence of the induced emf method. The other variant of deducing this expression is described in [13].

If we equate two analogous oscillating powers instead of complex powers: the power, passing through the closed surface is,

$$P_{K1} = -0.5\int_{-L}^{L} E_z J(z) dz \qquad (1.75)$$

and the power, created by the generator is,

$$P_{K2} = 0.5 e J(0) = 0.5 J^2(0) Z_A, \qquad (1.76)$$

where e is the emf of the generator, and we obtain

$$Z_{AII} = -\frac{1}{J^2(0)}\int_{-L}^{L} E_z J(z) dz. \qquad (1.77)$$

It follows from the obtained results, that the expression for oscillating power differs from the expression for reactive power by the absence of an asterisk denoting complex conjugate values. Therefore, the expression (1.77) for the input impedance of the radiator, derived from the equality of oscillating powers, differs from the expression (1.74), based on the equality of complex powers, by the absence of an asterisk.

After the appearance of an equality (1.77), equality (1.74) has been called the first formulation of the induced emf method. Expression (1.77), was called the second formulation of the induced emf method. This expression was first obtained on the basis of the theorem of reciprocity [14–16]. This theorem holds not only for two separate antennas but also for two points on the same antenna. Using that circumstance and applying the theorem to one radiator, one can obtain the expression (1.77). As it is shown here, if we are to use the concept of oscillating power, then this expression is easily deduced from the energy relations. But despite the fact that the expression (1.77) was obtained by means of the oscillating power theorem many years ago [17], most experts kept arguing that it is derived in accordance with the reciprocity theorem in contrast to the expression (1.74), which is obtained by proceeding from an equality of powers.

As can be seen from the above, the difference between the first and second formulations is caused by the fact that the first one is based on the equality of complex powers, consisting of the real and reactive component, and the second one—on the equality of the total (summation) powers, consisting of the active and oscillating components. Even here an advantage of the second formulation is obvious, since a reactive power unlike the oscillating power does not have physical meaning.

Indeed, the fact of the presence of oscillating powers is objectively substantiated in the book [6]. The oscillating power is a result of the existence of currents and voltages, the instantaneous values of which vary in time according to a sinusoidal law. By itself, the symbolic method that replaced the sinusoidal dependence of the current on time $i(t) = I \sin \omega t$ by the exponent $i(t) = I \exp(-j\omega t)$ was convenient in that it facilitated the calculations. At the same time, there was a temptation to use the second summand of the exponent as a voltage, and this unreasonable temptation turned into a constantly refuted and simultaneously used reactive power.

Analysis shows that the second formulation is stationary. This fact is its undoubted advantage. To verify this, one must show that, if the antenna current is changed by a value of the first order of smallness, then the input impedance will change by a magnitude of the second order. The input impedance Z_{AII} obtained from (1.77) does not change in the first approximation for any distribution of the trial current, which differs from the true current $J^0(z)$ by a small magnitude $\delta J(z)$. This means that if at $J(z) = J^0(z)$ a self-impedance of the radiator is equal to Z_{AII}, then for $J(z) = J^0(z) + \delta J(z)$ the self-impedance is also equal to Z_{AII}. The corresponding proof was given by J.E. Storer and is described in [18]. As shown in [12], the stationary property of the second formulation is due to the fact that the integral in the expression (1.77) is a rough functional of the current function, although an integrand is no rough functional of this function.

Let a straight perfectly conducting filament of a finite small radius a, whose axis coincides with z-axis, is located in a lossless medium and is used as a model of a vertical symmetrical dipole with arm length L (see Fig. 1.1a). The current along it is defined by the expression (1.26), i.e., a tangential component of the electric field of the filament along a radiator surface is determined by the expression (1.30). In this case both formulations of the induced emf method give the same result:

$$R_A = \frac{30}{\sin^2\alpha}[2(C + \ln\alpha - Ci2\alpha) + \sin 2\alpha(Si4\alpha - 2Si2\alpha) + \cos 2\alpha(C + \ln\alpha + Ci4\alpha - 2Ci2\alpha)],$$

$$X_A = \frac{30}{\sin^2\alpha}[\sin 2\alpha(C + \ln\alpha + Ci4\alpha - 2Ci2\alpha - 2\ln L/a) - \cos 2\alpha(Si4\alpha - 2Si2\alpha) + 2Si2\alpha].$$

(1.78)

Here $Six = \int_0^x \frac{\sin u}{u} du$ is a sine integral, $Cix = \int_\infty^x \frac{\cos u}{u} du$ is a cosine integral, $\alpha = kL$, $C = 0.5772...$ is the Euler's constant.

As can be seen from the expression for antenna reactance, X_A consists of terms of different order of smallness. The large member inside square brackets does not depend on the frequency. This member gives the summand

$$X_{A0} = -30\frac{\sin 2\alpha}{\sin^2 \alpha} \cdot 2(lnL/a - C/2) \approx -120 ln\frac{L}{1.335a} \cot \alpha. \quad (1.79)$$

The magnitude $\chi \approx 1/\Omega$ is called a small parameter of the thin antennas' theory (Ω is a parameter, used by Hallen). Let the parameter χ is equal to $\chi = 0.5/ln(L/a)$. Introducing the notation $W = 60/\chi$, we obtain an expression for the input reactance of an equivalent long line, open at the end:

$$X_{A0} = -W \cot \alpha.$$

The results of calculations using formulas (1.74) and (1.77) have shown that the difference between them occurs in the presence of losses in the radiator. Indeed, the current along an ideally conducting metal conductor is determined by the expression (1.26) and is a sine wave. Comparison of expressions (1.74) and (1.77) shows that in the case of a sinusoidal distribution, they are almost the same: in the numerator and denominator of one expression the product of complex conjugate magnitudes is present, and in the numerator and denominator of another expression the product of their real magnitudes is present. The difference was first observed while calculating the losses in antenna conductors (e.g., while calculating losses from the skin effect). In order to determine these losses, one must add a pure real propagation constant of a small imaginary magnitude. In this case the calculation in accordance with the second formulation gave a positive value for losses and the calculation in accordance with the first formulation gave a negative value for them. A similar situation arose while calculating the losses in the magneto-dielectric sheath of impedance antennas.

Calculation of a resistance loss to the ground also confirmed the correctness of results received by means of the oscillating power theorem (see Section 1.7). Thus, due to the presence of losses in the medium or in the antenna, application of the concept of reactive power is an obvious mistake. The correctness of the second formulation, based on the concept of oscillating power, became an accepted fact.

The coincidence of the results of calculations based on the use of different formulations, in the absence of losses in the antennas and the environment, shows that in these cases the results of the calculations performed in accordance with expression (1.74) are correct. The possibility of their use is beyond question. In the presence of losses, the calculation practically reduces to calculating the input impedances with the real propagation constant k (its result does not depend on the choice of formulation) and to replacing the real value of k by a complex quantity with the subsequent highlighting of an additional component of the input resistance.

The second formulation of the induced emf method was studied, when integral equations of Hallen [19] and Leontovich [20] for the current along a radiator axis were already written and solved. Solutions have been given in the form of expansions in the form of power series'. If the formulas, presented in [21] and [20] are used, one can show that the solutions of both equations are the same [22]. In this case, the coincidence of the results is not only numerical. The results were obtained in an explicit form (in the form of identical tabulated functions).

As already mentioned, solutions, obtained by induced emf method for the perfectly conducting filament, using different formulations, gave identical results. They coincide with the solutions of integral equations for different lengths of the radiator, if this length is not close to the parallel resonance when $J(0) \approx 0$. In the latter case, the input impedance, calculated by the induced emf method, becomes infinitely large, and the integral equations give finite results.

Summarizing, one can say that both formulations of the induced emf method are based on the same two theses. The first thesis assumes the sinusoidal character of the current distribution along the radiator. The second thesis signifies the equality of the power source and the power passing through the closed surface.

Both formulations are useful only in the case when the current distribution $J(z)$ along a radiator is known. The selection of the law of current distribution may be based only on a solution of integral equations for the current, i.e., on a rigorous solution to the problem. Physical basis for the selection of another distribution law does not exist. Hence there is no sense in speaking about the accuracy of the Levin, B.M. 2013 induced emf method, by artificially excluding the error caused by the inexact current definition. The accuracy of this method is the mutual accuracy of (1.26) and (1.74) or (1.77). The calculation experience has shown that (1.26) gives a reasonably acceptable approximation, until the value of α ceases to approach $\pi/2$.

As to the second thesis, used for derivation of the first formulation, its inapplicability is obvious, since the reactive power has no physical sense, and the input reactance of antenna is determined as a result of equating two quantities not having physical sense. It is difficult to justify the equality of two such quantities. The use of the second formulation significantly improves the results.

1.9 Generalized method of induced emf

A rigorous analysis of antennas' characteristics was executed only for a small number of simple variants of radiators. As a rule, in the problems under consideration, the characteristics of small radiators located in free space are analyzed. This is explained by the complexity of the problem. In this connection, in solving integral equations for the current, numerical methods became a frequent practice. These methods allow us to find the characteristics of complex antennas with large dimensions in comparison to a wavelength and takes into account the influence of neighboring antennas and metal bodies.

Numerical methods permit to reduce an integral equation to a set of algebraic equations with the help of Moments method. In the general case the integral equation for the current in a wire antenna has the form

$$\int_{(l)} J(\varsigma) K(z,\varsigma) d\varsigma = F(z), \qquad (1.80)$$

where $J(\varsigma)$ is the sought function (the current distribution along a wire), $K(z, \varsigma)$ is the kernel of the equation, which depends on coordinate z of the observation point and on coordinate ς of the integration point. $F(z)$ is a known function, it is determined by extraneous sources of a field. The terms proportional to the current may enter into this function, for example in the case of antenna with loads. Here this is of no great importance. The integral is taken over the entire wire length. It is easy to be convinced that the equations considered earlier are particular cases of the equation (1.80).

Let's present the unknown current $J(\varsigma)$ as a sum of linearly independent functions $f_n(\varsigma)$, which are named by the basis functions:

$$J(\varsigma) = \sum_{n=1}^{N} I_n f_n(\varsigma). \qquad (1.81)$$

Here I_n are unknown coefficients, which in the general case are complex. Substituting (1.81) into (1.80), we obtain:

$$\sum_{n=1}^{N} I_n \int_{(l)} f_n(\varsigma) K(z,\varsigma) d\varsigma = F(z). \qquad (1.82)$$

Often the second system of linearly independent functions $\varphi_p(z)$ is introduced. They are named by the weight functions. If we multiply both parts of equation (1.82) by $\varphi_p(z)$ and integrate it over the entire wire length and then repeat the operation at different values of p, we shall obtain the set of equations:

$$\sum_{n=1}^{N} I_n \int_{(l)} \varphi_p(z) \int_{(l)} f_n(\varsigma) K(z,\varsigma) d\varsigma dz = \int_{(l)} \varphi_p(z) F(z) dz, \; p=1,2\ldots N. \quad (1.83)$$

Obviously, number of equations N (1.83) must coincide with the number of unknown magnitudes N. The result of integrating each expression is its moment. The name of this method is a consequence of calculating the moments of functions.

If the system of weight functions coincides with the system of basis functions, such a variant of the Moments method is known as Galerkin's method. In this case

$$\sum_{n=1}^{N} I_n \int_{(l)} f_p(z) \int_{(l)} f_n(\varsigma) K(z,\varsigma) d\varsigma dz = \int_{(l)} f_p(z) F(z) dz, \ p = 1,2...N. \quad (1.84)$$

One can rewrite this set of equations as

$$\sum_{n=1}^{N} I_n Z_{np} = U_p, \ p = 1,2...N, \quad (1.85)$$

where

$$Z_{np} = \int_{(l)} f_p(z) \int_{(l)} f_n(\varsigma) K(z,\varsigma) d\varsigma dz, \ U_p = \int_{(l)} f_p(z) F(z) dz.$$

Equation (1.85) is also true for set of equations (1.83), if one replaces $f_p(z)$ with $\varphi_p(z)$ in formulas for Z_{np} and U_p.

Expression (1.85) is the set of linearly independent algebraic equations with N unknown I_n, having the dimensions of current. Coefficients Z_{np} and U_p have the dimensions of impedance and voltage; they can be calculated, e.g., by means of numerical integration. Accordingly, one can interpret the expression (1.85) as Kirchhoff's equation for the contour p with current I_p and emf U_p. The contour p enters into the system of N coupled contours. Here Z_{pp} is the self-impedance of the contour element, and Z_{np} is the mutual impedance of the contours n and p.

The set of equations (1.85) can be solved on the computer with the help of standard software. We write down this set in a matrix form:

$$[I][Z] = [U], \quad (1.86)$$

where $[Z]$ is the matrix of impedances, $[I]$ and $[U]$ are the current and voltage vectors, then the solution can be obtained by standard matrix inversion:

$$[I] = [Z]^{-1} [U]. \quad (1.87)$$

Substitution of obtained magnitudes I_n into (1.81) allows us to find a current distribution $J(\varsigma)$, and subsequently, all electrical characteristics of the radiator.

In practice the calculation of matrix elements Z_{np} may prove to be extremely difficult, since it is connected with double numerical integration. This difficulty often depends on the equation kernel. To alleviate these difficulties, δ-functions can be used in the capacity of weight functions: $\varphi_p(z) = \delta(z - z_p)$. Then, the double integral for the calculation of Z_{np} becomes a simple integral, the calculation of U_p does not requires integration, and the expression (1.83) takes the form

$$\sum_{n=1}^{N} I_n \int_{(l)} f_n(\varsigma) K(z_p, \varsigma) d\varsigma = F(z_p), \ p = 1, 2 \ldots N.$$

One can obtain this equality directly from (1.80) and (1.81), by equating the left and right parts of the equation (1.80) at isolated points. The number N of these points corresponds to that of obtained equations. For this reason, the given variant of the Moments method is known as the point-matching technique or the collocation method (see, e.g., [23]).

The collocation method ensures an exact equality of the left and right parts of the equation (1.80), at N points at least. In the intervals between the points, the difference between the two parts of the equation may increase sharply. When using the Moments method with weight functions of another type, the equality may disappear at all points of the interval where z is changing. But equating both moments of function (integration with some weight) minimizes the difference between the left and right parts of the whole interval where z is changing. This property in the final analysis is almost always more important than the exact equality at isolated points. Therefore, Galerkin's method allows providing, as a rule, an essentially more accurate solution than the collocation method. Yet, sometimes the collocation method is useful, especially to increase the number of points and to place them correctly.

The choice of basis functions is of great importance for using the Moments method, since the successful selection of functions permits to decrease the volume and the time of calculation under given accuracy or increases the accuracy under the same calculation time. With that end in focus, as a rule, the basis functions must correspond to the physical sense of the problem, i.e., must coincide, in the first approximation, with the actual distribution of the current along a radiator or its elements.

Basis functions are usually subdivided into two types: entire-domain functions, which are distinguished from zero along the entire radiator length, and functions of sub-domains, which are distinguished from zero along radiator segments. Members of Fourier series and polynomials of Tchebyscheff or Legendre can be used in the capacity of basis functions of the first type. Their field of application is limited mainly by solitary radiators of a simple shape. Basis functions of sub-domains are employed typically for an antenna of a complex shape. In particular, such an approach is expedient, if the antenna consists of arbitrarily situated segments of straight wires partially jointed with each other. A straight radiator may also consist of physically isolated segments of wire, if concentrated loads are located in the radiator at given distances from each other. Piecewise constant (impulse) functions (Fig. 1.10*a*), piecewise linear functions (Fig. 1.10*b*), and piecewise parabolic functions (Fig. 1.10*c*) are shown at Fig. 1.10 for illustration of sub-domains. These basis functions are special cases of a wider class of basis functions—

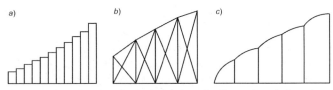

Fig. 1.10. The curve line as the sum of pulsed (*a*), piecewise linear (*b*) and piecewise parabolic (*c*) basis functions of sub-domains.

of polynomials. A simplest variant of approximation with the help of a polynomial is proposed in [24]:

$$J(\varsigma) = \sum_{m=0}^{M_n} I_{nm} (\varsigma - \varsigma_n)^m, \quad \sigma_n < \varsigma < \varsigma_{n+1}.$$

Here, M_n the selected degree of the polynomial on the segment n, and I_{nm} are unknown coefficients. Comparing this expression with (1.81), we obtain:

$$I_n = I_{n0}, \quad f_n(\varsigma) = \sum_{m=0}^{M_n} \frac{I_{nm}}{I_{n0}} (\varsigma - \varsigma_n)^m \text{ at } \sigma_n < \varsigma < \varsigma_{n+1} \text{ and 0 for other values}$$

of ς.

Similarly, members of Fourier series can be used as basis functions of sub-domains. A particular case of such functions are piecewise sinusoidal functions:

$$J(\varsigma) = \frac{I_p \sin k(\varsigma_{p+1} - \varsigma) + I_{p+1} \sin k(\varsigma - \varsigma_p)}{\sin k(\varsigma_{p+1} - \varsigma_p)}, \quad \varsigma_{p-1} \leq \varsigma \leq \varsigma_{p+1}. \quad (1.88)$$

Comparing this expression with (1.81) and choosing a simple variant, one can write:

$$f_p(\varsigma) = \begin{cases} \sin k(\varsigma - \varsigma_{p-1}) / \sin k(\varsigma_p - \varsigma_{p-1}), \varsigma_{p-1} \leq \varsigma \leq \varsigma_p, \\ \sin k(\varsigma_{p+1} - \varsigma_p) / \sin k(\varsigma_{p+1} - \varsigma_p), \varsigma_p \leq \varsigma \leq \varsigma_{p+1}, \\ 0 \text{ elsewhere.} \end{cases} \quad (1.89)$$

Application of expression (1.81) with the basis functions in the form (1.89) is equivalent to dividing the wire into short dipoles with overlapped arms and with centers at points ς_p, wherein the current I_p of the dipole *p* is maximal. Such a look at the resulting structure allows us to say that expressions (1.81) and (1.89) generalize the expression (1.26). When lengths of short dipoles are decreased, piecewise sinusoidal basis functions are converted to piecewise linear functions. Figure 1.10*b* permits the visualization of how the basis functions of sub-domains form the aggregate curved line corresponding to distribution of the current along an antenna.

In [25] the use of functions in the form (1.89) was proposed as the basis and weight functions. Such a variant of the Moments method has two advantages. First, a rapid convergence of results is ensured, i.e., dimension of the matrix [Z] is small in comparison with dimensions of the matrixes, when using different basis and weight functions. This means that application of piecewise sinusoidal functions as the basis and weight functions corresponds to the physical content of the problem. Second, closed expressions containing sine integrals and cosine integrals can be used to calculate many matrix elements.

The advantages of this variant of the Moments method are especially significant when calculating complex antennas. The integral equation for the current in the metal antenna is based on the fulfillment of the boundary condition on its surface. In accordance with (1.25), the boundary condition for a metal antenna of complicate shape has the form

$$\sum_{n=1}^{N}\int_{S_{n1}}^{S_{n2}} J_n(S_n)\left[k^2 G \vec{e}_{s_n}\vec{e}_p - \frac{\partial^2 G}{\partial p \partial s_n}\right]ds_n = -j\omega\varepsilon K_p(p). \quad (1.90)$$

Let's substitute the current distribution (1.81) with weight functions (1.89) into the equation (1.90) for a complicated wire radiator and multiply both parts of the equation to the weight function $f_p(z)$ in accordance with Galerkin's method. On integrating the result along the entire wire length, we obtain a set of N equations of type (1.85) with N unknown magnitudes I_n, on repeating this operation for different values of p. The coefficients in these equations are equal to

$$Z_{ps} = -\frac{1}{j\omega\varepsilon}\int_{Z_{S-1}}^{Z_{S+1}} f_s(z)\int_{\varsigma_{p-1}}^{\varsigma_{p+1}} f_p(\varsigma)\left[k^2 G_3 \vec{e}_\varsigma \vec{e}_z - \frac{\partial^2 G_3}{\partial z \partial \varsigma}\right]d\varsigma dz, \quad U_S = \int_{Z_{S-1}}^{Z_{S+1}} f_s(z)K_S(z)dz. \quad (1.91)$$

Comparing (1.91) with expression (1.77), where the magnitude E_{ps} is substituted from (1.24), it is easy to be convinced that the expression for Z_{ps} corresponds to the mutual impedance between dipoles p and n, calculated by the induced emf method. As seen from (1.91), the dipoles are considered as isolated, i.e., the current of each dipole is distributed in accordance with the sinusoidal law. Substituting extraneous field $K_s(z)$ into (1.91), we see that magnitude U_s is the emf of the generator connected at the center of the dipole s. Therefore, the set of equations (1.85) with coefficients Z_{np} and U_p is the set of Kirchhoff equations for the dipoles constituting the wire antenna.

Thus, the variant of Galerkin's method that was proposed by Richmond for calculating the current distribution in a complicated antenna, is equivalent to dividing of the radiator into isolated dipoles. Their own and mutual impedances are calculated by the induced emf method. For this reason, as Richmond pointed out, the method can be named as the generalized induced emf method. For calculating the wire antenna with concentrated loads, it is expedient to divide it into short dipoles and to place each load in the center of a dipole. It is useful to generalize the set of equations (1.85) and to write it in the form:

$$\sum_{p=1}^{P} I_p Z_{ps} = U_s - I_s Z_s, s = 1, 2 \ldots N, \qquad (1.92)$$

or in the matrix form

$$[I][Z] = [U - IZ]. \qquad (1.93)$$

The accuracy of the induced emf method for calculating a dipole decreases when the dipole length increases. The accuracy of calculation is acceptable at dipole arm length $L \leq 0.4\lambda$. The advantage of the generalized induced emf method consists in the fact that one can divide the long dipole onto several short dipoles, e.g., with the arm length no greater than 0.2λ. That allows ensuring the required exactness.

Calculation of the coefficients Z_{ps} requires double numerical integration. But the problem is simplified essentially, if the method described in [11] is used for calculating the mutual impedance of two arbitrarily situated dipoles. Here, the double integrals are reduced to simple integrals, and each integral is the sum of a series, consisting of members with successively changing signs. The components of series are calculated by means of recurrence formulas, almost as quickly as the components of the power series. This manner of calculating Z_{ps} is useful in general-purpose software.

In substance the generalized induced emf method makes it possible to ensure the accuracy of calculating complex antennas close to the accuracy of calculating the simplest radiators. Therefore, it is the basis for all calculation programs used in modern computers. This allows us to complete the story of a rigorous theory of thin linear radiators.

1.10 Multi-wire antennas and cables

Section 1.5 briefly describes a method of analyzing a system of horizontal wires parallel to the earth's surface. It was used in particular to calculate the complicated horizontal load of a vertical antenna. The wire load was made in the form of a meander and allowed us to substantially increase the effective height of the vertical antenna, on top of which it is installed, compared to the usual horizontal load consisting of wires of the same length connected in parallel with each other [26]. The method of calculating the load characteristics, including an input impedance, current in the wires and voltage at the nodes, depends on the number of wires and their geometrical dimensions, is described in [27]. In the same article a comparative results of tests are given.

The use of antenna with meandering load in a medium frequency range allows reduction of the size of the plot, which the antenna occupies, grounding area and the price of construction. The first specimen of this antenna was put into operation in 1983 in the town Pavlovo, on the radio center of the Baltic Shipping Company. In 1987, the radio center of a commercial sea port in Ventspils (Latvia) was equipped with three such antennas. The effectiveness of the new antennas was tested by measuring the magnitude of the fields, produced by them in comparison with the inverted L antennas. Both types were operating with the same transmitter. Tests showed that the field created by the antenna with a meandering load is greater by 1.5 to 1.6 times. This is equivalent to increasing the transmitter power by 2.3 to 2.6 times.

With time, the described method became applicable to calculations of the current distribution along the parallel wires located perpendicular to the ground surface, taking into account the mirror image from this conductive surface. This procedure has permitted the determination of the electrical characteristics of vertical multi-wire structures. One of these structures is the multi-folded radiator (Fig. 1.11), which consists of several folded radiators connected in series with each other. Parallel wires of multi-folded radiators are connected in pairs on the top and the bottom so that they form a system of elongated loops (of long lines) coupled and connected in series. This creates more opportunities for an analysis of new circuits and connections (series and parallel).

If the transverse dimensions of the multi-folded radiator are small in comparison with its height L and the wave length λ, then, as is shown in the

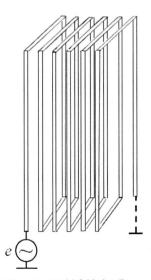

Fig. 1.11. Multi-folded radiator.

article [28], devoted to a research of electromagnetic oscillations in systems of parallel thin conductors, the currents in each wire of the system can be divided into in-phase and anti-phase components, and the entire system may be reduced to an aggregate of linear radiator and non-radiating long lines.

The other option of a parallel multi-conductor vertical structure is a multi-radiator antenna [27]. One possible variant of the multi-radiator antenna is given in Fig. 1.12a. The equivalent long line for this antenna is presented in Fig. 1.12b. The antenna consists of a central radiator 1 with a complex impedance load Z_0 and side radiators 2 located around the radiator 1 along the generatrixes of a cylinder and connected to the base of the radiator 1. The impedance load is a distinguishing feature of this antenna. It is included to provide a mode of operation close to the running wave mode on the section between the generator and the load. This allows us to increase the level of matching and to obtain the desired radiation pattern. An impedance load is fabricated in the form of a special power absorber. The multi-radiator antenna without a load has high characteristics in the vicinity of series resonances of individual radiators, but in order to obtain such characteristics in a continuous range, its diameter must be significantly increased in order to reduce the spatial coupling between these radiators. Combining the two principles of antenna construction, namely the inclusion of a complex load and the use of a multi-radiator structure, makes it possible to create an antenna of acceptable dimensions.

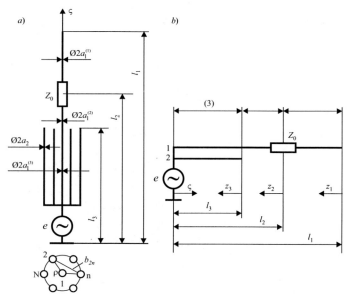

Fig. 1.12. Multi-radiator antenna with a complex impedance load (*a*) and an equivalent asymmetrical line (*b*).

The theory of coupled lines, described in Section 1.5, permitted to show, that the mutual coupling between lines in multi-conductor cables results in the emergence of the electromagnetic interference (crosstalk) in communication channels, and that an asymmetry of excitation and loads causes the emergence of the common mode currents in the lines.

In order to determine the signal magnitude at the end of a multi-conductor cable located inside a metal shield it is necessary to define the electrical characteristics of the lines. The magnitude of voltage across a load placed at the end of an adjacent line can be used as a measure of such a distortion [29]. The rigorous method of calculating the mutual coupling between lines enables the development of a simple and effective procedure of preventing interference.

Electromagnetic interference in communication channels (imbalance of a cable) is caused not only by the cable asymmetry, but also by asymmetry of excitation and load, which provokes the emergence of the in-phase currents in cables (common mode currents). The rigorous calculation method of the electrical characteristics of multi-conductor cables enables to determine these currents. Compensation of the in-phase currents allows us to decrease the EM radiation and susceptibility to the external fields. We employ a rigorous method—first for calculating characteristics of a two-wire line located inside a metal screen and then for mutual coupling between lines. It is considered that the lines are uniform, the electromagnetic waves are transverse (TEM), and the cable diameter is small in comparison with the wavelength.

A single pair of wires (twisted pair) inside a metal cylinder can be modeled as two wires of radius a, situated at a distance b from each other inside a metal cylinder with radius R and length L (Fig. 1.13). If wire radius a and distance b in multi-conductor cables are small in comparison with the cylinder radius R, i.e., $a, b \ll R$, then the wave impedance of the line is constant along its length (the wave impedance varies along the line, if the axial lines of the cylinder and the twisted pair do not coincide, and the inequality of b and R is not true).

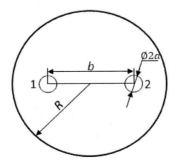

Fig. 1.13. Two wires inside a cylinder.

Also, we assume that the wires are straight and take the twisting into account by increasing length L of the equivalent line.

Because a pitch of a helix, along which each wire is located, is greater than the diameter b of the helix, inductance Λ per unit length undergoes a slight change on the replacement of a helical wire by a straight wire. The wire capacitance per unit length also varies only slightly from the helix pitch, i.e., twisting wires have no effect on the wave impedances of a structure.

But the asymmetrical location of the line in a real cable can cause a change of the wave impedance and the input impedance of the two-wire line. It is one of a causes of the cable asymmetry. The implementation of each two-wire line in the form of a twisted pair (helix), is another cause of the cable asymmetry. The twisted pair is the design, which causes a difference of the average distances between different wires and the mutual coupling (cross talk) between two two-wire lines surrounded by a common screen, even if the exciting emf of each line and the line load are symmetric.

The equivalent circuit for a single line inside the screen is shown in Fig. 1.14. The two-wire line is located above the ground (inside a metal cylinder). The current and the potential along the wire n of an asymmetrical line of N parallel wires situated above ground, in the general case, are determined by an expression (1.49). The boundary conditions for the currents and potentials in this circuit are

$$i_1(0) + i_2(0) = 0,\ u_1(0) = u_2(0) + i_1(0)Z,\ i_1(L) + i_2(L) = 0,\ u_1(L) = e + u_2(L). \quad (1.94)$$

Here, Z is the impedance of the line load. Sequentially substituting expressions (1.92) in the first and second equalities of the set (1.94), we find

$$I_2 = -I_1,\ U_2 = -I_1 Z.$$

Taking into account formulas for $1/W_{ik}$, presented in Section 1.5, we find from the third equation of the set (1.94) that

$$U_1 = I_1 Z \left(\frac{1}{W_{22}} - \frac{1}{W_{12}} \right) \bigg/ \left(\frac{1}{W_{11}} + \frac{1}{W_{22}} - \frac{2}{W_{12}} \right) = I_1 Z \frac{\rho_{11} - \rho_{12}}{\rho_{11} + \rho_{22} - 2\rho_{12}}.$$

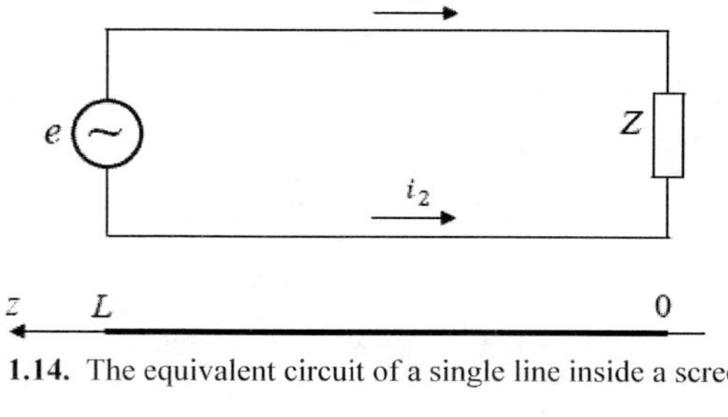

Fig. 1.14. The equivalent circuit of a single line inside a screen.

And from the fourth equation we obtain

$$I_1 = e/[Z \cos kL + j(\rho_{11} + \rho_{22} - 2\rho_{12}) \sin kL].$$

The input impedance of a two-wire line inside a metal screen (the load impedance of generator e) is $Z_1 = e/i_1(L)$. Substituting magnitude $i_1(L)$ from expression (1.49) and using the relationships between e, I_1, I_2, U_1, U_2, we find that

$$Z_1 = W \frac{Z + jW \tan kL}{W + jZ \tan kL}, \tag{1.95}$$

where $W = \rho_{11} + \rho_{22} - 2\rho_{12}$.

It is readily seen that expression (1.95) coincides with the expression for the input impedance of a two-wire long line that has no losses and is loaded at the end by impedance Z. The line is located in free space ant has the wave impedance W. The asymmetry of line leads to a difference of the electro-dynamics and electrostatic wave impedances of the wires ($\rho_{11} \neq \rho_{22}$, $W_{11} \neq W_{22}$). The calculation of currents $i_1(z)$ and $i_2(z)$ shows that the currents in a two-wire line are identical in magnitude and opposite in sign:

$$i_1(z) = -i_2(z) = I_1 \cos kz + jI_1(Z/W) \sin kz.$$

In a wire pair, there are only anti-phase currents. In-phase currents are absent in the wires, because the emf and load impedance are included between the line wires. The appearance of the in-phase currents can be caused by the connection of an additional emf or an additional load only between one of wires of this line and the screen.

In order to find the potential coefficients α_{ns}, one should take into account the following facts. If the system consists of two identical conductors (a wire and its image) and the structure is electrically neutral, the mutual partial capacitance coincides with the capacitance between the conductors [30] and is equal to

$$C = 1/[2(\alpha_{11} - \alpha_{12})],$$

where α_{11} is the self potential coefficient, and α_{12} is the potential coefficient of the image. The capacitance between the conductor and the ground is twice as much as the capacitance between two conductors: $C_l = 2C$. For two wires of radius a, located inside the metal cylinder of radius R at a distance b from each other, symmetrically with respect to the cylinder axes (see Fig. 1.13), we can write, using (4.20) from [30],

$$\alpha_{11} = \alpha_{22} = \frac{1}{2\pi\varepsilon} \cosh^{-1} \frac{R^2 + a^2 - b^2/4}{2Ra}.$$

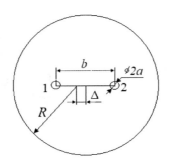

Fig. 1.15. Offset wires inside a cylinder.

Here ε is the permittivity of the medium inside the cable. If wire radius a and distance b are small in comparison with the cylinder radius R, then, in the air, $\rho_{11} = \rho_{22} \approx 60 \, ln(R/a)$. Similarly, using (4.22) from [30], we find: $\rho_{12} \approx 60 \, ln(R/\sqrt{ab})$, i.e., the wave impedance of a lossless two-wire line, symmetrically situated inside a metal cylinder, is a half of the wave impedance of the same line in the free space: $W_0 = \rho_{11} + \rho_{22} - 2\rho_{12} \approx 60 \, ln(b/a)$.

If the wires inside a metal cylinder of a radius R are located asymmetrically, e.g., they are displaced to the right by distance Δ (Fig. 1.15), then

$$\alpha_{11} = \frac{1}{2\pi\varepsilon} \cosh^{-1} \frac{R^2 + a^2 - (b/2 - \Delta)^2}{2Ra},$$

so, at $a, b \ll R$

$$\alpha_{11} = \frac{1}{2\pi\varepsilon} ln \left\{ \frac{R}{a} \left[1 + \frac{\Delta(b - \Delta)}{R^2} \right] \right\} = \frac{1}{2\pi\varepsilon} \left[ln \frac{R}{a} + \frac{\Delta(b - \Delta)}{R^2} \right], \quad \alpha_{22} = \frac{1}{2\pi\varepsilon} \left[ln \frac{R}{a} - \frac{\Delta(b + \Delta)}{R^2} \right].$$

In this case, the wave impedance of the line is

$$W = W_0 - \frac{120\Delta^2}{R^2}. \tag{1.96}$$

That is one of the possible causes of changing lines wave impedance inside the screen.

If the distance between the wires is increased by value Δ, then at $\Delta \ll b$

$$\rho_{12} \approx 60 ln \frac{R}{\sqrt{a(b + \Delta)}} \approx 60 \left(ln \frac{R}{\sqrt{ab}} - \frac{\Delta}{2b} \right). \tag{1.97}$$

This is the second cause. As can be seen from these results, a change of the distance between wires has a greater effect on the wave impedance of the line than a displacement of wire relative to the cylinder axis.

Figure 1.16 shows the equivalent circuit of two coupled two-wire lines inside a screen. One of the lines is excited by generator e and is loaded by

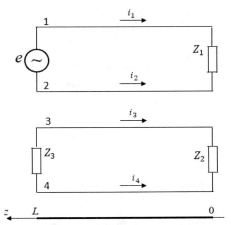

Fig. 1.16. The equivalent circuit of two coupled lines placed inside a screen.

complex impedance Z_1. The loads Z_2 and Z_3 are connected in the wires at both ends of the other line. It is necessary to emphasize that such a circuit has the most general nature. If, for example, generator e_1 is located at the end of the second line (at point $z = L$), the currents and voltages created by generator e are calculated considering that the input impedance of generator e_1 is equal to Z_3.

We take into account inequalities $a \ll b \ll d$, R (here d is the distance between the axes of twisted pairs). The wires inside the screen form the bunch, whose diameter in many cases is small in comparison with the diameter of the metal screen. But when the bunch consists of many wires, its diameter is close to the screen diameter. However, it is necessary to take into account that the maximal mutual coupling exists between adjacent lines. Therefore, by analyzing the mutual coupling between them it is possible to consider in the $d \ll R$ in the first approximation.

As was stated at the beginning of this Section, cable asymmetry leads to a mutual coupling between two two-wire lines, i.e., to interconnection and cross talk. The reason of such asymmetry is the structure of each two-wire line. This structure has a shape of a twisted pair (helix). In Fig. 1.17 the cross-section of two twisted-pairs is shown. This figure demonstrates the placement of the line conductors in different points of a given cross-section. That placement changes during winding and depends on the initial cross-section of the cable and on the variant of the winding. If, for example, in the initial cross-section the ends of helices 1 and 3 are located at the same point of cross-section (we shall call it the initial point) and the ends of helices 2 and 4 are displaced along the perimeter of cross-section by π from this point, it means that the distance between wires 1 and 3 (and also between wires 2 and 4) is

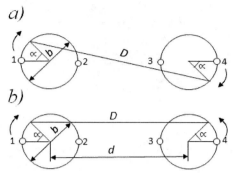

Fig. 1.17. Distance between wires 1 and 4 in the same (*a*) and opposite (*b*) direction winding of wire 4.

$D_{13} = D_{24} = d$ along the total length of the cable, whereas the distance between wires 1 and 4 (and also between wires 2 and 3) varies along wires from $d + b$ to $d - b$. For example, the distance between wires 1 and 4 (see Fig. 1.17*a*) is

$$D_{14} = \sqrt{(d + b\cos\alpha)^2 + b^2\sin^2\alpha} \approx d + b\cos\alpha + (b\sin\alpha)^2/(2d)$$

(here α is angular displacement of points 1 and 4 along the perimeter of cross-section), i.e., the average distance between these wires,

$$(D_{14})_0 = \frac{1}{\pi}\int_0^\pi D d\alpha = d + \frac{b^2}{4d}$$

differs from distance *d*. The potential coefficients as well as the electrodynamic and electrostatic wave impedances vary accordingly.

If, at the initial cross-section of the cable, the ends of helix 3 and 4 are displaced along the perimeter of cross-section by $\pi/2$ and $3\pi/2$ from the initial point, respectively, the distance between wire 1 and wire 3 (or wire 4) is

$$D_{13} \approx d + \frac{b}{2}(\cos\alpha + \sin\alpha) + \frac{b^2}{8d}(\sin\alpha - \cos\alpha)^2.$$

$$D_{14} \approx d + \frac{b}{2}(\cos\alpha - \sin\alpha) + \frac{b^2}{8d}(\sin\alpha + \cos\alpha)^2$$

The average distance between the wires in this case is

$$(D_{13})_0 \approx d + \frac{b}{\pi} + \frac{b^2}{8d}, \quad (D_{14})_0 \approx d - \frac{b}{\pi} + \frac{b^2}{8d},$$

i.e., the displacement of the helix ends of the two-wire line by $\pi/2$ changes the average distance between wires substantially. Difference between $(D_{13})_0$ and

$(D_{14})_0$ increases from value $\dfrac{b^2}{4d}$ to $2\dfrac{b}{\pi}$, where $b \ll d$. In order that the average distance D_0 between wires 1 and 4 does not differ from d, it is necessary to wind wire 4 in the opposite direction to the direction of winding of the other wires. In this case (see Fig. 1.17b)

$$D = d + b \cos \alpha,\ D_0 = d. \tag{1.98}$$

The electrodynamic wave impedances of the structure at the same direction of winding of both wires are

$$\rho_{11} = \rho_{22} = \rho_{33} = \rho_{44} = \rho_1 = 60\ ln(R/a),\ \rho_{12} = \rho_{34} = \rho_2 = 60\ ln(R/\sqrt{ab}),$$
$$\rho_{13} = \rho_{24} = \rho_3 = 60\ ln(R/\sqrt{ad}),\ \rho_{14} = \rho_{23} = \rho_4 = 60\ ln[R/\sqrt{a(d + b^2\pi/(4d))}\,].$$

In the case of lines located at a finite distance H from the cable axis, we find

$$\rho_1 = 60\,ln\dfrac{R(1 - H^2/R^2)}{a}.$$

These expressions for the other quantity ρ_n remain valid. This means that the wave impedance of a two-wire line, which has no losses and is situated inside a metal cylinder at a distance H from its axis, is in accordance with equation for W:

$$W = 60ln[b(1 - H^2/R^2)^2/a],$$

i.e., the wave impedance of this line decreases as the result of its displacement from the cable axis. When H is small and equal to Δ, we arrive at the expression (1.95).

Since the electrostatic wave impedances are

$$W_{ns} = \begin{cases} \Delta_N/\Delta_{ns}, & n = s, \\ -\Delta_N/\Delta_{ns}, & n \ne s, \end{cases}$$

where $\Delta_N = |\rho_{ns}|$ is the $N \times N$ determinant, and Δ_{ns} is the cofactor of the determinant Δ_N. For a structure made of four wires

$$W_{11} = W_{22} = W_{33} = W_{44} = W_1 = \Delta_4/\Delta_{11},\ W_{12} = W_{34} = W_2 = -\Delta_4/\Delta_{12},$$
$$W_{13} = W_{24} = W_3 = -\Delta_4/\Delta_{13},\ W_{14} = W_{23} = W_4 = -\Delta_4/\Delta_{14}.$$

The current and potential of wire n of an asymmetric line, consisting of N parallel wires, located above ground, is found from expressions (1.49). The boundary conditions for the currents and voltages in the circuit, shown in Fig. 1.16, are

$i_1(0) + i_2(0) = i_3(0) + i_4(0) = i_1(L) + i_2(L) = i_3(L) + i_4(L) = 0$, $u_1(0) = u_2(0) + i_1(0)Z_1$,
$u_3(0) = u_4(0) + i_3(0)Z_2$, $u_1(L) = u_2(L) + e$, $u_3(L) = u_4(L) + i_3(L)Z_3$.

Substituting expressions (1.49) for the currents and voltages in these equations, we find

$$I_1 = \frac{e}{Z_1 \cos kL + 2j[\rho_1 - \rho_2 + (\rho_3 - \rho_4)A]\sin kL}, I_3 = AI_1, \quad (1.99)$$

where

$$A = \frac{4(\rho_3 - \rho_4) + Z_1 Z_3 (1/W_3 - 1/W_4)}{-4(\rho_1 - \rho_2) + Z_2 Z_3 (1/W_1 + 1/W_2) + j2(Z_2 - Z_3)\cot kL},$$

If $\rho_3 = \rho_4$ (and, accordingly, $W = W_4$), then $A = 0$, the current at the beginning of the second long line is zero. In this case, the presence of the second two-wire long line has no effect on the first line. This result obviously corroborates the fact that the asymmetry of the cable leads to mutual coupling (crosstalk) between two two-wire lines. Knowing all parameters in expressions (1.49), one can calculate the load impedance of the generator e:

$$Z_l = \frac{e}{i_1(L)} = 2\frac{Z_1 + 2j[\rho_1 - \rho_2 + A(\rho_3 - \rho_4)]\tan kL}{2 + j[Z_1(1/W_1 + 1/W_2) - AZ_2(1/W_3 - 1/W_4)]\tan kL}$$

and the currents in the wires of the second (unexcited) line:

$i_3(z) = I_1 A \cos kz + jI_1/2[AZ_2(1/W_1 + 1/W_2) - Z_1(1/W_3 - 1/W_4)]\sin kz$, $i_4(z) = -i_3(z)$.

The sum of the currents is zero, i.e., as in the case of a single long line placed into the screen, the in-phase current is absent since the emf and the loads are located only between wires of each line. The voltages across passive loads are

$$V_1 = i_1(0)Z_1 = I_1 Z_1, V_2 = i_3(0)Z_2 = I_1 AZ_2,$$
$V_3 = i_3(L)Z_3 = I_1 Z_3 \{A \cos kL + j0.5[AZ_2(1/W_1 + 1/W_2) - Z_1(1/W_3 - 1/W_4)]\sin kL\}$.

As an example, consider a structure from two pairs of wires inside the screen with sizes (in millimeters): $a = 0.2$, $b = 0.5$, $d = 2$, $R = 2$. If the loads are identical and equal to $Z_1 = Z_2 = Z_3 = 100$ Ohm, ratio A of the currents at the beginning of the second and first line is equal to 0.13. If the load magnitudes are equal to the wave impedance of the single two-wire line inside the metal screen, i.e., $Z_1 = Z_2 = Z_3 = 55$ Ohm, the currents' ratio is substantially increased ($A = -0.76$). The absolute values of the currents depending on kz are plotted in Fig. 1.18. Here, k is the propagation constant of a wave in a medium, z is the coordinate along the line (see Fig. 1.16).

Consider the effect of loads placed between the wires and the screen, using a two-wire line as an example (Fig. 1.19). Unlike the circuit shown in

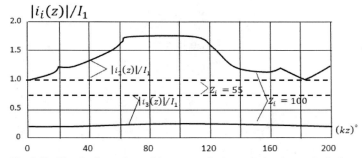

Fig. 1.18. The absolute values of the currents in the excited and unexcited wires.

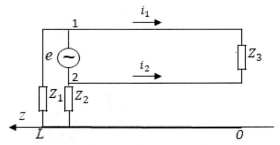

Fig. 1.19. The equivalent circuit of a single line with loads connected between the wires and the screen.

Fig. 1.14 near the generator each wire of the line is connected with the screen by a complex impedance Z_1 or Z_2. Their magnitudes depend on the circuit of a line excitation. In a real circuit the secondary winding of the transformer can act as emf e, exciting a two-wire line. In this case impedances Z_1 and Z_2 are stray capacitances between the secondary winding and a cable screen.

The boundary conditions for the currents and potentials in the circuit, shown in Fig. 1.19, have the form

$$i_1(0) + i_2(0) = 0, \quad u_1(0) = u_2(0) + i_1(0)Z,$$
$$i_1(L) + i_2(L) + u_1(L)/Z_1 + u_2(L)/Z_2 = 0, \quad u_1(L) = e + u_2(L). \quad (1.100)$$

Substituting expression (1.49) in the equations (1.100), we find the input impedance of a two-wire line

$$Z_l = \frac{e}{i_1(L) + \dfrac{u_1(L)}{Z_1}} = \frac{Z\cos kL + j(\rho_{11} + \rho_{22} - 2\rho_{12})}{1 + \dfrac{U_1}{I_1}\left[\dfrac{1}{Z_1} + j\left(\dfrac{1}{W_1} - \dfrac{1}{W_2}\right)\tan kL\right] + j\left[\dfrac{Z}{W_{12}} + \dfrac{\rho_{11} - \rho_{22}}{Z_1}\right]\tan kL},$$

(1.101)

and the sum of the currents in the line wires is

$$i_s(z) = i_1(z) + i_2(z) = jI_1\left[Z\left(\frac{1}{W_{12}} - \frac{1}{W_{22}}\right) + \frac{U_1}{I_1}\left(\frac{1}{W_{11}} + \frac{1}{W_{22}} - \frac{2}{W_{12}}\right)\right]\sin kz.$$
(1.102)

Therefore, the placement of the loads between the wires and the screen leads to the appearance of the in-phase current in the wires and the current along the inner surface of the cable screen, equal in magnitude but opposite in direction. For two wires of the same radius situated symmetrically to the cylinder axis

$$Z_l = [Z + 2j(\rho_1 - \rho_2)\tan kL]/B,$$

where

$$B = 1 + j\frac{z\rho_2}{\rho_1^2 - \rho_2^2}\tan kL + j\frac{1}{2(\rho_1 + \rho_2)}[Z + (\rho_1 + \rho_2)C] + \frac{1}{2Z_1}[Z + (\rho_1 + \rho_2)C + 2j(\rho_1 - \rho_2)\tan kL],$$

and

$$i_s(z) = j\frac{eC\sin kz}{Z\cos kL + j2(\rho_1 - \rho_2)\sin kL},$$

It is not difficult to convince that, if $\frac{1}{Z_1} = \frac{1}{Z_2} = 0$, quantity C is zero and the expressions for U_1 and Z_l coincide with the similar expressions for the circuit without loads. From the presented results it is also easy to obtain the expressions for the circuit with one load, for example, when, $\frac{1}{Z_1} = 0$.

The above analysis confirms that the cause of the appearance of in-phase currents in wires of the line is the asymmetry of its excitation produced in particular by the secondary transformer winding. Asymmetry of loads at $z = 0$, i.e., at the far end of the line, produces similar results. The in-phase currents in the excited line induce the in-phase currents in wires of the adjacent unexcited line, even if it is totally symmetric (with respect to ground and the excited line). Removal of the excitation and load asymmetry in the excited line results in the disappearance of these currents in wires of excited and unexcited lines.

In order to reduce or eliminate the in-phase currents, it is necessary to violate the asymmetry, e.g., to neutralize the effect of stray capacitances to ground (to the cable screen). To this end, in [31] it was proposed to compensate the current through stray capacitance with the current equal in magnitude and opposite in direction, which is created by an additional transformer winding.

As is shown in Section 1.5, the theory of electrically coupled lines is based on the telegraph equations and on the relationship between the potential coefficients and the coefficients of electrostatic induction. In the case, when the wires and the medium have no losses, electrostatic W_{ns} and electrodynamic ρ_{ns} wave impedances between wires n and s are real-valued quantities, and k is the propagation constant of the wave in the medium.

As follows from the above, electrodynamic wave impedance ρ_{ns} is proportional to the self or mutual inductance of the wires segment, i.e., is proportional to the reactance, which is connected in series with wires. Electrostatic wave impedance W_{ns} is proportional to the mutual capacitance between wires, i.e., to the susceptance between them. Therefore, it is expedient to connect the loss resistance in a wire (e.g., the skin-effect loss) in series with the inductance and to connect the leakage conductance in parallel with the mutual capacitance.

In order to take into account, the loss in the medium and in the wires, one must consider that wave impedances W_{ns} and ρ_{ns} and the propagation constant k are complex values. If the inductance of the wire n per unit of its length is Λ_0 and its active resistance is R_0, its impedance per unit length is $j\rho_{nn} = j\omega\Lambda_0 + R_0$, i.e., the self electrodynamics wave impedance of a loss wire is equal to

$$\rho_{nn} = \rho_0(1 - jR_0/\rho_0),$$

where $\rho_0 = \omega\Lambda_0$ is the electrodynamic wave impedance in the absence of losses, R_0 is total loss resistance in the wire n and in the metal screen per unit length.

For the mutual electrodynamic wave impedance between wires n and s, we obtain

$$\rho_{ns} = \rho_{ns0}(1 - jR_{ns0}/\rho_{ns0}),$$

where $\rho_{ns0} = \omega M_{ns0}$, M_{ns0} is the mutual inductance between wires n and s per unit length, and R_{ns0} is the loss resistance in both wires per unit length.

Similarly, for the admittance between wires n and s per unit length we find: $jW_{ns} = j\omega C_{ns0} + G_{ns0}$, i.e., the electrostatic wave impedance in a medium with losses is

$$W_{ns} = W_{ns0}(1 - jG_{ns0}/W_{ns0})$$

where $W_{ns0} = \omega C_{ns0}$, C_{ns0} is the mutual partial capacitance between wires n and s per unit length, and G_{ns0} is the leakage conductance per unit length.

Thus, the evaluation of the electrical performances of the coupled lines with losses can use the results previously received for the lossless lines by substitution of the complex wave impedances into the earlier expressions. Here, the losses in wires and losses in an imperfectly conducting metallic tube (screen) are taken into account.

A rigorous method for the calculation of characteristics of two-wire lines inside a metal screen allows revising the mechanism of mutual coupling between lines in multi-conductor cables. It permits the determination of the values of voltage (interferences) across impedances located at the beginning and the end of the adjacent line at the given power in the main line. The reason for cross talks is the asymmetry of the cable structure (the different average distance between wires) and, accordingly, the asymmetric wave impedances. An asymmetry decrease must reduce cross talks in multi-conductor cables, i.e., will allow increasing the carrying capacity of a channel. This is true also for multi-conductor connectors.

The reason for the emergence of the in-phase currents in the lines of a multi-conductor cable is the asymmetry of excitation and loads. As it was noted, the compensation of this asymmetry will allow us to decrease the *EM* radiation and to reduce its susceptibility to external fields.

Chapter 2
Integral Equation of Leontovich

2.1 Integral equations for linear metal radiators

Chapter 1 describes the approximate methods of calculating radiators. These methods have been applied on the eve of a broad wave of theoretical works devoted to rigorous methods of calculation. New methods were based on the derivation and solution of integral equations. It should be recalled that these theoretical works were carried out against the background of a pessimistic attitude towards the possibility of solving the set task. The book [1] is in particular evidence of this attitude. In this book, as a rigorous method for analyzing a direct linear radiator, it is proposed using calculation of an elongated ellipsoid (spheroid) with an eccentricity slightly different from unity, based on the computation of its eigenfunctions.

Subsequently, this method was brought to the final results and then was forgotten due to the limited possibilities of its application. Such a prospect of an alternative method of analysis was predicted into a commentary on the translation of a book published in the Soviet Union in 1948 (editor S.M. Rytov). True, this translation was published 7 years later than the book itself. The strict method based on the calculation of the eigenfunctions of a spheroid is now only of historical interest and lies outside the scope of the present book.

As shown in Chapter 1, knowledge of the current distribution along a linear radiator allows us to find the electromagnetic field and all electrical characteristics of the radiator. For this reason, determination of the current distribution is the important problem of the antenna theory. The current distribution must satisfy the condition of the lack of current on the radiator's ends:

$$J(\pm L) = 0. \tag{2.1}$$

The radiator current $J(z)$ must create electromagnetic field $E_z(J)$, satisfying the boundary condition on its surface

$$E_z(a, z)|_{-L < Z < L} + K(z) = 0. \tag{2.2}$$

Here the cylindrical coordinate system is used, a and L are the radius and the arm length of a dipole respectively, and $K(z)$ is an extraneous emf.

Expression (2.2) is the mathematical record of the fact that the total field, which is a sum of the extraneous field and the current field, is zero on the surface of a perfectly conducting radiator. The extraneous field is specified usually as the product of potentials difference e between edges of the gap into the δ-function. Magnitude $K_1(z) = e\delta(z)$ corresponds to connecting the generator at the middle of the radiator, at point $z = 0$, and $K_2(z) = e\delta(z - h)$ corresponds to its shift, i.e., to placement of the generator at point $z = h$.

Equality (2.2) contains all integral equations of the theory of thin antennas as in an embryo. The external appearance of the equations depends mostly on the selection of a radiator model, because this model determines the type of function $E_z(J)$. For example, if the radiator model is made in the form of a conductive filament, we use the expression (1.30) for the field and obtain Hallen's integral equation for the filament

$$\int_{-L}^{L} J(\varsigma) G_1 d\varsigma = -j\frac{1}{Z_0}\left(C\cos kz + \frac{e}{2}\sin k|z|\right), \qquad (2.3)$$

where

$$G_1 = exp(-jkR_1)/4\pi R_1, \quad R_1 = |z - \varsigma|.$$

For the model of a straight thin-wall metal cylinder we use the expression (1.28) for the field and obtain Hallen's integral equation with exact kernel

$$\frac{1}{2\pi}\int_{-L}^{L} J(\varsigma)\int_0^{2\pi} G_2 d\varphi d\varsigma = -j\frac{1}{Z_0}\left(C\cos kz + \frac{e}{2}\sin k|z|\right), \qquad (2.4)$$

Here

$$G_2 = exp(-jkR_2)/4\pi R_2, \quad R_2 = \sqrt{(z-\varsigma)^2 + 4a^2\sin^2\varphi/2}.$$

For the current along a filament of a finite radius we use the expression (1.27) and obtain the often-used integral equation with approximate kernel

$$\int_{-L}^{L} J(\varsigma) G_3 d\varsigma = -j\frac{1}{Z_0}\left(C\cos kz + \frac{e}{2}\sin k|z|\right), \qquad (2.5)$$

where

$$G_3 = exp(-jkR_3)/4\pi R_3, \quad R_3 = \sqrt{(z-\varsigma)^2 + a^2}.$$

For the current along a filament of a finite radius we also can use the expression (1.27), but with replacement of R_1 by R_3. In this case we obtain Pocklington's integral equation [32]

$$\int_{-L}^{L} J(\varsigma)\left(k^2 + \frac{\partial^2 G_3}{\partial z^2}\right)d\varsigma = j\omega\varepsilon_0 K(z). \qquad (2.6)$$

Constant *C* in each equation is found from condition (2.1).

The named equations and methods for their solution have much in common with each other. The first solution of the Hallen's equation with an approximate kernel was given by Hallen itself and described in detail in [21]. During the process of finding the solution, the function of the current $J(z)$ was presented in the form of a series in the inverse powers of a parameter $\Omega = 2ln(2L/a)$. By means of the method of successive approximations (iterative procedure), an expression for the current was obtained, whose members are expressed in sines and cosines integrals.

The iterative procedure proposed by King and Middleton [33] gave a more accurate result. But the common and most important feature of different equations was the complexity of finding their solution. In the expansion of the function for the current into a series, as a rule, it was not possible to determine more than the first two terms of this series because of the existing singularities. The iterative process slowly converged, and it was not obvious that it even converges at all. The search of a numerical solution also did not demonstrate a smooth process. In this regard, the opinion has arisen that the proposed equations either do not have the solution, or the solution is not one only (*several solutions*).

Because of the particular importance of this question, it was considered in [34]. The author has comprehensively analyzed the equation

$$\left(\frac{\partial^2}{\partial z^2} + k^2\right)\int_{-1}^{1} J(t)K(z-t)dt = f(z),$$

where

$$K(u) = \frac{1}{2\pi}\int_{0}^{2\pi} \frac{\exp\left\{-jk\left(u^2 + 4a^2 sin^2\frac{\varphi}{2}\right)^{1/2}\right\}}{\left(u^2 + 4a^2 sin^2\frac{\varphi}{2}\right)^{1/2}} d\theta.$$

Here $J(t)$ is the sought current, and $f(z)$ is the known exciting field. This field is proportional to the tangential component of the electric field and can be considered as a continuous function. The integral equation is complemented by the specific property of the current at the wire ends, namely $J(\pm 1) = 0$.

The analysis confirmed that the equation has a solution, and this is the only solution.

Further, Leontovich's integral equation [20] and the methods of its solution as applied to various problems are considered in detail, since the severity of this equation and the possibility of its use, in the author's opinion, exceed the similar qualities of other equations (although the author of the book is not the co-author and or a relative of the co-author of this article, but only his namesake).

Leontovich's equation played an important role in the progress of the theory of thin antennas and rightfully occupies the main place in this book.

2.2 Derivation of Leontovich's equation

Assume that a source of electromagnetic field are electrical currents parallel to the z-axis and having a circular symmetry:

$$\vec{J} = j_z e_z, j_z = j_z(z) = \text{const}(\varphi). \tag{2.7}$$

Then a vector potential A of a field has only component A_z, which on the surface of the radiator model in a shape of a thin-wall straight metal cylinder with circular cross-section of a radius a is equal to

$$A_z(\rho,z) = \frac{\mu}{8\pi^2} \int_0^{2\pi} T(z,\varphi) d\varphi, \tag{2.8}$$

where $T(z, \varphi) = \int_{-L}^{L} J(z) \frac{\exp(-jkR)}{R} d\varsigma$, $R = \sqrt{(z-\varsigma)^2 + \rho_1^2}$, $\rho_1 = \sqrt{\rho^2 + a^2 - 2a\rho \cos \varphi}$.

If we integrate $T(z, \varphi)$ by parts and successively use the circumstance that the radiator radius is small in comparison with its length and the wavelength, i.e., to neglect the summands of the order of a/L and ka and to keep the summands proportional to the logarithm of these quantities, we obtain:

$$T(z,\varphi) = -2J(z)\ln\rho_1 - \int_{-L}^{L} \exp(-jk|\varsigma - z|)\ln 2p|\varsigma - z|\left[\frac{dJ(\varsigma)}{d\varsigma} - jkJ(\varsigma)\text{sign}(\varsigma - z)\right] d\varsigma.$$

Here p is some constant having the dimensions of inverse length. Since at $\rho > a$

$$\int_0^{2\pi} \ln p \rho_1 d\varphi \equiv \int_0^{2\pi} \ln\left(p\sqrt{\rho^2 + a^2 - 2a\rho\cos\varphi}\right) d\varphi = 2\pi \ln p\rho,$$

then

$$A_z(\rho, z) = (\mu/4\pi)[-2J(z)\ln p\rho + V(J, z)], \tag{2.9}$$

where

$$V(J,z) = \int_{-L}^{L} \exp(-jk|z - \varsigma|)\ln 2p|\varsigma - z|\left[jkJ(\varsigma) + \text{sign}(z - \varsigma)\frac{dJ(\varsigma)}{d\varsigma}\right] d\varsigma.$$

The tangential component of the electric field of the antenna is

$$E_{zA}(\rho,z) = -j\frac{\omega}{k^2}\left(k^2 A_z + \frac{\partial^2 A_z}{\partial z^2}\right). \tag{2.10}$$

Substituting (2.9) into (2.10) and setting ρ equal to a, we find:

$$E_{zA}(a,z) = \frac{1}{j4\pi\omega\varepsilon_0}\left[-\chi_1^{-1}\left(\frac{d^2J}{dz^2}+k^2J\right)+\frac{d^2V}{dz^2}+k^2V\right]. \quad (2.11)$$

Here $\chi_1 = \dfrac{1}{(2 ln pa)}$ is a small parameter of the thin antenna theory, used in [20]. As is shown in [3], in the capacity of constant $1/p$ one should choose the distance to the nearest inhomogeneity, i.e., the smallest of three magnitudes: wavelength λ, antenna length $2L$ and the radius R_C of its curvature. In case of a straight radiator, the whose length does not exceed the wavelength, one can consider that $1/p = 2L$, i.e.,

$$\chi = -\chi_1 = 1/[2 ln(2L/a)]. \quad (2.12)$$

From (2.10) and (2.11), we obtain with allowance (2.2) the desired equation

$$\frac{d^2J(z)}{dz^2}+k^2J(z) = -j4\pi\omega\varepsilon_0\chi K(z) - \chi W(J). \quad (2.13)$$

Here $K(z) = -E_{zA}(a, z)$ is the field, creating the current in the antenna, $-j4\pi\omega\varepsilon_0\chi K(z)$ is the extraneous emf, and $-\chi W(J)$ is emf, created by the antenna currents. As can be seen from (2.13), this equation together with the components, which contain the extraneous emf, the current and the current derivative, also has the element $W(J) = \dfrac{d^2V}{dz^2}+k^2V$, incorporating the integral $V(J, z)$ and its derivatives. In Section 1.4 it is shown that one concentrated emf cannot create the sinusoidal current along the dipole. The mentioned element is the additional emf, which depends on the current of the antenna. This emf takes the radiation into account and distributes it along the antenna.

The meaning of manipulations performed during derivation of (2.13), first, is that a logarithmic singularity was highlighted and isolated. The function A_z in expression (2.9) including integral $V(J, z)$ is a continuous function everywhere in contrast to the original integral (2.8). Another important advantage of the equation (2.13) is the absence of an argument φ, since the integration with respect to φ has been executed. Nevertheless, this equation is derived for the current of a straight thin-wall cylindrical antenna, and so the equation (2.13) is equivalent to Hallen's equation with exact kernel.

In order to solve the equation (2.13), in [20] the perturbation method is used, i.e., the current is sought in the form of expansion into a series of powers of the small parameter χ:

$$J(z) = \chi J_1(z) + \chi^2 J_2(z) + \chi^3 J_3(z) + \ldots$$

Further we will use another form for the expansion of the current into a series, which differs little from this form, but, as it will become clear later on, it is more convenient for use:

$$J(z) = J_0(z) + \chi J_1(z) + \chi^2 J_2(z) + \chi^3 J_3(z) + \ldots \quad (2.14)$$

Substituting this series into the equation (2.13) and equating coefficients for equal powers of χ, in the case of an untuned radiator, when $J(\pm L) = 0$, we obtain, the set of equations:

$$\frac{d^2 J_0(z)}{dz^2} + k^2 J_0(z) = -j4\pi\omega\varepsilon_0 \chi K(z), J_0(\pm L) = 0,$$

$$\frac{d^2 J_1(z)}{dz^2} + k^2 J_1(z) = -\chi W(J_0), J_1(\pm L) = 0, \quad (2.15)$$

$$\frac{d^2 J_n(z)}{dz^2} + k^2 J_n(z) = -\chi W(J_{n-1}), J_n(\pm L) = 0, n > 1,$$

where $W(J_{n-1}) = \dfrac{d^2 V(J_{n-1}, z)}{dz^2} + k^2 V(J_{n-1}, z)$. Since the solution is sought in the form of a series in powers of a small parameter χ, these expressions must take into account that the functional $W(J, z)$ is linear, i.e.,

$$W(J, z) = \sum_{n=0}^{\infty} \chi^n W(J_n).$$

As follows from (2.15), in calculating the terms of the series (2.14), the radiator current can be considered concentrated on its axis. At the same time, the degree of accuracy adopted in deriving the equation is preserved. This circumstance greatly simplifies calculations, including calculations using recurrent formulas.

From the first equation, we obtain that, in the zero approximation the current is distributed according to a sinusoidal law and at an arbitrary point of the radiator is equal to

$$J_0(z) = j \frac{e\chi}{60 \cos kL} \sin k(L - |z|). \quad (2.16)$$

It is easy to make sure that the input impedance of the antenna in this approximation is

$$Z_{A0} = -j60\chi^{-1} \cot kL.$$

It has only the reactive component, which coincides with the input impedance of the equivalent long line, whose wave impedance is equal to

$60/\chi$. The active component appears in the first approximation. As shown in [20], where the tuned and untuned radiators are first considered separately, the expression for the current in the untuned radiator is general in nature and can be used in both cases.

The next member at the feeding point $z = 0$ is obtained in [20] from the second equation of the set in the form

$$J_1(0) = e(B + jC), \qquad (2.17)$$

where $B = \dfrac{1}{30\cos^2\alpha}\left\{\cos\alpha(\cos\alpha \mathrm{Di}2\alpha - \sin\alpha \mathrm{Si}2\alpha) - \dfrac{1}{4}(\cos 2\alpha \mathrm{Di}4\alpha - \sin 2\alpha \mathrm{Si}4\alpha)\right\}$,

$C = \dfrac{1}{30\cos^2\alpha}\left\{\cos\alpha(\sin\alpha \mathrm{Di}2\alpha + \cos\alpha \mathrm{Si}2\alpha) - \dfrac{1}{4}(\sin 2\alpha \mathrm{Di}4\alpha + \cos 2\alpha \mathrm{Si}4\alpha) - \dfrac{1}{2}\sin 2\alpha \ln\alpha\right\}$.

Here $\alpha = kL$ is electric arm length of a symmetric radiator, $\mathrm{Di}2\alpha = C + \ln 2\alpha - \mathrm{Ci}2\alpha$, $\mathrm{Di}4\alpha = C + \ln 4\alpha - \mathrm{Ci}4\alpha$, $C = 0.5772\ldots$ is Euler's constant, functions Six и Cix are sine and cosine integrals of the argument x.

The obtained result allows us to find the value of the active component of the radiator input impedance by means of dividing B to the square of the value $A = \dfrac{J_1(0)}{je} = \dfrac{\tan\alpha}{60}$ (i.e., with the help of the expression $R_A = B/A^2$) and to determine the additive to the reactive component by means of the expression $\Delta X = C/A^2$. Hence, in particular, one can to find the input resistance and the input reactance of the cylindrical dipole with the arm length $\lambda/4$, $\lambda/2$, $3\lambda/4$ respectively (while changing the shape of the dipole, the magnitude of reactance changes slightly).

Since prior to the development of methods for calculating input impedances were based on the use of integral equations, numerical values were obtained only by using the induced emf method, and a natural desire arose to compare new results with them. Unfortunately, this comparison turned out to be a difficult task. The results, based on the substitution of the values of the known functions, were close to each other in the greater part of the range, except for the parallel resonance region, when the length of the radiator arm was approaching to the half of wavelength. Near a parallel resonance, the integral equation gave finite values, but the input impedance, calculated by the method of induced emf, increased infinitely, since the denominator of the corresponding expression tended to zero.

The new option of solving the system of equations (2.15) was based on the use of the mathematical method of variation of constants. The first attempts to use this method were described in [17]. This decision allowed us to obtain new results and to answer separate questions. The reason for the different results in the vicinity of parallel resonance is considered in Section 2.3.

2.3 Method of variation of constants

As is known, in accordance with the method of the constants' variation the solution of the non-uniform differential equation is sought as follows. First, we find the solution of an auxiliary uniform equation, whose right hand side is zero. Then the constant magnitude used in this solution is replaced by an unknown magnitude, and the resulting function is substituted into the left side of the equation, which equates with the right side, allowing us to obtain a solution to the non-uniform equation.

If this equation is of a higher order than the first one, such a technique can meet with insurmountable difficulties. In our case, we are talking about the equation for J_n from the system of equations (2.15). Although this is a second-order equation, the application of this technique is facilitated by the similarity of the left and right sides of the equation. Each of them contains a function and its second derivative with the same coefficients, i.e., their ratio can be determined using one of the members. This makes it easy to go from solving a known uniform equation to solving a non-uniform one.

In the capacity of a right side of the non-uniform equation for the current $\chi^n J_n$ we should take the extraneous emf. The variant of the corresponding expression is chosen depending on the symmetry of the radiator. For definiteness, we will consider that the radiator is symmetric. In this case, in accordance with the equations set (2.15) the extraneous emfs are concentrated at the point $z = 0$.

As stated in Section 2.1, the total field is the sum of the extraneous field and the field of currents. This means that when calculating the next member of the series for the current, we must take in the capacity of the extraneous emf the field created by the previous member of the series, i.e., the current of a smaller order of smallness flowing along the entire length of the antenna. As can be seen from (2.15), the external field created by the zero component of the current is

$$E_\varsigma(J_0) = -W(\chi J_0). \tag{2.18}$$

The current J_0 is continuous along the radiator and has a constant amplitude. The field $E_z(J_0)$ of this current on the surface of the radiator is also continuous. It is obvious, that this field depends not only on the distance between the points of radiation and reception, but also on the phase of the current. Therefore, the field of the entire radiator is determined by integrating the product of the field to the magnitude $\sin k(L - z)$ along the length of the radiator. In accordance with (2.13), the external emf created by the field $E_1(J_0)$ is χ times larger than the field, i.e., the negative extraneous emf exciting the second component of the current is

$$e_1 = -\chi \int_{-L}^{L} E_\varsigma(J_0) \sin k(L - |\varsigma|) d\varsigma. \tag{2.19}$$

Let's write the *nth* member of the series (2.14) as

$$J_{(n)}(\varsigma)\sin k(L-\varsigma) = \chi^n J_n(\varsigma)\sin k(L-\varsigma). \quad (2.20)$$

Here, $J_{(n)}(\varsigma)$ is the amplitude of the *nth* member of the series for the current. Each previous member of the series is the current $J_{(n-1)}(\varsigma)$. This current creates the field $\int_{-L}^{L} E_\varsigma(\chi^{n-1}J_1)\sin k(L-|\varsigma|)d\varsigma$ in the form of an integral along the entire length of the antenna. The field is multiplied to the small parameter χ and creates emf e_n, which excites the current $J_{(n)}(\varsigma)$—the next member of the series. This means that the emf e_n, exciting the *nth* member of the series, starting from the second, is obtained by sequentially multiplying the field of the previous series member to χ. Accordingly, the external emf, exciting the *nth* component of the current, is equal to

$$e_n = -\chi^{n-1}\int_{-L}^{L} E_\varsigma[\chi J_1(\varsigma)]\sin k(L-|\varsigma|)d\varsigma. \quad (2.21)$$

The field magnitude $\int_{-L}^{L} E_\varsigma[J_{(n)}(\varsigma)]d\varsigma$ is defined in accordance with (1.30). Substituting this value into (2.21), we obtain an expression, which resembles the formula (1.77) for the input impedance of the dipole. This formula is known from the method of induced emf. According to (1.26) $J(z) = J(0)\sin k(L-|z|)$. In accordance with (2.20) the current $J_{(n)}(z)$ has the same form. If we replace $J(z)$ in (1.77) by $J_{(n)}(\varsigma)$, then

$$\int_{-L}^{L} E_\varsigma[J_{(n)}(\varsigma)]J_{(n)}(\varsigma)d\varsigma = -J_{(n)}^2(0)Z_{All}. \quad (2.22)$$

In the left part of this expression the sum of powers in elementary antenna segments of length $d\varsigma$ is recorded. In the right part the total power is recorded. Their equality is the meaning of this expression. A similar equality can be written for any member of the series (2.14).

If in the expression (2.22) we replace each power by emf, which excites elementary segments and the whole antenna, then we obtain

$$\int_{-L}^{L} E_\varsigma[J_{(n)}(\varsigma)]\sin k(L-|\varsigma|)d\varsigma = -J_{(n)}(0)Z_{All}. \quad (2.23)$$

Here the left part of expression (2.23) differs from the left part of expression (2.22) by replacing the multiplier $J_{(n)}(\varsigma)$ with $\sin k(L-\varsigma)$. The right part of the equality describes the position at the antenna input—at the point $\varsigma = 0$. It shows that the indicated emf is equal to the product of the current $J_{(n)}(0)$ and the antenna impedance. We assume that the input impedance is the same for all currents, starting from $J_{(1)}(\varsigma) = \chi J_1(\varsigma)$, and is equal to Z_{All}, because for the indicated currents each emf and the corresponding impedance are measured at the same point.

72   ANTENNAS: Rigorous Methods of Analysis and Synthesis

In accordance with (2.23)

$$e_n = -J_{(n)}(0)Z_{AII} = -\chi^n J_n(0)Z_{AII}.$$

On the other hand, in accordance with (2.21)

$$e_n = -\chi^{n-1}J_{(1)}(0)Z_{AII} = -\chi^n J_1(0)Z_{AII}.$$

where from

$$J_n(0) = J_1(0), \quad J_n = J_1,$$

i.e.,

$$J_{(n)}(\varsigma) = \chi^n J_1(\varsigma). \tag{2.24}$$

The same result is obtained, if to use the method of constant variation and compare two equations of set (2.15)—for the current $J_1(z)$ and for the current $J_n(z)$:

$$\frac{d^2 J_{(1)}(z)}{dz^2} + k^2 J_{(1)}(z) = e, \quad \frac{d^2 J_{(n)}(z)}{dz^2} + k^2 J_{(n)}(z) = e_n \frac{Z_{A0}}{Z_{AN}}.$$

While checking the ratio of currents in the left parts of equations the right side of each equation must correspond to the current. Therefore, during checkout one must multiply the emf on the right side of the second equation to the inverse ratio of impedances. Since $e_1 = \chi J_1(0)$, $e_n = -\chi^n J_1(0) Z_{AII}$, then $\dfrac{e_n}{e_1} = -\dfrac{\chi^n J_1(0)}{\chi J_1(0)} = -\chi^{n-1}$. From here $J_{(n)}(\varsigma) = -\chi^{n-1} J_{(1)}(\varsigma)$, i.e., $J_{(n)}(\varsigma) = \chi^n J_n(\varsigma) = \chi^n J_1(\varsigma)$ in accordance with (2.24). Thus, the verification confirms that the ratio of currents in the left parts of arbitrarily chosen equations coincides with the ratio of the emf in the right parts of the same equations.

As follows from the obtained results, each emf, starting from e_1, creates a new current in the same load Z_{AII}. Equality (2.22), obtained in accordance with the expression known as the basic formula of the induced emf method, allows us find the relationship between J_n and J_1. The solution, obtained by means of Leontovich's integral equation, coincides with the solution found by means of the second formulation of the induced emf method. In this solution the asterisk, i.e., a sign for using the concept of reactive power, is absent.

It is interesting to note that the reactive component of the input impedance Z_{AII} contains a summand, whose magnitude is substantially greater than the other summands. This summand is equal to

$$jX_{A1} = -j\frac{30}{\sin^2 \alpha} \sin 2\alpha \cdot 2\ln(L/a).$$

Replacing the last factor of this expression by $1/\chi$, we find

$$jX_{A1} = -j\frac{60}{\chi \tan \alpha} = \frac{1}{\chi J_1(0)}.$$

It is easy to verify that this magnitude is equal to the input impedance of a two-wire long line, open at its end as well as the input impedance of a symmetric thin radiator with the length L, whose wave impedance in the first approximation is equal to $W = \dfrac{60}{\chi}$. This term is positive, and therefore, as can be seen from (2.23), the second emf differs in sign from the first emf in accordance with the general order of the sequential change of signs. As a result, the signs of the series members also sequentially change.

Using (2.24), we obtain from (2.14)

$$J(z) = J_0(z) + \chi J_1(z) - \chi^2 J_1(z) + \chi^3 J_1(z) - \ldots \qquad (2.25)$$

The expression (2.25) allows us to find the current of the n*th* order of smallness. The ease, with which these calculations are performed, indicates the advantages of the equation, which suggests that the radiator is located on the axis of the coordinate system. This makes it possible to determine the n*th* term of the series (2.14) for the current at an arbitrary point z of the radiator and, accordingly, the n*th* approximation if previous approximations are known, both in the general case and for the symmetric radiator. To find the input current of the antenna, we need to put $z = 0$ in the expression for the current.

The obtained results permit us to construct a set of equations for the members of the series (2.14) and to find the amplitudes of the currents. The found order on changing the members of the series allows us to conclude that the members of the series (2.14), starting from $\chi J_1(z)$ form a geometric progression. The progression begins with the first term, equal to $a_1 = \chi J_1(z)$. The denominator of the progression is constant and equal to $q = -\chi$. This is an important attestation of the usefulness of condition (2.24). If the number N of members of such a progression grows endlessly, then its sum tends to the limit

$$s = a_1/(1 - q).$$

In our case, the sum of the members of the progression is $\sum_{n=1}^{\infty} \chi^n J_1(z) = \chi J_1(z)/(1+\chi)$. Hence the total current of the radiator in accordance with (2.25) is equal to

$$J(z) = J_0(z) + \chi J_1(z)/(1+\chi), \qquad (2.26)$$

where according to (2.16) $J_0(z) = e/Z_{A0}$. In accordance with (2.19) and (2.23) $\chi J_1(0) = e_1/Z_{A1}$, where similarly (2.23) $e_1 = -\chi J_0(0)Z_{A1}$, i.e., $\chi J_1(0) = -\chi J_0(0)$. Hence the input impedance is

$$Z_A = \frac{e}{J_0(0) - \frac{\chi J_0(0)}{(1+\chi)}} = \frac{e}{J_0(0)\left(1 - \frac{\chi}{1+\chi}\right)}.$$

As can be seen from the description of the solution technique, which made it possible to obtain expression (2.26), the transition to the series (2.14) for the current allows us to isolate the first member, which differs significantly from the rest of the members, because it does not have a real component. At the same time, the isolation of the first member permits to simplify greatly the description of using geometric progression.

The obtained result is approximate, since it takes into account only the reactive component of the input impedance. In this expression, an active component of the input impedance is absent. To find it, we compare this result with the known magnitude Z_{All} in the second approximation. To do this, we calculate the value Z_A, taking into account two first members of the expression (2.25):

$$\frac{e}{Z_{All}} = J_0(0) + \chi J_1(0),$$

i.e., $\chi J_1(0) = \frac{e}{Z_{All}} - J_0(0)$. As it follows from (2.26), the total current at the point $z = 0$ is equal to

$$J(0) = J_0(0) + \chi J_1(0)/(1+\chi) = J_0(0) + \frac{e}{Z_{All}(1+\chi)} - \frac{J_0(0)}{(1+\chi)}.$$

The second component of this sum is equal to product of the current $\frac{e}{Z_{All}}$ in the second approximation to $\frac{1}{1+\chi}$. In fact, it is the total current at the point $z = 0$, since the emf and the load are the same for the total current and the current in the second approximation. The sum of the first and third members in the right part of this equality is equal to zero, the first member of the total current is changed and compensates the third member. Thus,

$$Z_A = (1+\chi)Z_{All}. \qquad (2.27)$$

This expression allows us to calculate the total sum of the series for the dipole current and take into account the active component of the current and the input impedance.

It is necessary to emphasize that the results of calculating the input impedance of a linear symmetric radiator Z_{All} in the second approximation, obtained by three methods: by the induced emf method in the second formulation, by the method, described in [20] and by the method, presented in the given chapter—are identical. The expression given in [20] for the input impedance of the untuned radiator, presented as a combination of integral and trigonometric sines and cosines, completely coincides with the similar expression that the induced emf method gives as the result of integrating expression (1.77). The difference in the results arises when calculating the input impedance near the parallel resonance, since in this case the induced emf method is forced to use the same general expression, and the result tends to infinity. In [20], the input impedance at parallel resonance is determined by solving the equation for the tuned radiator, and that allows us to obtain the final result. In fact, these components are given for the tuned cylindrical dipoles with a circular cross-section and arm length $\frac{\pi}{4}, \frac{\pi}{2}$ and $\frac{3\pi}{4}$ in a second approximation: $R_A = 73.2, 93.5$ and 105.4; $X_A = 42.5, 44.8$ and 45.5.

These values are given for the components of the input impedance Z_A of the antenna (for the generator load) in the case when the generator is included in the antinodes of the current. If the generator is placed on a distance z_0 from the antinode, then the impedance of its load changes. For a dipole length, equal to $\lambda/2$, it is equal to $Z_A = (73.2 + j42.5)/\cos\left(\frac{\pi z_0}{2L}\right)$, for a length λ, $Z_A = (73.2 + j42.5)/\cos\left(\frac{\pi z_0}{2L}\right)$. Multiplying these magnitudes to the corresponding factor of the Table 2.1 allows us to find the total components of the input impedance.

One can consider that the active component of the input impedance of a linear symmetric radiator is equal to the active component of the impedance Z_{AII}. But a simpler and more convenient method of calculating the active component is based on the analogy of a symmetric radiator with a Hertz dipole. In accordance with this analogy

$$R_A(h) = 20k^2 h_e^2, \qquad (2.28)$$

where $h_e = \frac{2}{k}\tan\frac{kL}{2}$ is the effective length of the symmetric metal radiator, equal to a ratio of an area under a current curve to the input current $J(0)$. From here

$$R_A(h) = 80\tan^2\frac{kL}{2}.$$

The result would be more accurate, if we take into account the phase of a dipole current, which is not constant as in the case of the Hertz dipole,

Table 2.1. Relation of components.

L/a	5	20	50	200	500	2000	5000
χ	0.311	0.167	0.128	0.0944	0.0805	0.0658	0.0587
$1+\chi$	1.311	1.167	1.128	1.0944	1.0805	1.0658	1.0587

but grows in the direction of a generator. This phase at the point located at a distance x from the free end of the dipole arm is equal in the first approximation to $\tan^{-1}\dfrac{X_A}{R_A}$, where X_A and R_A are the reactive and active components of the input impedance for the dipole with the arm length x.

It should be said that the formula for Z_{AII} is widely known, and an attempt to redetermine it does not give any practically useful result, i.e., it did not allow us to obtain a significantly more precise result compared to elementary calculations, which were ignoring small magnitudes of the order of wire radius. As can be seen from expressions (1.30) and (2.22), in the calculating process it is necessary to determine the integrals of the form

$$Y_m = \int_0^L \frac{exp(-jkR_m)}{R_m} \sin k(L-\zeta) d\zeta,$$

where $R_1 = \sqrt{(z-L)^2 + \rho^2}$, $R_2 = \sqrt{(z+L)^2 + \rho^2}$, $R_0 = \sqrt{z^2 + \rho^2}$. In particular the integral

$$Y_1 = \frac{1}{2j} \int_{-L}^{L} \left\{ \frac{exp[jk(-R_1 + L - \zeta)]}{R_1} - \frac{exp[jk(-R_1 - L + \zeta)]}{R_1} \right\} d\zeta$$

consists of two integrals. Applying in the first and second integrals the substitutions $t = -R_1 + L - \zeta$ and $u = -R_1 - L + \zeta$, we obtain

$$Y_1 = \frac{1}{j}\left[\int_{L-\sqrt{L^2+a^2}}^{-a} \frac{exp(jkt)}{t} dt + \int_{-L-\sqrt{L^2+a^2}}^{-a} \frac{exp(jku)}{u} du \right] =$$
$$- j\left[2Ei(-jka) - Ei\left(-\frac{jka^2}{2L}\right) - E_i(-j2kL) \right].$$

Results of the integration contains terms with arguments of the order of kL, ka and smaller. If an argument x is small,

$$Ei(jx) = Cix + jSix = \ln\gamma x + jx.$$

If we neglect small quantities, then summarizing the integration results gives

$$\frac{e}{\chi J_1(0)} = Z_A(\alpha) + \Delta(\alpha),$$

where $Z_A = e^{2j\alpha}[Ei(-4j\alpha) - 2Ei(-2j\alpha) + \ln\gamma\alpha] + 2\ln 2\gamma\alpha - 2Ei(-2j\alpha) - 2j\sin 2\alpha \ln(L/a)$,

$\Delta = j\{-4ka + (ka^2/(2L)) - [2ka - (3ka^2/(4L)] \cos 2\alpha - 3ka^2/(4L)\}$,

and

$$\chi J_1(z) = e/Z_A.$$

This result is almost equal to the result of the calculation, which begins with neglecting small magnitudes.

The following figures demonstrate active and reactive components of the radiator input impedance in the second approximation and with allowance of the total sum of the series. These figures also allow us to compare the obtained results for different dipole sizes with the results of an application of the induced emf method.

Figure 2.1 gives the reactance of dipoles with the ratio L/a equal to 5 and 50, and Fig. 2.2 shows the reactance of dipoles with the ratio L/a equal to 500 and 5000, respectively. The total reactance X_A is given by a solid curve, the reactance in the second approximation $X_{A2} = X_{AII}$—by a dotted line. In accordance with (2.27) the total reactance (absolute magnitude) is greater than the reactance in the second approximation by $(1 + \chi)$ times. The magnitudes χ and $1 + \chi$ depending on L/a are given in Table 2.1.

From the figures it can be seen that the difference between the curves for the total reactance and the reactance in the second approximation is small and

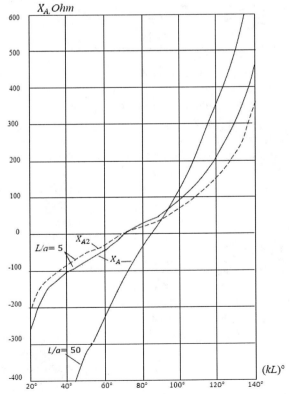

Fig. 2.1. Reactive component of the input impedance for the dipole with $L/a = 5$ and 50.

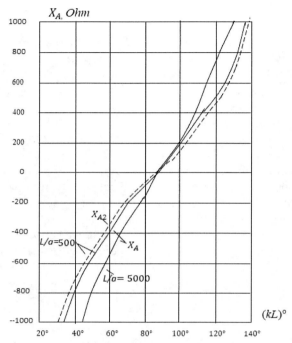

Fig. 2.2. Reactive component of the input impedance for the dipole with $L/a = 500$ and 5000.

decreases with decreasing antenna thickness. A similar relation is valid also for the active components of the input impedance.

The active component R_A of the input impedance is shown in Figs. 2.3 and 2.4. It has a similar character, but the resistance in the second approximation does not depends on the magnitude of L/a. The total resistance depends on L/a, it is proportional to $(1 + \chi)$. In Fig. 2.3 there are curves for the active component R_{A2} in the second approximation, and also the total resistance R_A for $L/a = 5$ and 50. The total resistance R_A for $L/a = 500$ is given in Fig. 2.4. Also, in Fig. 2.4 it is given the resistance $R_A(h)$, which is calculated in accordance with (2.28).

This result can be defined more precise, if to take into account that the current is not constant, as in the case of the Hertz dipole, but decreases as a result of approach to the generator, since the radiation losses cause an attenuation of the current in an equivalent long line.

As it follows from the above results and graphs, illustrating these results, the sum of the series (2.14) is close to the sum of the first two or three terms. L.A. Vainshtein, solving the so-called key equation and analyzing the numerical results, came to similar conclusions. He briefly spoke about these results in 1964 in Kharkov, at a conference on diffraction and propagation of radio

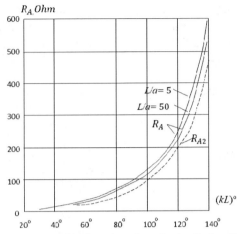

Fig. 2.3. Active component of the input impedance for the dipole with $L/a = 5$ and 50.

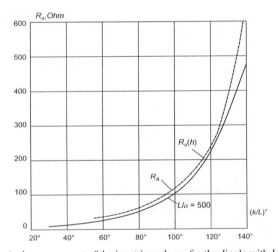

Fig. 2.4. Active component of the input impedance for the dipole with $L/a = 500$.

waves. In this connection, it is necessary to point out that the calculation of the sum of the series (2.14) has not so practical as fundamental importance. The calculation shows that the result obtained in [20] (the second approximation for the current and the input impedance) is very close to the exact solution.

2.4 Integral equation for two metal radiators

In the analysis of the structure, consisting of several radiators forming the antenna array, we begin with a special case. Lets consider two parallel symmetrical radiators of different lengths that are axially shifted relative to each other (Fig. 2.5). According to (1.16), if electric currents $J_I(\sigma)$ and $J_{II}(\varsigma)$ with circular frequency ω flow along the radiators, then the field created by these currents is

$$E_z = -j\frac{\omega}{k^2}\left(k^2 A_z + \frac{\partial^2 A_z}{\partial z^2}\right), \qquad (2.29)$$

and in accordance with the principle of superposition

$$A_z = A_{1z} + A_{2z}. \qquad (2.30)$$

Using a model of two radiators in the form of two straight thin-walled cylinders with length L_1 and L_2, with radius a_1 and a_2, respectively, we write down according to (1.20)

$$A_{1z}(\rho,z) = \frac{\mu}{8\pi^2}\int_{-L_1}^{L_1} J_1(\sigma)\int_0^{2\pi} \frac{e^{-jkR_1}}{R_1} d\varphi d\sigma,$$

$$A_{2z}(\rho,z) = \frac{\mu}{8\pi^2}\int_{h-L_2}^{h+L_2} J_1(\varsigma)\int_0^{2\pi} \frac{e^{-jkR_2}}{R_2} d\psi d\varsigma. \qquad (2.31)$$

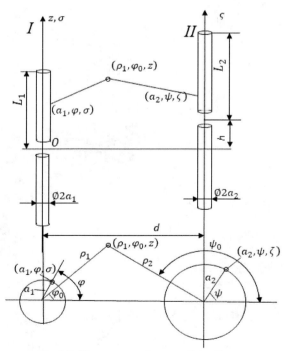

Fig. 2.5. The structure of two parallel radiators.

Here R_1 and R_2 are the distances from the observation point with coordinates (ρ_1, φ_0, z) to integration points (a_1, φ, σ) and (a_2, ψ, ς) on the surface of the first and second radiators, respectively, i.e.,

$$R_1 = \sqrt{(z-\sigma)^2 + \rho_1^2 + a_1^2 - 2a_1\rho_1 \cos(\varphi - \varphi_0)},$$
$$R_2 = \sqrt{(z-\varsigma)^2 + \rho_2^2 + a_2^2 - 2a_2\rho_2 \cos(\psi - \psi_0)}.$$

For the current in each radiator one can write an equation similar to equation (2.13). Let the observation point be located near the surface of the first radiator. Then, if

$$a_1, a_2 \ll d, \tag{2.32}$$

where d is the distance between the axes of the radiators, one can consider that the vector potential $A_{1z}(\rho, z)$ has a logarithmic feature near the first radiator, the isolation of which similarly (2.9) gives

$$A_{1z}(a_1, z) = (\mu/4\pi)[-2J(z)ln\rho\rho + V(J, z)].$$

The vector potential A_{2z} near the second radiator does not have such a feature, since if condition (2.32) is satisfied, the magnitude R_2 is not small for any ς: $R_2 \geq d - \rho_1 - a_1$. Accordingly the tangential component of the electric field created by the current $J_1(z)$ is written as expression (2.11), and it has a large magnitude of the order χ^{-1}. The field $E_z(J_{II})$, created by the current of the second radiator on the surface of the first one, does not contain a large summand.

Similar to (2.2), the boundary condition on the surface of the first radiator has the form

$$E_z(J_I, a_1, z) + E_z(J_{II}, a_1, z) + K_I(z) = 0, -L \leq z \leq L. \tag{2.33}$$

where $K_I(z)$ is the extraneous field, $E_z(J_I, a_1, z)$ and $E_z(J_{II}, a_1, z)$ are the fields of the first and second radiators on the surface of the first radiator. Substituting (2.11) into (2.33), we obtain the integral equation for the current $J_I(z)$:

$$\frac{d^2J_I(z)}{dz^2} + k^2 J_I(z) = -j4\pi\omega\varepsilon_0\chi[K_I(z) + E_z(J_{II}, a_1, z)] - \chi W(J_I), \tag{2.34}$$

where, $J_I(\pm L_1) = 0$. Here, $\chi = 1/[2ln(2L_1/a_1)]$. Equation (2.34) generalizes equation (2.13) written for the current in a single radiator.

The right part of this expression contains three components. The first component is the exciting emf, the second component is emf caused by influence of the second radiator, and the third component is emf, which takes the radiation into account.

82   ANTENNAS: Rigorous Methods of Analysis and Synthesis

While solving the equation (2.34), we represent the currents $J_I(z)$ and $J_{II}(\varsigma)$ in the form of a series in powers of small parameters χ_1 and χ_2, respectively:

$$J_I(z) = \sum_{n=0}^{\infty} \chi_1^n J_{In}(z), J_{II}(\varsigma) = \sum_{n=0}^{\infty} \chi_2^n J_{IIn}(\varsigma). \qquad (2.35)$$

where $\chi_2 = 1/[2ln(2L_2/a_2)]$. It means that the solution of equation (2.34) is sought as a series for the current in powers of a small parameter, similar to the series (2.14). Since functionals $W(J_1, z)$ and $E_z(J_{II}, a_1, z)$ are linear, they can also be presented in the form of similar series. If χ_1 and χ_2 have the same order of smallness:

$$\chi_1 \sim \chi_2, \qquad (2.36)$$

then the equation (2.34) for a nonresonant radiator reduces to the set of equations for $J_{In}(z)$, the generalized set of equations (2.15), written for a single radiator. In the right side of these equations, a new summand $\left(\dfrac{\chi_2}{\chi_1}\right)^n E_z(J_{II,n})$ appears next to $W(J_{In})$.

From the first equation it follows that when one of the radiators is excited by the concentrated emf, the current in it has a sinusoidal character in the first approximation which is also valid in the presence of the second radiator. Using the method of constants variation, it is easy to be convinced that the second radiator creates the field $E_o(\chi_2^{n-1} J_{II,n-1})$ involved in excitation of the following current component $J_{In}(z)$. The second radiator introduces a new summand into the integrand of the expression (2.21) for the extraneous emf located in the center of the first radiator and excited the nth component of the current.

In the left part of this expression, the first term in the bracket is the extraneous emf for the nth component of the current produced by the first radiator, and the second term is the extraneous emf for the same component produced by the second radiator.

$$e_n = -\chi^{n-1}\int_{-L}^{L} E_\varsigma[J_{I(1)}(\varsigma)]sink(L-|\varsigma|)d\varsigma + \int_{-L}^{L} E_\varsigma[J_{II(n)}(\varsigma)]sink(L-|\varsigma|)d\varsigma.$$

From here it is easy to pass a generalization for the expression of (2.22):

$$\int_{-L}^{L} \{-E[J_{I(n)}] + E[J_{II(n)}]\} J_{I(n)}(\varsigma)d\varsigma = -J_{I(n)}^2(0)Z_{AII}. \qquad (2.37)$$

During the analysis of radiators, located in a structure of two connected antennas, it is necessary to take into account the conditions (2.32) and (2.36). If these conditions are not met, the calculation may have a significant error. This circumstance must be taken into account when applying the method of calculation to folded, multi-wire and multi-radiator antennas, whose wires are located parallel to each other for distances, that are small compared to their length. The parameters χ of the coupled radiators, as a rule, have the same order of smallness. However, one must remember that the possibility of their use requires verification, if they are significantly different from each other.

2.5 Radiators with distributed loads

The next section is devoted to the application of Leontovich's equation to current in radiators with distributed surface impedance (or in other words, with distributed loads). The boundary conditions on the surface of the wire with distributed surface impedance, i.e., at $-L \le z \le L$, are given in Section 1.5 in the form of expression (1.36). An example of such a radiator is an antenna in the form of a metal rod covered with a magnetodielectric shell (Fig. 2.6). The antenna has the shape of a circular cylinder, with radius "a" which is small compared to its length $2L$ and wavelength λ.

In accordance with the equivalence theorem, when considering electromagnetic processes in the free space surrounding the radiator, one should replace an antenna by a field on its surface and then operate only with fields. But for clarity and simplicity of reasoning, it is expedient to mentally

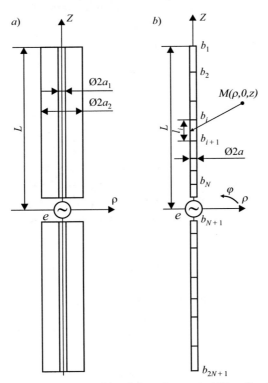

Fig. 2.6. Radiators with constant (a) and piecewise constant (b) surface impedance.

metallize the radiator surface, replacing the magnetic field strength $\vec{H}$ by the surface electric current density $\vec{J}_s$ in accordance with the equality

$$\vec{J}_s = [\vec{e}_\rho, \vec{H}], \qquad (2.38)$$

where $\vec{e}_\rho$ is the unit vector directed along the axis ρ. In that case

$$H_\varphi(a, z) = j_z(z) = J(z)/(2\pi a). \qquad (2.39)$$

Here, $J(z)$ is the linear current along the metallized antenna (the total current of the radiator).

Substituting into the boundary condition (1.36) the expression (2.39) for the magnetic field and the expression (2.11) for the tangential component of the electric field created by the current $J(z)$, we obtain the equation for the current in the impedance radiator, similar to equation (2.13):

$$\frac{d^2 J(z)}{dz^2} + k^2 J(z) = -j4\pi\omega\varepsilon_0 \chi[K(z) - J(z)Z(z)/(2\pi a)] - \chi W(J), \qquad (2.40)$$

which must satisfy the condition of zero current at the radiator ends. The right side of the equation consists of three terms. The first term is the exciting emf, the second term is a distributed load—the surface impedance $Z(z)$, the third term takes into account the radiation. Assuming that the surface impedance is constant along the length of the radiator, we introduce the designation

$$j\frac{k\chi Z}{60\pi a} = U. \qquad (2.41)$$

The equation for the current will take the form

$$\frac{d^2 J}{dz^2} + k_1^2 J = -j4\pi\omega\varepsilon_0 \chi K(z) - \chi W(J), \qquad (2.42)$$

where the magnitude $k_1 = \sqrt{k^2 - U}$ is constant in the case of an antenna with a constant surface impedance. If the order of the magnitude of U is close to the order of magnitude of k^2, then the surface impedance essentially affects the current distribution along the radiator, and the magnitude of

$$k_1 = \sqrt{k^2 - U} = \sqrt{k^2 - jk\chi Z/(60\pi a)} \qquad (2.43)$$

has the sense of the new propagation constant of a wave along the antenna. The relation k_1/k is usually called a slowdown.

We will seek the solution of equation (2.42) in the form of a power series of a small parameter, similar to the series (2.14). In the first approximation, the current distribution along the radiator has a sinusoidal character:

$$\chi J_1(z) = j\frac{k\chi e}{60 k_1 \cos k_1 L} \sin k_1(L - |z|). \qquad (2.44)$$

The input impedance of the radiator in this approximation is purely reactive and is equal to the input impedance of a uniform long line open at the end, in which the wave propagation velocity is k_1/k times less than in air, and the characteristic impedance k_1/k times greater.

The second equation based on equation (2.42) allows us to determine the current in the second approximation, to calculate the magnitude of the active component of the input impedance and to revise the magnitude of its reactive component. For a variant of the antenna excited by a concentrated emf located in its middle, we obtain:

$$\chi^2 J_2(z) = j \frac{\chi^2 e}{120 \cos^2 \beta} \frac{\alpha^2}{\beta^2} \Theta(\beta, \alpha) \sin k_1(L - |z|), \qquad (2.45)$$

where

$$\Theta(\beta, \alpha) = \frac{1}{2}\left(\frac{\beta}{\alpha} + \frac{\alpha}{\beta} + j\frac{2ml}{\alpha}\right) e^{j2\beta} Ei(-j2m) - \frac{1}{2}\left(\frac{\beta}{\alpha} + \frac{\alpha}{\beta} - j\frac{2ml}{\alpha}\right) e^{-j2\beta} Ei(j2l) -$$

$$\left(\frac{\beta}{\alpha} + \frac{\alpha}{\beta} + j\frac{ml}{\alpha}\right)(1 + e^{j2\beta}) Ei(-jm) + \left(\frac{\beta}{\alpha} + \frac{\alpha}{\beta} - j\frac{ml}{\alpha}\right)(1 + e^{-j2\beta}) Ei(jl) +$$

$$\left[\left(1 + \frac{e^{j2\beta}}{2}\right)\left(\frac{\beta}{\alpha} + \frac{\alpha}{\beta}\right) + j\frac{ml}{\alpha}\right]\ln m - \left[\left(1 + \frac{e^{-j2\beta}}{2}\right)\left(\frac{\beta}{\alpha} + \frac{\alpha}{\beta}\right) - \frac{ml}{\alpha}\right]\ln l +$$

$$j\left(\left(\frac{\beta}{\alpha} + \frac{\alpha}{\beta}\right)\sin 2\beta + \frac{2ml}{\alpha}\right)\ln\frac{\gamma}{2\alpha} - (1 + e^{-j2\alpha} + 2\cos^2\beta) + 4e^{-j\alpha}\cos\beta + j\frac{\beta}{\alpha}\sin 2\beta.$$

Here the notations are used: $Ei(jx) = Cix + jSix$, $m = \beta + \alpha$, $l = \beta - \alpha$, $\beta = k_1 L$, $\alpha = kL$.

Because the current in the second approximation at the emf application point is equal to

$$J(0) = \chi J_1(0) + \chi^2 J_2(0),$$

then far from the parallel resonance

$$Z_{All} = e\left[\frac{1}{\chi J_1(0)} - \frac{J_2(0)}{J_1^2(0)}\right] = -j\frac{60\beta}{\chi\alpha}\cot\beta + \frac{30}{\sin^2\beta}\Theta(\beta,\alpha),$$

and near the parallel resonance

$$Z_{All} = \frac{e}{\chi^2 J_2(0)} = 120\frac{\beta^2}{\alpha^2}\chi^{-2}\Theta^{-1}(\beta,\alpha).$$

The presence of surface impedance significantly changes the characteristics of the antenna. When calculating the field created by the current of the impedance radiator, it must be taken into account that the voltage drop on the antenna itself (on the impedance included in the antenna) does not contribute to the radiated field, and therefore in the expression (1.77) for the input impedance, the magnitude E_z should be replaced by the difference $E_{z1} = E_z - H_\varphi Z$. Then instead of (1.77) we get

$$Z_{All} = -\frac{1}{\chi^2 J_1^2(0)} \int_{-L}^{L} \left[E_z(\chi J_1) - \frac{z\chi J_1(\varsigma)}{2\pi a} \right] \chi J_1(\varsigma) d\varsigma. \qquad (2.46)$$

One must remember that the surface impedance not only adds a summand to this expression, but also changes the current distribution along the radiator, since the propagation constant is changed. The field $E_z(J)$ of an impedance antenna, like the field of a metal radiator, is calculated in accordance with expression (1.28). Despite the fact that in this case, the propagation constant is changed in the integral of the right side of expression (1.28), the first multiplier of the integrand is as before equal to zero. For the longitudinal component of the electric field of the antenna at an arbitrary point M with coordinates $(\rho, 0, z)$ located at a distance d from the antenna axis (see Fig. 1.2a) we obtain:

$$E_z = -j\frac{30J(0)}{\varepsilon_0}\frac{k_1}{k}\left[\frac{\exp(-jkR_1)}{R_1} + \frac{\exp(-jkR_2)}{R_2} - \frac{\exp(-jkR_0)}{R_0}\cos k_1 L\right]. \qquad (2.47)$$

Here $R = \sqrt{(z-\varsigma)^2 + d^2}$, $R_+ = \sqrt{(z+\varsigma)^2 + d^2}$, and $R_1 = \sqrt{(z-L)^2 + d^2}$, $R_2 = \sqrt{(z+L)^2 + d^2}$, $R_0 = \sqrt{z^2 + d^2}$ are the distances from the observation point M to the upper and lower ends of the radiator, as well as its middle (see Fig. 1.2a). Substitution of this expression in (2.46) gives

$$Z_A = j60\int_0^L \left(\frac{e^{-jkR_1}}{R_1} + \frac{e^{-jkR_2}}{R_2} - 2\frac{e^{-jkR_0}}{R_0}\cos k_1 L\right)\sin k_1(L-z)dz + \frac{Z}{4\pi a}\left(2L - \frac{\sin 2k_1 L}{k_1}\right). \qquad (2.48)$$

This expression differs from the similar expression (1.77) for a metal radiator by a term proportional to the surface impedance Z. If d is the distance between two parallel radiators of the same dimensions, then the magnitude of Z_A is a mutual impedance between two radiators. If $d = a$, where a is the radius of the antenna, then Z_A is the self-impedance of the antenna. It is impossible to reduce expression (2.47) to tabulated functions in the usual way, since the integrand depends on two propagation constants, k and k_1. This can only be done for calculating the self-impedance of a thin radiator with a radius $a \ll L, \lambda$.

To do this, we introduce a magnitude satisfying the inequality $a \ll \delta \ll L, \lambda$. Then for $|\varsigma - z| > \delta$ $R = |\varsigma - z|$, for $|\varsigma - z| < \delta$ $e^{-jkR} = 1$. Dividing the integration interval into segments from $-L$ to $-\delta$, from $-\delta$ to 0, from 0 to δ and from δ to L and consistently using appointed conditions, we obtain:

$$R_A = \frac{30}{\sin^2\beta}\left\{\left(\frac{\beta}{\alpha}+\frac{\alpha}{\beta}\right)\left[\ln\frac{m}{l}-Cim+Cil+\frac{\cos 2\beta}{2}\left(\ln\frac{m}{l}-2Cim+2Cil+Ci2m-Ci2l\right)\right.\right.$$
$$\left.+\frac{\sin 2\beta}{2}(Si2m-Si2l-2Sim+2Sil)\right]-2(\cos\beta-\cos\alpha)^2 -$$
$$\frac{\beta^2-\alpha^2}{\alpha}[Sim-Sil+\cos 2\beta(Sim-Sil-Si2m+Si2l)+\sin 2\beta(Ci2m-Ci2l-Cim+Cil)]\},$$

$$X_A = -\frac{60\beta}{\chi\alpha}\cot\beta+\frac{30}{\sin^2\beta}\left\{\left(\frac{\beta}{\alpha}+\frac{\alpha}{\beta}\right)\left[Sim+Sil+\frac{\cos 2\beta}{2}(2Sim+2Sil-Si2m-Si2l)+\frac{\sin 2\beta}{2}\right.\right.$$
$$\left.\left(\ln\frac{\gamma^2 ml}{4\alpha^2}+Ci2m+Ci2l-2Cim-2Cil\right)\right]-2\sin\alpha(2\cos\beta-\cos\alpha)+\frac{\beta}{2}\sin 2\beta-\frac{\beta^2-\alpha^2}{\alpha}*$$
$$\left[Cim+Cil-\ln\frac{\gamma^2 ml}{4\alpha^2}+\cos 2\beta(Cim+Cil-Ci2m-Ci2l)+\sin 2\beta(Sim+Sil-Si2m-Si2l)\right].$$
(2.49)

Here the notation adopted in (1.78) and (2.45) are used.

The field of the impedance radiator in the far zone and its radiation pattern can be calculated by considering the radiator as the sum of the elementary dipoles in accordance with the method described in Section 1.6 and using the expression (2.44) as the distribution of the current along the radiator. This calculation for the field of a symmetric radiator with an emf in the center, gives,

$$E_\theta = j60J(0)\frac{\exp(-jkR)}{R}\frac{k_1 k\sin\theta}{(k_1^2-k^2\cos^2\theta)}[\cos(kL\cos\theta)-\cos k_1 L]. \quad (2.50)$$

Using E_θ permits to find by means of expressions of Section 1.6 the radiation power. If we divide it to the square of the first component of the generator current, we get the antenna resistance of radiation, which coincides with the magnitude of R_A, presented in (2.49).

When the length of an electric radiator is small ($kL \ll 1$), then by means of (2.49), and replacing functions with the earlier members of the series expansion in, we arrive at the expression

$$R_A = R_\Sigma = 20k^2L^2. \quad (2.51)$$

This expression shows that in this case the radiation resistance of the impedance and the metal antenna coincide in magnitude. For short antennas whose length is comparable to the wavelength ($L < 0.3\, k\lambda/k_1$), one can use the formula

$$R_\Sigma = 20k^2 h_e^2, \quad (2.52)$$

where $h_e = \dfrac{2}{k_1}\tan\dfrac{k_1 L}{2}$ is the effective length of the symmetric impedance radiator. From here

$$R_\Sigma = 80\dfrac{k^2}{k_1^2}\tan^2\dfrac{k_1 L}{2}.$$

The properties of radiators with constant surface impedance which are considered in this section do not depend on the method of impedance creation. Let us, for example calculate the surface impedance of an antenna made in the form of a metal rod with radius a_1, surrounded by a ferrite shell with radius a_2 with absolute magnetic permeability μ and dielectric permittivity ε (see Fig. 2.7a).

The field structure inside the antenna will not change if we cut the ferrite shell into parts in the shape of rings and place between them equidistantly infinitely thin and perfectly conducting metal disks connected to the central rod. If the distances between the discs are small compared to the wavelength in the ferrite, such a structure can be considered as a set of short-circuited radial lines.

Using the telegraph equations for the radial line, we find the input impedance of the line, taking into account that at $\rho = a_1$ the voltage is zero. For the surface impedance we get

$$Z = \dfrac{E_z}{H_\varphi}\bigg|_{\rho=a_2} = Z_r\dfrac{2\pi a_2}{h}.$$

Here Z_r is input impedance of the radial line, h is the distance between adjacent discs. If the diameter of the antenna is small compared to the wavelength in the shell material, then

$$Z = j120\pi\mu_r k a_2 \ln(a_2/a_1),$$

and in accordance with (2.43)

$$k_1 = k\sqrt{1 + \mu_r\dfrac{\ln(a_2/a_1)}{\ln(2L/a_2)}}. \qquad (2.53)$$

From (2.53) it can be seen that the surface impedance of thin antennas is directly proportional to the magnitude μ_r and does not depend on ε_r. Thus, in order to obtain a significant slowing down in a thin impedance antenna, it is necessary to choose a magnetodielectric with a high μ_r as a material for the shell. The propagation constant of the wave k_1 in the antenna, made in the form of a metal rod surrounded by a ferrite shell, is always greater than propagation constant of the wave in the air.

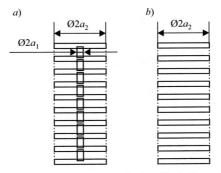

Fig. 2.7. Antennas in the form of a rod with a ferrite shell (*a*) and in the form of a set of parallel disks (*b*).

Such an antenna, in addition to ordinary losses, has losses in the shell, too. They can be taken into account by introducing complex magnetic permeability

$$\mu_r = \mu_1 - j\mu_2. \tag{2.54}$$

Substituting (2.54) into (2.53), we obtain the complex propagation constant

$$k_1 = k_{11} - jk_{12}$$

and complex magnitudes α and β in the expression (2.49). Additional active components in the input impedance created by reactive components α and β and proportional to these magnitudes is an additional resistance loss of the radiator. The expression for calculating this resistance loss in the radiator is given in [35] and is not mentioned here, since its bulkiness is unlikely to allow us to use it.

Figure 2.8 shows the calculated curves and experimental values of the active and reactive components of the input impedance for the three variants of asymmetric radiators, whose parameters are summarized in Table 2.2.

It can be seen from the figures that the experimental values agree quite well with the calculated ones. Some discrepancy in the value of R_A at high frequencies is explained by an increase of μ_2 with an increase in frequency f (in the calculation the magnitude of μ_2 is taken constant). The antenna efficiency depends on the frequency, increases with increasing f and is equal on average to about 0.5 for option 1 and about 0.8 for the variant 2.

As a second option, let us consider a thin antenna consisting of a set of equidistant parallel disks (Fig. 2.7*b*). The absence of an internal metal rod significantly changes the characteristics of the radiator. In this case the current at $\rho = 0$ is zero and the surface impedance Z is equal to

$$Z = -j\frac{240\pi}{ka_2\varepsilon_r}, \tag{2.55}$$

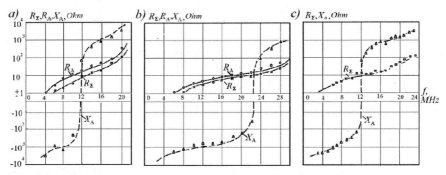

Fig. 2.8. Input impedance of the first (*a*), second (*b*) and third (*c*) variants of antennas with ferrite shell.

Table 2.2. Parameters of impedance radiators.

Variant of the radiator	L, m	a_2, m	a_1, m	μ_1	ε_r	k_1/k
1	2.0	0.021	0.007	40	10	3.0
2	1.5	0.15	0.007	25	10	2.3
3	1.0	0.021	0.007	100	10	5

i.e., $k_1^2 = k^2 - j4\chi/(\varepsilon_r a_2^2)$. In this case, in contrast to the previous one, the resonant frequency does not decrease, but rather sharply increases (except for very large dielectric permittivity).

The presented results of the analysis of electrical characteristics of antennas with constant surface impedance are based on the articles [3] and [20]. They were obtained by a group of authors led by E.A. Glushkovsky, and were first published in 1967 (in the journal "Radiotechnics") and were described in detail in [35]. The authors tried to simplify the cumbersome expressions obtained for the field, input impedance, resistance loss in the earth and others using the method of induced emf and the theory of integral equations. Unfortunately, these attempts did not help.

A new method for solving Leontovich's equation for the metal radiator allows us to avoid the difficulties of calculating characteristics of impedance antennas. As shown above, the integral equation for the current along the antenna with a constant surface impedance has the form (2.42). The current is distributed over a sinusoid with a new propagation constant k_1, and its magnitude in the first approximation is determined by the expression (2.44). If we use the perturbation method and look for the solution of equation in the form of a series in powers of a small parameter χ, similar to series (2.14), then in the general case of an untuned radiator we get a set of equations, which differs from (2.15) only by the replacement of k by k_1.

Since when calculating the next member of the series for the current, we must take in the capacity of the extraneous emf the field created by the previous member of the series, i.e., the current of a smaller order of smallness flowing along the entire length of the antenna, then as can be seen from (2.15), the extraneous emf created the second component of the current is equal to

$$E_2(\chi J_1) = -W(\chi J_1).$$

In this case, the magnitude $\chi J_1(z)$ is taken from expression (2.44), and the field $E_2(\chi J_1)$ in principle must be calculated in accordance with expression (2.47), which should be used in this case instead of (1.30). Replacing the magnitude $J(z)$ by $\chi J_1(z)$ in the expression (1.77), we can find, by a formula similar to (2.23), the load impedance Z_{AIII} for all current components, starting from the second member. For this, it is useful to know the value of $E_2(\chi J_1)$. However, as follows from the results of calculating currents and the input impedance of a metal radiator, it is not necessary to calculate this magnitude.

The extraneous emf, exciting the nth component of the current, is determined from (2.21). Introducing the notation of (2.22) and using the formula type (2.23), we come to equality (2.24) and expression (2.25) for the series. The expression (2.25) allows us to calculate the current of the nth order of smallness. The found order of changing the members of the series allows us to conclude that the members of the series (2.14) form a geometric progression. Similar to the case of a metal antenna without distributed load, the progression is endless, where the first term is equal to $a_1 = \chi J_1(0)$, the denominator of the progression is constant and equal to $q = -\chi$, and the sum of its terms tends to the limit $s = a_1/(1-q)$, i.e.,

$$Z_A = \frac{e}{J(0)} = \frac{e[1+\chi]}{\chi J_1(0)} = -j\frac{60 k_1[1+\chi]}{k\chi}\cot k_1 L.$$

In this expression, an active component of the input impedance is absent. To determine it, we calculate this component in the first approximation using the formula (2.52)

$$R_{A1} = 20 k^2 h_e^2.$$

In the general case an effective length h_e of an antenna is equal to the ratio of an area under a current curve to the input current $J(0)$. In this case $h_e = \frac{2}{k_1}\tan\frac{k_1 L}{2}$, i.e.,

$$R_{A1} = 80\frac{k^2}{k_1^2}\tan^2\frac{k_1 L}{2}.$$

Because $Z_A = Z_{A1}(1 + \chi)$, where Z_{A1} is the input impedance of the antenna in the first approximation, then

$$Z_A = (1 + \chi)\left[80\frac{k^2}{k_1^2}\tan^2\frac{k_1 L}{2} - j\frac{60k_1}{k\chi}\cot k_1 L\right]. \qquad (2.56)$$

The results of calculating the input impedance in accordance with (2.56) are close to the previously obtained results, but the calculation itself is much simpler.

The radiator with a constant impedance along its surface length is a special case of a radiator whose impedance varies along the antenna. Let, for example, the radiator consists of $2N$ sections of length l_i, and on each of these the surface impedance Z_i is constant (Fig. 2.6b). If the radius of the radiator a is much smaller than the wavelength λ and the length of the radiator $2L$, then the current distribution J_i along each segment has a sinusoidal character in the first approximation. The current is continuous along the antenna and absent at its ends:

$$J_i(b_i) = J_{i-1}(b_i), J_1(b_1) = J_{2N}(b_{2N+1}) = 0.$$

Lets isolate a logarithmic singularity from the total vector-potential of electric field. The vector-potential is obtained by means of summing up the vector potentials created by the currents of individual segments. Substitute the result in (1.16) and adopt, $\rho = a$. Then the integral equation for the current $J_m(z)$ of each segment of the radiator in accordance with the boundary condition (1.36) can be written in the form:

$$\frac{d^2 J_m(z)}{dz^2} + k^2 J_m(z) = -j4\pi\omega\varepsilon_0\chi[K(z) - J(z)Z_m/(2\pi a)] - \chi\sum_{i=1}^{2N}W(J_i,z), b_{m+1} \le z \le b_m.$$

Here Z_m is the magnitude of the surface impedance at the mth segment.

As before, we consider such an impedance, which is significantly affecting the distribution of the current. In this case

$$\frac{d^2 J_m(z)}{dz^2} + k_m^2 J_m(z) = -j4\pi\omega\varepsilon_0\chi K(z) - \chi\sum_{i=1}^{2N}W(J_i,z), b_{m+1} \le z \le b_m. \qquad (2.57)$$

Here $k_m = \sqrt{k^2 - jk\chi Z_m/(60\pi a)}$ is the wave propagation constant along the mth segment of the antenna. We search for the solution of equation (2.56) in the form of a series in powers of the small parameter χ:

$$J_m(z) = \sum_{n=0}^{\infty}\chi^n J_{mn}(z). \qquad (2.58)$$

If the values $J_{m1}(b_m)$ and $J_{m1}(b_{m+1})$ of currents are given at the ends of the mth section, and for simplicity it is assumed that the emf is located at the boundary of two segments, then from the equation for $J_{m1}(z)$ we find

$$\chi J_{m1}(z) = \chi J_{m1}(b_m) \frac{\sin k_m(z - b_{m+1})}{\sin k_m l_m} + \chi J_{m1}(b_{m+1}) \frac{\sin k_m(b_m - z)}{\sin k_m l_m},$$

where

$$I_m = \frac{\chi}{\sin k_m l_m} \sqrt{J_{m1}^2(b_m) + J_{m1}^2(b_{m+1}) - 2 J_{m1}(b_m) J_{m1}(b_{m+1}) \cos k_m l_m},$$

$$\varphi_m = \tan^{-1} \frac{J_{m1}(b_m) \sin k_m l_m}{J_{m1}(b_{m+1}) - J_{m1}(b_m) \cos k_m l_m}, z_m = b_m - z.$$

In order to find the current distribution along the entire radiator, it is necessary for the current boundary conditions to add the conditions of charge continuity at the boundaries of the segments, i.e., put

$$\frac{dJ_i(b_i)}{dz} = \frac{dJ_{i-1}(b_i)}{dz}.$$

This equality is valid for all *i*, excluding the point $z = h$ of generator placement, where

$$\frac{\partial J_{m1}(h+0)}{\partial z} - \frac{\partial J_{m1}(h-0)}{\partial z} = 2 \frac{\partial J_{m1}(h+0)}{\partial z} = -j \frac{ke}{30}.$$

Here $\dfrac{\partial J_{m1}(h+0)}{\partial z}$, and $\dfrac{\partial J_{m1}(h-0)}{\partial z}$ are the values of the derivatives to the right and left of the point $z = h$.

Further, we will consider that the radiator is symmetric, and the emf is included in its center between the N*th* and (N+1)*th* segments. The relations found make it possible to express the amplitude and phase of the current in any segment through the amplitude and phase of the current of the previous segment, and therefore through the parameters of the segments and one of the currents. These expressions demonstrate the law of current distribution along the radiator and coincide with expressions (1.44)–(1.47), which in the first chapter are given for currents in a stepped impedance long line equivalent to the radiator.

To find the input impedance of the antenna with piecewise constant surface impedance, we first determine the vector potential. To do this, we substitute the current distribution (1.47) into expression (1.19) for the longitudinal tangential component of the vector potential

$$A_z(a,z) = \frac{\mu J(0)}{4\pi} \sum_{i=1}^{N} A_i \int_{b_{i+1}}^{b_i} \left[\frac{e^{-jkR}}{R} + \frac{e^{-jkR_+}}{R_+} \right] \sin[k_i(b_i - \varsigma) + \varphi_i] d\varsigma,$$

where $A_i = I_i / J(0)$.

From (1.28), using (1.44) and (1.45) for mutual compensation of the summands, we get

$$E_z = -j30J(0)\left\{\frac{A_1 k_1}{k}\left(\frac{e^{-jkR_1}}{R_1} + \frac{e^{-jkR_2}}{R_2}\right) - \frac{2A_N k_N}{k}\cos(k_N l_N + \varphi_N)\frac{e^{-jkR_0}}{R_0}\right\}. \quad (2.59)$$

This expression generalizes expression (2.47), and it uses the same notation for R, R_+, R_1, R_2 and R_0.

Substituting (2.59) into (2.46), one can find the input impedance of the antenna:

$$Z_A = j60\left\{\left[\frac{A_1 k_1}{k}\left(\frac{e^{-jkR_1}}{R_1} + \frac{e^{-jkR_2}}{R_2}\right) - \frac{2A_N k_N}{k}\cos(k_N l_N + \varphi_N)\frac{e^{-jkR_0}}{R_0}\right]\sin[k_m(b_m - z) + \varphi_m]dz\right\} +$$

$$\frac{1}{2\pi a}\sum_{m=1}^{N} A_m^2 Z_m \left[l_m - \frac{1}{k_m}\sin k_m l_m \cos(k_m l_m + 2\varphi_m)\right].$$

$$(2.60)$$

Using the new method for solving the integral equation, we find similar to (2.56)

$$Z_A = (1 + \chi)(R_A + jX_A) \quad (2.61)$$

where

$$R_A = 20k^2 h_e^2 = 80k^2 \left(\sum_{m=1}^{N} h_{em}\right)^2, \quad X_A = -W_N \cot(k_N l_N + \varphi_N).$$

Here h_{em} is effective length of the segment m, which is equal to the ratio of the area under the current curve $J_m(z)$ to the input current $J(0)$. N is the segment number near the generator.

An expression for the field of the radiator in the far zone, obtained using (1.47) and (1.56), has the form

$$E_\theta = j\frac{30kJ(0)\exp(-jkR_0)}{\varepsilon_r R_0} F(\theta), \quad (2.62)$$

where

$$F(\theta) = \sin\theta \sum_{m=1}^{N} A_m \left[\frac{\cos(\varphi_m - kb_m \cos\theta) - \cos(\varphi_m + k_m l_m - kb_{m+1}\cos\theta)}{k_m + k\cos\theta} + \frac{\cos(\varphi_m + kb_m \cos\theta) - \cos(\varphi_m + k_m l_m - kb_{m+1}\cos\theta)}{k_m - k\cos\theta}\right].$$

The above expressions clearly demonstrate the simplicity of the new method.

2.6 Radiators with resistive impedance

The antennas considered in Section 2.5 are radiators with reactive surface impedances. Their main difference from metal radiators is in a change in the wave propagation constant along the antenna. Another variant of a surface impedance is a resistive coating. This coating slightly effects the magnitude of the propagation constant, but leads to a smooth attenuation of the current amplitude along the antenna. As an example, let us consider transparent antennas. Analysis of the properties of such antennas, performed by means of Leontovich's equation, significantly helped to determine the capabilities of antennas and to speed up the process of their development and application.

Let's start with the necessary information about transparent antennas. The creation of such antennas became possible thanks to thin transparent and conductive films working out. Such antennas have unconditional merits. First, they can be made invisible. Second, they can be used in the capacity of screens for projecting different images—both still (photos) and moving (for example, *TV*). This additional use option is especially important for small devices where antennas are installed, that is, for operation at high radio frequencies.

Thin films of ITO (Indium-Tin-Oxide), placed on high-quality glass substrates, are electrically conductive and optically transparent at ultrahigh and superhigh frequencies. This allows us to use them as flat antennas for mobile communications and other applications. A check reveals that the transmission coefficient increases with increasing film resistivity and provides a sufficiently high transparency (about 95%) if the film resistivity is greater than 5 Ω/square.

To better understand the constraints imposed by the low conductivity of the *ITO* film, we should consider the surface resistivity of the film as a function of its thickness d. This resistivity is denoted as R_{sq1}. According to Leontovich's boundary condition, if the thickness d of the metal film is greater than the penetration depth, the film resistivity is equal (in Ohms) to,

$$R_{sq1} = R_{sq} = 1/(\sigma\delta), \tag{2.63}$$

where σ is its specific conductivity with respect to constant current (in S/m). The penetration depth δ is given by the formula

$$\delta = 1/\sqrt{\pi f \mu \sigma}, \tag{2.64}$$

where f is frequency (in Hz), $\mu = \mu_0 = 4\pi \cdot 10^{-7}$ F/m is the absolute permeability. If the thickness d of the metal film is much smaller than the penetration depth δ, the film's sheet resistivity is equal to

$$R_{sq1} = R_{sq}\delta/d = 1/(\sigma d). \tag{2.65}$$

The specific resistivity of *ITO* films is substantially greater than the specific resistivity of printed cards and metal antennas, where copper and aluminum are used. For example, the specific resistivity R_{sq1} of the transparent film CEC005P is equal to 4.5 Ω/square. The specific conductivities of copper and aluminum are respectively 5.8×10^7 and 3.5×10^7 S/m, and hence, in accordance with (2.63) and (2.64), the specific resistivity of a copper plate, whose thickness is greater than the penetration depth, at frequencies 1 and 5 GHz, is equal to 6.9×10^{-3} S/m and 18.4×10^{-3} S/m respectively (the specific resistivity of an aluminum plate with analogous thickness at these frequencies is equal to 4.2×10^{-3} S/m and 11.1×10^{-3} S/m respectively). That means, the resistivity of an *ITO* transparent film is greater by several orders than the resistivity of copper and aluminum, i.e., the conductive films are different from materials, commonly used in antennas (copper, aluminum), by a decrease of conductivity, and that changes the properties of radiators.

For comparative analysis of antennas made of materials with high and low conductivity, it is necessary to apply methods for solving the corresponding boundary value problems of electrodynamics. These methods are divided into direct numerical and approximate analytical. The conclusion of specialists on the results of the application of these methods in prolonged researches is clearly formulated in [36]: "The undoubted advantage of analytical methods is that they are physically clearer in comparison with numerical methods. Analytical methods allow us to determinate the effect of device parameters on its individual characteristics."

In recent years, transparent antennas have been the subject of many works. However, these works were devoted only to definition and improvement of material properties. Physical processes in transparent antennas, their electrical characteristics and their difference from characteristics of metal antennas with a high conductivity were not considered as a rule.

For understanding physical processes in antennas knowledge of the current distribution law along their axes has great importance. This knowledge allows defining all main characteristics of antennas and serves as the basis for the analysis of any antenna. This postulate remains valid in spite of elaboration of such calculation programs as the program CST, since primarily these programs basically allow calculating input characteristics (and characteristics dependent on them). Calculation of a current distribution by means of these programs is a difficult problem. Second, such a program doesn't allow us to understand the reasons for distribution change, to take them into account and to use in designing antennas. Unfortunately, the distribution of currents along transparent antennas has not been considered in the published papers.

It should also be recognized that the method of analyzing the characteristics of a transparent antenna described in book [10] turned out to be very approximate and did not lead to a reliable agreement in the calculated and experimental results. The calculation method described below, in spite of its apparent simplicity, made it possible to significantly improve the agreement with the experimental results. An equation for a current in a transparent antenna is an integral equation with non-zero (impedance) boundary conditions. It is a variant of equation (2.40). This equation is valid, if a surface impedance of an antenna is large enough to change a wave propagation constant along the antenna in the first approximation. In the case of a transparent antenna, it is necessary first to take into account the surface impedance R_{sq1}, created by losses in the transparent film. The width b of the transparent antenna is less important and has a smaller effect on the characteristics of the antenna.

The presence of a distributed resistive load means that the current along the radiator has a sinusoidal character with a slowly decreasing amplitude

$$J(z) = J(0)e^{-\alpha z} \sin kz. \tag{2.66}$$

For the current in the mode of natural oscillations we can write in accordance with (2.40), substituting $Z = jR_{sq1}$:

$$\frac{d^2 J(z)}{dz^2} + k^2 J(z) = \frac{k\chi J(z) R_{sq1}}{60\pi a}.$$

Substituting $J(z)$ into the left side of this equation gives

$$J(z)(\alpha^2 - 2k\alpha \cot kz) = \frac{k\chi J(z) R_{sq1}}{60\pi a},$$

where from

$$\alpha = \frac{k}{\tan kz} \pm \sqrt{\left(\frac{k}{\tan kz}\right)^2 + \frac{k\chi R_{sq1}}{60\pi a}}. \tag{2.67}$$

Let, for example, $R_{sq1} = 4.5$ Ом, $a = 0.006$ м, $\chi \approx 0.12$. At the frequency $f = 5$ GHz the wave length λ is equal to 0.06 m and the propagation constant $k \cong 105$, i.e., $\tan kz$ tends to infinity, and

$$\alpha = \sqrt{\frac{k\chi R_{sq1}}{60\pi a}} \cong 7.1.$$

In Fig. 2.9 the current distributions of two radiators with the same dimensions are compared. One radiator is made from a metal with perfect conductivity, and another radiator is made from film *CEC005P*. The dimensions of the radiators in millimeters are shown in Fig. 2.9*a*. The curve of the current (see Fig. 2.9*b*) along a metal antenna is denoted by number 1, and the curve of the current along a transparent antenna is denoted by number 2. As is seen from Fig. 2.9*b,* the current decay in the metal antenna with perfect

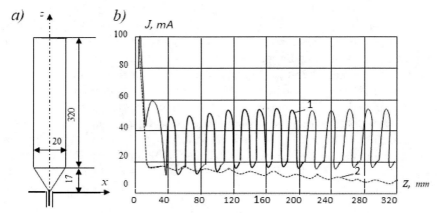

Fig. 2.9. Dimensions of the antenna (a) and current distribution along the metal (1) and transparent (2) radiators on a frequency 5 GHz (b).

conductivity is absent. The current of the transparent antenna decays rapidly. It decreases 5 times at a distance 0.23 m from the source and 10 times at a distance 0.33 m from the source. These distances are close to the calculation results.

Therefore, the current distribution along the transparent antenna axis is sinusoidal in nature, but the sinusoid amplitude decreases in accordance with exponential law. If $\alpha L = 2$, where L is its arm length, the amplitude of the current is decreased by e^2 times, where e is the base of the natural logarithms, i.e., current is almost zero. If $\alpha L > 2$, the operating length of the antenna is even shorter. The performed analysis leads to an important conclusion: the length of the radiating segment of the antenna in a first approximation is independent of frequency. This means that increasing the antenna length in order to ensure operation at lower frequencies is useless. The input impedance of such an antenna does not have a sharp resonance, and the effective length of the antenna is small.

Losses in a transparent antenna is one of the reasons for its low efficiency. The low level of matching of the transparent antenna with the cable leads to an additional reduction in efficiency of the antenna. Therefore, to increase the antenna's efficiency, it is first necessary to ensure a smooth transition from the central conductor of the cable to the transparent film forming the antenna. For example, one should place between them a metal plate with a variable width. But, as a rule, this is not enough.

One way to improve antenna matching with a cable or generator is to use complementary and self-complementary structures. The complementarity principle is connected with the duality principle based on the symmetry of Maxwell's equations with respect to the quantities $\sqrt{\varepsilon_0}E$ and $\sqrt{-\mu_0}H$. From this symmetry follows the existence of metallic (electrical) and slotted (magnetic) antennas. If the metal and slot antennas are placed on a flat metal surface of infinite size next to each other and occupy the entire surface, they are called complementary. As known, the input impedances of complementary antennas are related to each other by the expression

$$Z_s = (60\pi)^2/Z_e. \tag{2.68}$$

Here, the magnitude Z_s is the input impedance of the slot radiator, equal to the input impedance of the metal radiator, which occupies the rest of the endless sheet. The magnitude Z_e is the input impedance of the metal radiator, the shape and dimensions of which coincide with the shape and dimensions of the slot. If the antennas located on an endless sheet have the same shape and dimensions, then they are called self-complementary, and

$$Z_s = Z_e = 60\pi,$$

i.e., the input impedance of each radiator is purely active, and does not depend on the frequency and is equal to the wave impedance 60π. The special properties of such structures make them unique and are partially preserved at finite dimensions of the metal sheet.

The first of the studied structures of this type is shown in Fig. 2.10. It consists of triangular metal and slot radiators of the same shape and dimensions. An antenna with such properties does not surely have to be flat. For example, an antenna located on the surface of a circular cone, which is shown in Fig. 2.11, has the same features. If the angular width of the metal strip and the slot in the form of the conical helix is the same, we can speak of the self-complementary structure consisting of identical conical surfaces, one of which is covered with a metal shell, and the other is free from the shell.

In the general case, expression (2.68) is valid for an input impedance of a symmetric slot antenna of arbitrary shape and size, located on a circular metal cone of infinite length, excited on its top (Fig. 2.12). To derive this

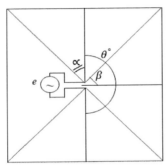

Fig. 2.10. Flat symmetric self-complementary radiator.

Fig. 2.11. Symmetric self-complementary radiator on a conic surface.

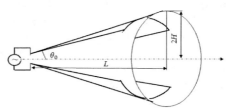

Fig. 2.12. Symmetric slot on a surface of a circular cone.

expression, it is necessary to consider a symmetric magnetic V-radiator located in free space, each arm of which is divided by a conical metallic surface passing through its axis, and to use the principle of duality. The performed analysis showed that the choice of the surface for the placement of a self-complementary radiating structure is not accidental. This is a surface of revolution, in particular a circular cone, which in the limit turns into a plane. The corresponding result was first obtained in [38] and described in [10]. The shape of a metal and slot radiator located on a circular cone may be different—similar to the shape of radiator located on a plane. In the simplest case, the boundary of the radiator coincides with the generatrix of the cone.

The magnetic radiator used in the analysis of the self-complementary conical structure is located on the surface along which the magnetic field lines pass. These lines may be curvilinear, for example, may have a parabolic shape. In this case, the metallic surface that coincides with the surface along which the magnetic field lines are located has the shape of a circular paraboloid. The point of excitation coincides with the top of the paraboloid (Fig. 2.13).

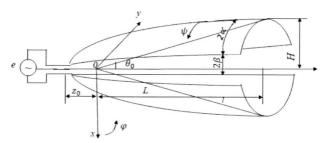

Fig. 2.13. Symmetric slot on a surface of a circular paraboloid.

In a general case, it is a surface of revolution with a generatrix in the shape of curve line. On such a surface, as on a circular cone, which represents this particular case, it is also possible to place complementary radiators, to which expression (2.68) is applicable.

The book [27] describes the generalized results of the analysis of complementary and self-complementary antennas of various shapes and sizes. It is shown that the characteristics of the antennas placed along the faces of the pyramid lying horizontally are close to the characteristics of the antennas located on the surfaces of rotation. Self-complementary antennas, consisting of several metal and slot radiators, and various schemes for connecting individual radiators to generator poles are considered. It has been shown that the wave impedance of a self-complementary antenna depends on the number of metal radiators and the circuit connection of these radiators to the poles of the generator.

Based on these results, the following tables show the main characteristics of radiators of various structures, shapes and sizes.

Table 2.3 for symmetrical radiators located on different surfaces (flat, conical, parabolic) shows the maximal gain and the radiation efficiency. The radiation efficiency is equal to signal magnitude under angle θ_0 with respect to main lobe magnitude. The angle θ_0 is indicated on the figures and depends on the antenna height H.

Figure 2.14 shows the shape and dimensions of a symmetric antenna located on the faces of the pyramid. The shape of the pyramid is determined by the angles γ and δ. Table 2.4 gives the gain and the signal magnitude (compared to the main lobe) for two variants of this antenna (placed on a pyramid with square and rectangular cross-section) at different angles θ_0, which depend on the height of the antenna H.

Figures 2.15–2.17 shows the shapes of the flat self-complementary antennas with rotational symmetry with one, two and four symmetrical metal radiators. Antennas with one metal dipole are presented in Fig. 2.15. They have different widths depending on the angle at the flat cone vertex:

Table 2.3. Gain and efficiency of different symmetrical radiators.

λ/H	Gain at θ_0					Radiation efficiency at θ_0				
	flat	cone		paraboloid		flat	cone		paraboloid	
	90°	30°	15°	30°	15°	90°	30°	15°	30°	15°
1	2.8	16.1	17.8	11	12.4	0.97	0.96	0.98	0.98	0.98
2	1.9	4.7	4.9	4.2	6.8	0.91	0.96	0.98	0.96	0.97
2.5	2.3	3.4	3.4	3.5	5.3	0.92	0.98	0.95	0.95	0.97
3	1.9	2.5	2.5	2.8	4.4	0.93	0.90	0.91	0.95	0.95

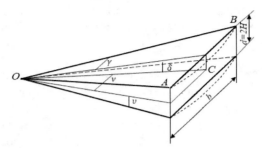

Fig. 2.14. Symmetric antenna located on the faces of the pyramid.

Table 2.4. Gain and efficiency of symmetrical antennas located on pyramid's sides.

λ/H	Gain		Radiation efficiency	
	$\gamma = \delta = 15°$	$\gamma = 30°, \delta = 15°$	$\gamma = \delta = 15°$	$\gamma = 30°, \delta = 15°$
1	23.3	17.6	0.97	0.98
2	6.56	10.2	0.98	0.95
2.5	5.21	6.87	0.98	0.95
3	4.08	5.5	0.93	0.94

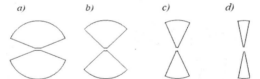

Fig. 2.15. Flat antenna with one symmetric metal radiator.

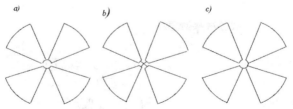

Fig. 2.16. Flat antenna with two symmetric metal radiators.

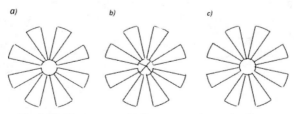

Fig. 2.17. Flat antenna with four symmetric metal radiators.

135° (*a*), 90° (*b*), 45° (*c*), 22.5° (*d*). Antennas with two metal dipoles are given in Fig. 2.16. They differ from each other in their circuit connection with the generator poles. The cones coupled with each other by a thin line are attached to one and the same pole: adjacent cones in variant *a* are connected together; in variant *b* one cone is connected with the other cone through a cone, in variant *c*—three cones are connected with one pole and the fourth cone is connected with another pole. The properties of diverse variants are different. In Fig. 2.17 different connection circuits to the generator are demonstrated for an antenna with four metal dipoles.

Accordingly, Table 2.5 gives the figures in numbers for the each antenna and the main parameters of these antennas: the angle θ_0 between the axes of metal dipoles, the angular half-width α of a metal dipole and the angular half-width $\beta = \theta_0/2 - \alpha$ of a slot, and also the antenna wave impedance W_1 and the ratio C_l of the capacitance per unit length of each antenna for the absolute permittivity ε_0 of air. The magnitude W_2 is the wave impedance of a three-dimensional antenna. In Fig. 2.18 as an example, a general view of a three-dimensional conical antenna with two metal radiators is presented.

The proximity of the values of wave impedances of cables and antennas is a necessary condition for the efficiency of antennas. The known variants of self-complementary antennas do not satisfy this condition, since their wave impedances are greater than the wave impedances of standard cables. The antennas considered here have different wave impedances, including a significantly smaller one. This makes it much easier to reconcile in a wide frequency range and expand the scope of self-complementary antennas. The efficiency of three-dimensional radiators is significantly higher than the efficiency of flat antennas of the same height.

Table 2.5. Parameters and wave impedances of antennas with rotation symmetry.

Number of dipoles	Fig.	θ_0	α	β	C_l/ε_0	W_1	W_2
1	2.15a	180°	67.5°	22.5°	2.94	40.8π	40.8π
	2.15b	180°	45°	45°	2	60π	60π
	2.15c	180°	22.5°	67.5°	1.35	88.7π	88.7π
	2.15d	90°	11.25°	78.75°	1.04	115.4π	
2	2.16a	90°	22.5°	22.5°	4	17.8π	20.4π
	2.16b	90°	22.5°	22.5°	8	15π	19π
	2.16c	90°	22.5°	22.5°	5.35	22.4π	26.5π
4	2.17a	45°	11.25°	11.25°	12	9.9π	8.4π
	2.17b	45°	11.25°	11.25°	24	7.5π	9.1π
	2.17c	45°	11.25°	11.25°	13.04	10.6π	10.6π

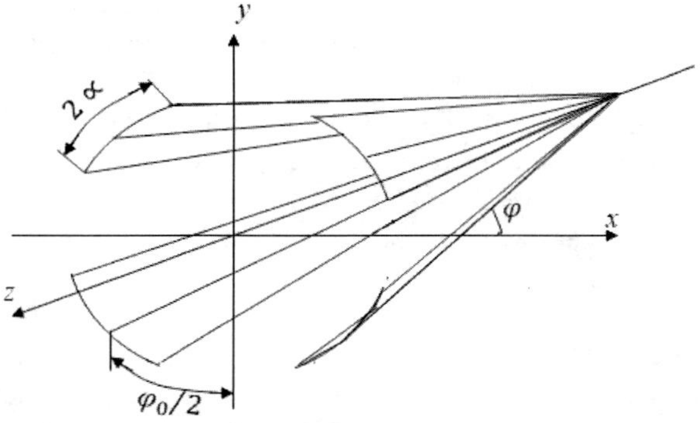

104   ANTENNAS: Rigorous Methods of Analysis and Synthesis

Fig. 2.18. Three-dimensional self-complementary antenna on the conic surface.

The considered antenna variants were checked in particular from the point of view of their use as transparent antennas. Two variants of asymmetrical radiators were selected coming from the basis of the values of their wave impedances (see Fig. 2.19). An antenna comprising of two connected with each other metal radiator (monopoles) with an angular width of 45° has the wave impedance, equal to 15π (47 Ohms). By adjusting the angular width of each radiator one can provide exact equality of wave impedances of the antenna and the standard cable.

In Fig. 2.20 the reflectivity of three antennas is compared: the curve 1 shows the reflectivity of antenna with the triangular metal base (Fig. 2.9), the curve 2 demonstrates the reflectivity of antenna with one transparent cone of length $L = 0.45$ m (Fig. 2.19a) and the curve 3 is the reflectivity of antenna with two transparent cones of the same length (Fig. 2.19b). As it is seen from (Fig. 2.20), the antenna with two metal radiators provides a smooth change in reflectivity in a wide frequency range. This result shows the significant advantage of the antenna in the shape of self-complementary radiator with one or two flat cones.

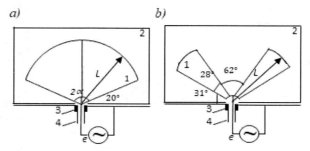

Fig. 2.19. Asymmetrical self-complementary antennas with one (a) and two (b) flat cones. 1 – transparent film, 2 – glass substrate, 3 – connector, 4 – cable.

Integral Equation of Leontovich 105

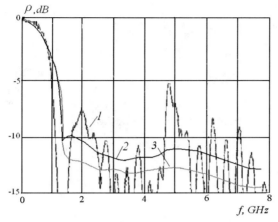

Fig. 2.20. Reflectivity of the first (1), second (2) and third (3) variants of the transparent antennas.

2.7 Radiators with concentrated loads

As an example of a radiator with concentrated loads we consider the antenna with complex impedances shown in Fig. 2.21a. The integral equation for the current in such an antenna can be easily obtained from the integral equation for the current in the metal radiator. The inclusion of the concentrated complex

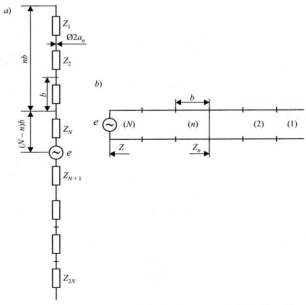

Fig. 2.21. Radiator with loads (a); equivalent stepped long line (b).

impedance Z_n at the point $z = z_n$ is equivalent to the inclusion of concentrated emf $e_n = -J(z_n)Z_n$ at this point. Accordingly, instead of equation (2.13) we will obtain

$$\frac{d^2 J}{dz^2} + k^2 J(z) = e_1 - \sum_{n=1}^{2N} J(z_n)Z_n - \chi W(J). \tag{2.69}$$

If the radiator is symmetrical and the loads Z_n in both its arms are the same and placed at equal distances z_n from the coordinates' origin, then

$$\frac{d^2 J}{dz^2} + k^2 J(z) = e_1 - \sum_{n=1}^{N} [J(z_n) - J(-z_n)]Z_n - \chi W(J).$$

If only one load Z_1 is installed in the radiator—at the point z_1, then

$$\frac{d^2 J}{dz^2} + k^2 J = e_1 - J(z_1)Z_1 - \chi W(J). \tag{2.70}$$

If the exciting emf is located in the radiator center, then the currents along the antenna are equal respectively to

$$\chi J_{11}(z) = j \frac{e_1}{60 \cos kL} \sin k(L - |z|),$$

$$\chi J_{12}(z) = -\chi J_{11}(z_1) Z_1 \begin{cases} \sin k(L - z_1 - |z|), & z \geq z_1 \\ \sin k(L + z_1 - |z|), & z \leq z_1 \end{cases}. \tag{2.71}$$

Thus, as one would expect, the current along the radiator with one concentrated load contains two sinusoidal components: the first is created by the generator, the second is related to the presence of load. If $Z_1 = 0$, the second component disappears.

Applying the method of variation of constants for solution of the integral equation, we use the extraneous emf as the right-hand side of the non-uniform equation. As a result, the fields of the located near adjacent radiators and the surface impedance of the antenna under study (the impedance, constant along the antenna, and the impedance, changing stepwise) are successively included in the right side of these equations. As the considered variant of the antenna with the concentrated load shows, the inclusion of this impedance in the antenna actually results in the appearance of the free term proportional to the load magnitude in the solution of the equation. Adding such a component does not contradict the logic of solving the integral equation, nor the logic of the method of induced emf. The oscillating power of the generator, first, is spent on radiation, and second, is allocated in the complex load.

It should be emphasized that changing the input impedance of the antenna, caused by the inclusion of loads, leads not only to the appearance of a free term in the expressions for the current and the input impedance, but also to changing the integral term, since the current distribution along the antenna and the field of current change. Inclusion of the load and changing the current distribution significantly change all the electrical characteristics of the antenna: the input impedance, the radiation pattern, and quantities that depend on them. Therefore, the load can be used similarly with extraneous fields (emf) to change the current distribution and the electrical characteristics of the antenna in the required directions.

If there are a lot of loads and they are located at small electrical distances from each other, then we can consider that the loads are connected continuously along the entire length of the antenna. This means a transition from a radiator with a finite number of loads to an impedance radiator. The impedance radiator, in contrast to a simple metallic one (without loads), has additional degrees of freedom. These degrees of freedom allow us to create antennas with the required electrical characteristics, in other words, to solve the problems of antenna synthesis.

The most important of these tasks is developing a wide-range radiator and obtaining high directivity in a wide frequency range. They are discussed in the next chapter.

2.8 Curvilinear radiators

Antennas, which are used today, have a wide variety of shapes and dimensions. An antenna located on the surface of a cone, paraboloid or pyramidial, is a wide metal plate, flat or bending along a cross-section. In the axial direction, the antenna can be straight or consists of separate segments of different shape located under different angles to each other.

The characteristics of different antennas are calculated with a varying degree of accuracy. The equivalent model for flat and three-dimensional antennas consists of intersecting wires. An intricate wire antenna may have a shape of a continuous stepped line. By reducing the lengths of the segments of this line, one can in the limit go to smooth curve. Unfortunately, these solutions are usually approximate.

A thorough analysis of the characteristics of the antenna is based on integral equations written and solved for thin direct radiators. Even the accuracy of calculating an inclined asymmetrical antenna of thin wire (monopole) often is insufficient. The complexity of this calculation is due to the complexity of the problem. On the one hand, the conduction current gradually decreases when distance from the ground increases, turning into a displacement current. On the other hand, the field of the current consists of different components located under different angles.

In Fig. 2.22*a* the structure of the *V*-radiator of thin wires is shown as an example. In Fig. 2.22*b* radiator has a similar, but more complex structure.

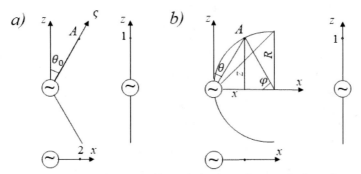

Fig. 2.22. *V*-antenna (*a*) and antenna with arm in the shape of an arc of a circumference (*b*).

Its arm is made in the form of an arc of a circumference. For instance, a slot antenna cut in a metal cylinder (tube) of large radius in a plane perpendicular to the cylinder axis has such a shape. Both structures differ from the direct radiator, which is considered by rigorous methods based on solving integral equations for the current. In the general case, the antenna arm may consist of several segments of various shapes.

In the process of analysis it is expedient to replace such an antenna with several direct radiators located along the axes of a rectangular coordinate system and to solve the equation for the current in each radiator. The points of the new radiators should coincide with the projections of the original antenna on the axes of the new coordinates. This is a condition of close coincidence of the currents and fields of the new radiators with the currents and the fields of the original antenna. The figures show radiators that replace the original antennas. The numbers 1 and 2 indicate, for example, the points of new radiators corresponding to the starting point *A*. As can be seen from the figures, the currents of the new radiators at points 1 and 2 must be added together in accordance with the rule of the addition of vectors. In this case their sum coincides with the current in the original antenna. This equality allows us to determine the magnitude of the currents in the new radiators.

To determine the horizontal and vertical components of the antenna field, it is necessary to divide the current of the wire into the components. Since the current amplitude is proportional not to the diameter of the flow, but to the square of its diameter, it should be considered that the ratio of the vertical and horizontal components of the current of an inclined wire is proportional to $\cot^2 \theta$ (and not to $\cot \theta$). Here θ is the angle of inclination of the wire. Accordingly, the sum of the vertical current component of the vertical wire and the horizontal current component of the horizontal wire is equal to the current of the inclined wire.

Of course, the described method of calculation is not very strict. With increasing antenna size, an error increases inevitably. But the accuracy of the proposed method is quite high.

In the case of the V-antenna, it is expedient to replace the original antenna with two radiators: vertical and horizontal (see Fig. 2.22a). The current along the vertical radiator

$$J_1(z) = J(0)\cos^2\theta_0 \sin k(L_1 - |z|), \tag{2.72}$$

where $J(0)$ is the current at the antenna input, $L_1 = L\cos\theta_0$ is the length of the vertical arm of the radiator, L and θ_0 are the length and the inclination angle of the V-antenna arm. In this case, for the first and second derivatives of the radiator current we get

$$\frac{dJ_1(z)}{dz} = -kJ(0)\cos^2\theta_0 \cos k(L_1-z), \quad \frac{d^2J_1(z)}{dz^2} = -k^2 J(0)\cos^2\theta_0 \sin k(L_1-|z|), \tag{2.73}$$

i.e., the left side of Leontovich's equation for the vertical radiator is

$$\frac{d^2J_1(z)}{dz^2} + k^2 J_1(z) = 0. \tag{2.74}$$

The current of horizontal radiator with arm length $L_2 = L\sin\theta_0$ is

$$J_2(x) = J(0)\sin^2\theta_0 \sin k(L_2 - |x|). \tag{2.75}$$

It is easy to see that in this case we get a similar result.

$$\frac{d^2J_2(x)}{dx^2} + k^2 J_2(x) = 0. \tag{2.76}$$

The sum of expressions (2.74) and (2.76) is equal to

$$\frac{d^2J(\varsigma)}{d\varsigma^2} + k^2 J(\varsigma) = 0, \tag{2.77}$$

where ς is the coordinate along the original antenna. Expression (2.77) coincides with the left-hand side of Leontovich's equation for the original antenna.

From that it follows that when the integral equation is divided into two, the sum of the left sides in the new integral equations is equal to the left side of the original equation. Similar equality is valid for the right sides of equations. This means that the shapes of the curves for the currents of both radiators is the same. Either of the new radiators is direct, that is, the shape of the currents of all radiators coincides with the shape of the current of the direct radiator, and all currents are distributed in accordance with a sinusoidal law. Only the magnitudes of the currents change. They depend on the angle of inclination of the antenna arms.

The case of a symmetric antenna with inclined straight arms of the same length is the simplest. As a second example, let's consider the more complicated case when the antenna has the shape of an arc (Fig. 2.22b). In this case, the share of current in each of the new radiators is continuously changing. It should be emphasized that this is a fraction of the current, and not its magnitude. The magnitude of the current in the previous case also changes, since in the original antenna it varies in length in accordance with the current distribution. The new antenna has the shape of an arc, and the proportions of the horizontal and vertical currents vary along these arcs from 0 to 1 and vice versa. The reason for these changes is that the arc is moving and pivoting, and its elements are in different directions. Therefore, for example, the projections of the segments of the same length on the x-axis change from segment to segment due to turning.

Let the vertical radiator be located along the z axis with the center at the origin of coordinates, the horizontal radiator be located along the x axis with the center at the same point, and the length of each arm be L. Then the current in the point z of the vertical radiator should be multiplied by $\cos^2 \theta$, and the current at the point x of the horizontal radiator should be multiplied by $\sin^2 \theta$. As can be seen from Fig. 2.22b, $\theta = \varphi/2$. The projections of the arc segment OA on the z-axis and x-axis are $z = R\sin \varphi = R\sin 2\theta$ and $x = R(1 - \cos \varphi) = R(1 - 2\sin^2\theta)$ respectively. The currents along the new vertical and horizontal radiators at points 1 and 2 are equal respectively to

$$J_1 = J(0) \cos^2 \theta \sin k(L - z) = J(0) \cos^2 \theta \sin [kR(1 - \sin 2\theta)],$$
$$J_2 = J(0) \sin^2 \theta \sin k(L - x) = J(0) \sin^2 \theta \sin [kR(1 - 2\sin^2 \theta)]. \quad (2.78)$$

Then, for each new radiator, we can write expressions similar to expressions (2.74) and (2.76) for the first and second derivatives and for the left side of the Leontovich equation. As in the previous case, the sum of these expressions is equal to

$$\frac{d^2 J(\varsigma)}{d\varsigma^2} + k^2 J(\varsigma), \quad (2.79)$$

where ς is the coordinate along the source radiator, and expression (2.79) coincides with the left-hand side of Leontovich's equation for the original radiator.

The obtained results significantly expand the possibilities of using Leontovich's equation for solving new problems.

2.9 Slot radiators

An ideal slot antenna is a slot cut in an infinitely large, ideally conducting, infinitely thin plane. The calculation of such antennas is performed using the duality principle [39]. In accordance with this principle, a slot antenna is a practical embodiment of a magnetic radiator, and the characteristics of a slot antenna can be determined based on the analogy between an electric and a magnetic radiator, if the characteristics of a metal (electric) radiator of similar size and shape are known. As shown in [40], if the slot is cut in a non-planar metal screen, the integral equation for voltage U between the edges of the slot coincides in appearance with the equation for current in an equivalent metal antenna of analogous shape and dimensions. However, the operator $G(U)$, coming into this equation, which is linear with respect to U, depends on the shape of the screen, i.e., in a general case, it differs from the operator $G(J)$ for a rectilinear metal antenna located in free space. Therefore, the question of the electrical characteristics of the slots cut in non-flat screens of different shape remains open.

Analysis of the characteristics of slot antennas located in three-dimensional space on circular metal structures [38] created new perspectives in solving the described problem, by means of detection of slots in non-flat screens, whose characteristics coincide with the characteristics of metal radiators. These are slotted antennas, complementary to metal ones. For example, it is a symmetrical two-sided slot antenna, located on a circular metal cone of infinite length and excited on the cone top (Fig. 2.23a).

In accordance with the duality principle, the radiation resistance of a two-way slot antenna cut in a flat perfectly conducting metal screen of unlimited dimensions and infinitesimal thickness is equal to

$$R_{\Sigma M} = \frac{(120\pi)^2}{R_{\Sigma e}}. \qquad (2.80)$$

Here $R_{\Sigma e}$ is the radiation resistance of a metal radiator of a similar shape and dimensions. This expression was obtained by comparing the power radiated by both antennas. To compare the oscillating power created by these radiators with each other, one can get a similar expression for the input impedances of the radiators:

$$Z_M = \frac{(120\pi)^2}{Z_e} \qquad (2.81)$$

One of them (slotted) will be called magnetic in accordance with the nature of its current, and the other radiator (metal), electric.

Let's move from the V-shaped magnetic radiator (Fig. 2.23b) to the slot antenna (Fig. 2.23c). To do this, we divide each an arm of the magnetic radiator

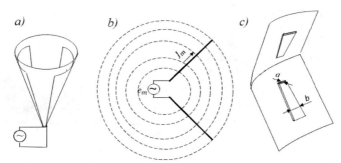

Fig. 2.23. Slot antenna on a circular metal cone (*a*) and transition from magnetic radiator (*b*) to slot antenna (*c*).

by means of a metal surface of a circular conic passing through the arm axis. Since the magnetic field lines of the radiator coincide with the surface having the shape of a circular cone, the field of the radiator does not change. The metal surface divides the magnetic radiator into two radiators. They are located on different sides of this surface, inside and outside the conic solid angle. Since in this case the magnetomotive force e_m, exciting the radiator, and oscillating power $P = e_M J_M$, created by the radiator, do not change, the fraction of magnetic current J_M in each of the new radiators is equal to the fraction of power into their part of space.

Let m be the fraction of power inside the solid angle, and $(1-m)$ be the fraction outside it. Then the input admittance of the magnetic radiator located inside the cone is equal to

$$Y_1 = e_M/(mJ_M) = Y/m, \qquad (2.82)$$

where $Y = 1/Z_M$ is the total admittance of the original radiator. As can be seen from Fig. 2.23c, the cross-section of the magnetic radiator with a current mJ_M has the trapezoidal shape with sides b and a, and $b \gg a$. Such a radiator is equivalent to a one-sided slot of width b. The second radiator is equivalent to a similar one-way slot with a current $(1-m)J_M$. Its input admittance is

$$Y_2 = \frac{e_M}{(1-m)J_M} = Y/(1-m). \qquad (2.83)$$

If both slots are replaced by one double-sided slot and we assume that its admittance is equal to the sum of the admittances of both slots, then

$$Y_s = Y_1 + Y_2 = Y/[m(1-m)]. \qquad (2.84)$$

The input impedance of the double-sided slot using (2.81) and (2.84) is equal to

$$Z_s = 1/Y_s = m(1-m)Z_M = (120\pi)^2 \, m(1-m)Z_e. \qquad (2.85)$$

The input impedance of the slot is actually the input impedance of the metal radiator located near the slot, or the input impedance of a uniform two-wire line. When the length of the two-wire line increases, its input impedance tends to the wave impedance, which for the same width of the slot and the metal radiator is equal to 60π. Comparison of this fact with the expression (2.84) shows that the magnitude m is equal to 0.5 for any angle of the cone aperture. This means that the power radiated into the solid angle is equal to the power radiated outside it.

A special case of the considered transition is the transition from a straight vertical magnetic radiator to a vertical slot. This slot and a flat metal radiator are the complementary antennas. In this case, the slot antenna is located in an infinite vertical metal plane, and the coincidence of the operators G(U) and G(J) is beyond doubt. Both operators are linear. In the general case considered here the operators are also the same, but the straight arms of the magnetic V-radiator are located at an angle to each other, and the operator G(J) of such metal radiator cannot be called linear. Such a slot antenna, without doubt, is similar to metal radiators, discussed in the previous section. The difference between the radiators shown in Figs. 2.22a and 2.23b is only that one of them is electric, and the other is magnetic.

If the slot antenna is located on the faces of the pyramid, its operator differs from both mentioned operators, only because in this case the metal and slot radiators are not complementary.

As mentioned earlier, slot antennas located on circular metal surfaces (cut in the indicated surfaces) can operate as slot antennas, complementary to metal ones. Such surfaces are the surfaces of circular cones and paraboloids, as well as the surfaces of circular cylinders. In this regard, we will consider the structure of two identical magnetic radiators located on the surface of a metal cylinder excited in anti-phase (Fig. 2.24a). Magnetic radiators are implemented in the form of slot antennas. In this case, the operators of slot antennas are linear and coincide with the operators of linear metal radiators. Half the power of each antenna is radiated into the internal volume of the cylinder.

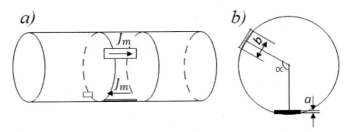

Fig. 2.24. Two parallel magnetic radiators (a) and two slot antennas (b).

114 ANTENNAS: Rigorous Methods of Analysis and Synthesis

A similar situation, as shown in [38], takes place in the case of two parallel infinitely long filaments, whose linear charge densities are the same in magnitude and opposite in sign. Half of the energy flow propagating along a line of two parallel wires is concentrated inside an imaginary circular cylinder passing through the wires axes. The second half passes outside. With a great deal of confidence, it can be argued that a similar situation holds true for wires of finite length, despite the fact that the current varies along the wire. More on this can be found in Section 4.2.

The slot antennas shown in Fig. 2.24 create the system of two anti-phase radiators located along the generatrix of a circular cylinder. It can be any generatrix, with an arbitrary arc length between radiators around the circumference of the cylinder. The input impedance of each radiator is equal to

$$Z_A = Z_{11} - Z_{12},$$

where Z_{11} is the self-impedance of each radiator, and Z_{12} is the mutual impedance between the radiators. The directional pattern of the structure of two thin radiators is determined by the distance between the radiators and does not depend on the arc length between the axes of the slots, i.e., a metal surface manifests itself only as a structure, on which a slot is placed.

2.10 Impedance magnetic antennas

As stated in the previous section, a slot antenna is a practical embodiment of a magnetic radiator, and the characteristics of a slot antenna can be determined in accordance with the duality principle, based on the analogy between an electric and magnetic radiator. The impedance magnetic radiator should be considered the same analogue of an impedance electric radiator. On the surface of this antenna non-zero boundary conditions are satisfied. As is shown in [3], these are conditions of the type

$$\frac{H_z(a,z) + K_1(z)}{E_\varphi(a,z)} = -\frac{1}{Z_1(z)}, \quad -L \leq z \leq L. \tag{2.86}$$

Here $E_\varphi(a, z)$ and $H_z(a, z)$ are the azimuthal component of the electric field and the longitudinal component of the magnetic field, respectively, $K_1(z)$ is the extraneous emf, and $Z_1(z)$ is the surface impedance, which in a general case depends on coordinate z.

The equation for the current in an impedance electric radiator is written in (2.40).

The slot antenna has the shape of a strip. The cross-section of the electric radiator is the cross-section of the round wire. The shape of the cross-section affects the small parameter χ, but does not affect the general form of the equation for the current. Therefore, the integral equation for the total magnetic current $J_m(z)$, flowing along one side of a strip of an impedance magnetic radiator, taking into account the duality principle, is written in [3] as

$$\frac{d^2 J_M}{dz^2} + k_1^2 J_M = -j2\pi k \chi Z_0 \left[K_1(z) - J_M \frac{1}{2aZ_1} \right] - \chi W(J_M), \quad (2.87)$$

where $k_1 = \sqrt{k^2 - jk\pi\chi Z_0/(aZ)}$. Here μ_0 is the magnetic permeability, $Z_0 = 120\pi\ \Omega$ is the wave impedance of free space. The magnitude of k_1, as usual, make the sense of a propagation constant for the current (in this case magnetic) along the radiator. From the expression (2.87) it follows that if the magnitudes k^2 and $k\pi\chi Z_0/(aZ)$ have the same order of smallness, then the surface impedance already affects the current distribution in the first approximation. If the impedance is of a purely reactive nature, and $jk\pi\chi Z_0/(aZ) > k^2$, then the magnitude of k_1 will be pure imagination, i.e., the magnetic current along the antenna will not propagate (will attenuate quickly).

Using the results obtained for the impedance of electric radiators, one can write expressions for the current distribution J_M and the input admittance Y_M of the magnetic radiator, excited by the magnetomotive force e_M, applied in its middle:

$$J_M = -j\chi \frac{k}{k_1} \cdot \frac{2\pi Z_0}{\cos k_1 L} e_M \sin k_1(L - |z|), \quad Y_M = -jW_1 \cot k_1 L, \quad (2.88)$$

where $W_1 = \frac{60}{Z_0^2 \chi} \cdot \frac{k_1}{k}$ is the wave impedance. Input admittance in the second approximation is equal to

$$Y_M = \frac{e_M}{J_M(0)} = \frac{Z_e}{Z_0^2}, \quad (2.89)$$

where Z_e is the input impedance of an electric radiator having the same dimensions and the same magnitude of the propagation constant.

The field strength E_z of a conventional metal radiator is determined by the expression (1.30). The radiation pattern of an electric and magnetic radiator with a constant surface impedance is calculated in accordance with the expression

$$F(\varphi, \theta) = \frac{\sin\theta}{\frac{k_1^2}{k^2} - \cos^2\theta} [\cos(kL \cos\theta) - \cos k_1 L]. \quad (2.90)$$

As is shown in [3], in order to get the equation for the current in the unloaded slot, one must put $Z_1 = \infty$ in the expression (2.87) and write a similar equation for the other half-space, and then demand the continuity of voltage

116 *ANTENNAS: Rigorous Methods of Analysis and Synthesis*

V on two sides of the slot. As a result, we come to the basic equation of the theory of slot antennas, derived by Pheld [40].

Of essential interest is the case when the surface impedance is large enough to change the current distribution along the antenna and the equivalent line in the first approximation in χ, where χ is a small parameter in the theory of thin antennas. Such an example is the problem of natural oscillations of a narrow rectangular trough [3, 41]. The trough is considered as a narrow slot in a flat screen, uniformly loaded with distributed impedance Z_1, with magnitude equal to the ratio E/H at the input of the short-circuited section of a rectangular waveguide (Fig. 2.25a). Since in the first approximation in χ the current distribution is sinusoidal, the impedance is created by fields of the type TE_0, i.e., $E_y/H_z = -jZ_1 \dfrac{k}{\gamma}\tan\gamma d$, where γ is the wave propagation constant with allowance for the surface impedance, and d is the depth of the trough. If we write an equation for the natural frequencies of the trough oscillation, then, as a result, one can interpret a uniformly loaded slot as a segment of the long line with a capacitor connected in parallel to this line (Fig. 2.25b).

Impedance slot antennas include the antenna in the form of a magnetic rod excited by a loop (Fig. 2.26a) and the antenna in the form of a circular

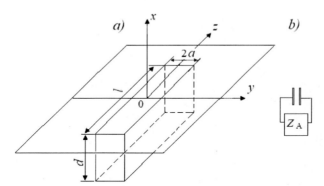

Fig. 2.25. Slotted antenna in the form of a loaded slot (a) and its equivalent circuit (b).

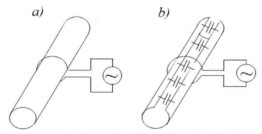

Fig. 2.26. The impedance magnetic radiators in the form of a magnetic rod excited by a loop (a) and a circular metal cylinder with a narrow longitudinal slot (b).

metal cylinder with a narrow longitudinal slot (Fig. 2.26b). The magnetic radiator, as a rule, is excited by a wire loop located around its central section. As a result, the loop is loaded by an impedance equal to

$$Z_M = 1/Y = U_{ind}/J(0) = Z_0^2/Z_e. \qquad (2.91)$$

Here U_{ind} is the induced voltage, and, $J(0)$ is the loop current. The loop current is the magnetomotive force e_M, applied at the point of excitation, and the induced voltage is the magnetic current of the radiator. The input impedance Z_e of relatively short radiator with a total length $2L < \lambda/2$ and a radius a can be calculated by means the expression

$$Z_e = 80\left(\frac{k}{k_1}\right)^2 tan^2 \frac{k_1 L}{2} - j120\left(ln\frac{2L}{a} - 1\right)\cot k_1 L. \qquad (2.92)$$

Analysis of the characteristics of specific antennas [42] begins with the calculation of the surface impedance. The surface impedance of a magnetic rod fabricated of ferrite can be determined, as shown in Section 2.5, by representing the rod as a set of radial lines, or by solving the problem of diffraction of a cylindrical wave converging to the surface of the rod. As a result, we find

$$Z = \frac{E_\varphi}{H_z}\bigg|_{\rho=a} = j120\pi\sqrt{\mu_r/\varepsilon_r}\,\frac{J_1(ma)}{J_0(ma)} \qquad (2.93)$$

where m is the wave propagation constant in the ferrite, $J_1(ma)$ and $J_0(ma)$ are the Bessel functions. Hence, for thin rods ($ma \ll 1$) the surface impedance and propagation constant are equal to

$$Z = j60\pi\mu_r ka, \quad k_1 = \sqrt{k^2 - 2\bigg/\left(\mu_r a^2 ln\frac{2L}{a}\right)}. \qquad (2.94)$$

From the last expression it follows that the condition of propagation of magnetic current along the rod is,

$$\mu_r \geq \frac{k^2 a^2}{2} ln\frac{2L}{a}.$$

The equal sign in this inequality corresponds to the critical wavelength λ_0, where, $\lambda_0 = 2\pi a\sqrt{ln\frac{2L}{a}\bigg/(2\mu_r)}$.

At $\lambda > \lambda_0$ the magnetic current is damped, at $\lambda < \lambda_0$ it spreads along the antenna. If μ_r increases infinitely, the ferrite rod becomes an ideal magnetic radiator. Unfortunately, such ferrites do not exist, and the magnetic permeability of the rod is additionally reduced due to the demagnetizing factor.

If we increase the diameter of the ferrite rod, then the expressions for the surface impedance and the wave propagation constant along the rod will change, but the properties of the antenna as a whole will change slightly. As shown in [43], when $\lambda = \lambda_0$, the directional pattern of the ferrite rod, excited by the loop, changes (at low frequencies and it coincides with the directional pattern of the loop, and for larger frequencies it coincides with the directional pattern of the impedance radiator).

The second example of an impedance magnetic radiator is a narrow slot cut in a metal circular cylinder. From the general expression for k_1 it follows that the propagation constant of the magnetic current along the radiator is real, if the surface impedance is capacitive in nature. To do this, along a narrow slot of the cylinder we need to include several capacitors. Then the surface impedance will be created by parallel connection of the capacitors and the cylinder inductance.

Inductive impedance of a metal cylinder with a length L and a radius ρ in accordance with (2.93) is

$$j\omega\Lambda = Z\frac{2\pi\rho}{L} = \frac{j120\pi^2 k\rho^2 \mu_r}{L}.$$

The total impedance (inductance plus capacitance) is equal to

$$Z_+ = j\omega\Lambda \Big/ \left(1 - \frac{f^2}{f_0^2}\right),$$

where f_0 is the frequency of a parallel resonance. The cross-section perimeter of the slot antenna with a width a is equal to $2a$. Because the total load of the slot antenna is Z_+, its surface impedance is

$$Z = Z_+ \cdot L/a.$$

Accordingly, the propagation constant is

$$k_1 = \sqrt{k^2 - j\pi k \chi Z_0/(aZ)} = \sqrt{k^2 - \chi\left(1 - \frac{f^2}{f_0^2}\right)\Big/(\rho^2 \mu_r)}.$$

Equating the radicand (expression under root) to zero, we obtain an equation that allows us to determine the critical frequency for a wave, propagating along the slot. This frequency is equal to f_0. If the frequency is less than f_0, then the magnetic current does not propagate along such a slot antenna. When $f = f_0$ and accordingly $k_1 = k$, the slot represents the ideal magnetic radiator. Increasing the frequency leads to a sharp increase of the propagation constant, when a large number of half-waves of the current are placed along the antenna, and their fields cancel each other. When the operating frequency is close to the critical one f_0, i.e., in the narrow frequency range near f_0, the antenna has high efficiency.

The input impedance introduced by the rod antenna into the exciting loop is defined by the expression (2.91), where Z_e is taken from (2.92).

The impedance magnetic radiator in the form of a circular metal cylinder with a narrow longitudinal slot and capacitors connected along the slot has found use as a receiving antenna for the *VHF* range [44]. The author considers the antenna developed by him as a loop and calculates the active component of its input impedance at the resonance point in accordance with expression

$$R = Z_0^2/(80k^2L^2),$$

i.e., he believes that the resistance introduced into the loop is equal to the resistance of the Hertz magnetic dipole. This insertion resistance is $\pi^2/4$ times smaller than that given in (2.92) and can be considered as a first approximation that does not take into account the sinusoidal distribution of the magnetic current and the resonant nature of the input impedance of the radiator.

Figure 2.27 shows the results of calculations and measurements for the input impedance of a metal cylinder with a narrow longitudinal slot and capacitors. The main dimensions (in meters): $2L = 2.0$, $\rho = 0.11$, $2a = 0.03$. The capacity was created by capacitors with the capacitance 51 pF located at a distance $b = 0.025$ m from each other. Curve 1 is the measurement result for the loop, curves 2 and 3 are the results of the calculation and measurement for the antenna, respectively. In Fig. 2.28 curves for the loop and the antenna

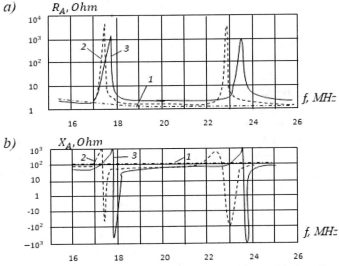

Fig. 2.27. Active (*a*) and reactive (*b*) components of the input impedance of the loop (1 – experiment) and a slot antenna cut in a circular metal cylinder (2 – calculation, 3 – experiment).

120   ANTENNAS: Rigorous Methods of Analysis and Synthesis

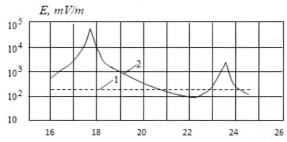

Fig. 2.28. The experimental results for the field of loop (1) and slotted (2) antenna, cut in the circular metal cylinder.

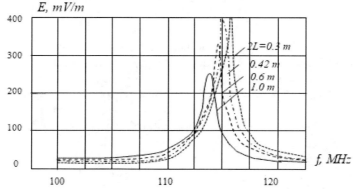

Fig. 2.29. The dependence of the first resonance frequency of the slot antenna, cut into a circular metal cylinder from the antenna length $2L$.

fields in the far zone are shown at the same magnitude of current in the loop depending on the frequency. In Fig. 2.29 experimental results are given for the antenna with dimensions (in meters) $2a = 0.01$, $\rho = 0.021$, $b = 0.02$ and capacitors with the capacitance 25 pF. The results are showing, how the frequency of the first resonance changes with changing the antenna length $2L$.

Returning to [44], it should be said that the author created a new antenna without being carried away by its features and without wasting time on useless theorizing. In this case, it is impossible not to recognize the fairness of this approach, given the venerable age of the antenna. But this topic needs comments. Any radiator can be considered, using different level of understanding of the problem. The simplest approach to calculating the input impedance of the antenna is to consider it as a circuit consisting of a capacitance, an inductance and a resistance, the magnitude of which is related to radiation and losses. The second approach considers the antenna as a structure of infinite length. Of course, in many cases this approach is simpler.

But a result close to the truth requires analyzing the antenna as a radiator of finite length.

Speaking frankly about such a controversial issue at the risk of running into harsh criticism, I want to quote King, who wrote in [45]:

"Since the time of Hertz, the infinitely small dipole replaces physically feasible antennas in studies associated with the location of sources in different environments and is too complex for analysis ... It should be noted that this method bypasses the main issues of antenna problems and does not solve them."

As an example of an overly cautious approach to the analysis of not too complicated questions, it is expedient to recall how we usually perceive the phrase that the current does not propagate along any conductor. This does not mean a total absence of radiation. This fact only means that the current exponentially decays. The length of the section with decaying current may be sufficient to create a signal of the desired magnitude.

2.11 Distribution of the current over the antenna wire surface

The name of the next section coincides with the content of a topic, which often interests specialists: how significant is the transverse change in the surface current density of radiators located in parallel and how does it affect the characteristics of these radiators. This question is considered in [46]. There, in particular, it is shown that the transverse changes in current are significant, even if the distances between the radiators are larger than several wire diameters.

A fundamental approximation in the theory of thin wire antennas is that the surface current density is constant along the perimeter of the wire cross section. To correspond to this approximation, the wires must satisfy two conditions: the radius of the wire should be much less than the wavelength in free space, and no two wires should run parallel to each other within a distance equal to the sum of several diameters of the wire.

The mentioned article discusses wires of finite dimensions with a piecewise sinusoidal current distribution. To account for changes in the axial component of the current along the perimeter of the cross-section, a Fourier series is used. The finite dimensions and use of the Fourier series are the principal features of the analysis. Other details can be attributed to the traditional conditions for wire antennas: the wire radius is considered small, the transverse component of the current is missing, the surfaces at the wire ends are not considered. The wire in accordance with the Shchelkunov's equivalence theorem is replaced by free space with electric and magnetic currents of a given density. In this case, the wire is considered ideally conductive, and the field inside it is absent.

The equation for the current is solved by the Moments method. The basis functions along the length l of the radiator are piecewise sinusoidal functions, the basis functions along the angle φ (along the wire circumference) are the Fourier series. Weight functions are taken to coincide with the basis ones, i.e., Galerkin's method is used.

In Fig. 2.30 numerical examples for the surface current density $J(\varsigma)$ and its phase φ are given for two parallel half-wave dipoles with radii $a = 10^{-3}\lambda$. The dipoles are excited at their central points. The results are shown for two magnitudes of c/a: $c/a = 1.1$ (Fig. 2.30a) and $c/a = 50$ (Fig. 2.30b). The density of surface current is normalized with respect to the isolated radiator. The results obtained by the described method are given in the form of a continuous curve 1. They are compared with the results obtained by the method based on the replacement of the metal radiator surface by the mesh model [47]. The latter are given by a dotted line and are designated by the number 2. In the grid model each radiator is modeled by sixteen ultra-thin wires evenly spaced around its perimeter, which are excited in the center by in-phase δ-generators.

Analysis of the results obtained using the Fourier series showed that parallel wires can have a significant mutual influence on the distribution of current density along the perimeter of each of them. This circumstance may affect the antenna efficiency, for example, the efficiency of the multi-turn loop. When analyzing radiators using the theory of thin antennas, including mesh models, this effect does not affect the obtained input impedance and the field

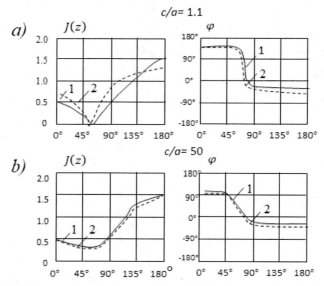

Fig. 2.30. The distribution of the amplitude and phase of the surface current in two half-wave dipoles with $c/a = 1.1$ (a) и 50 (b).

in the far zone. The results obtained in this case are correct, since the change in current density along the perimeter of the antenna affects its input impedance and the field in the far zone, negligibly. It can be assumed that the results of calculations using a mesh model were affected by the absence of transverse wires. We can conclude that the external influence on the distribution of the surface current density along the wire perimeter is undoubted.

Chapter 3
Inverse Problems of Antenna Theory

3.1 Wide-range linear radiator and impedance line

The inverse problem of the antennas' theory, or the problem of synthesis, is the problem of creating antennas with the required electrical characteristics. A particular case of such a task is the creation of a wide-range radiator, i.e., creation of an antenna providing a high level of matching with a transmitter or a cable in a wide frequency range and a maximal radiation in a plane, perpendicular to the antenna axis. This task is of great practical importance. A promising method for solving this problem is the use of concentrated loads.

A typical linear radiator (thin, without loads) fails to meet these requirements. The reactive component of its input impedance is great everywhere, except in an area of the series resonance. That results in deteriorated matching of antenna with cable. If the radiator arm is larger than 0.7λ, the radiation in the plane, perpendicular to antenna axis, decreases, since the current distribution along a thin linear radiator without loads (Fig. 3.1a) is close to being sinusoidal, and at high frequencies anti-phase segments are formed on the current curve (Fig. 3.1b, curve 1).

Electrical characteristics of radiators, i.e., its input impedance, directional pattern, etc. depend on current distribution along them. In order to change this distribution, one can use extraneous fields (exciters) or loads—both distributed and concentrated. Even one load can significantly change the current distribution, and hence the electrical characteristics of the antenna. If a great number of loads are placed along the antenna at small electrical distances from each other, we can consider that they are included continuously along the entire antenna length, and that means a transition from the radiator with a finite number of loads to the radiator with distributed load, i.e., to the impedance radiator. The boundary condition on the surface of this radiator, located between the points $z = -L$ and $z = L$ along z axis of the cylindrical coordinates system, are given by (1.36). Such a boundary condition is valid,

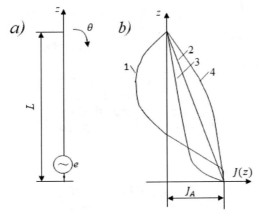

Fig. 3.1. A linear monopole (a) and the laws of current distribution along it (b).

if the structure of the field inside one medium (e.g., inside an antenna cover) is independent of a field structure in another medium (e.g., in ambient space).

By including concentrated loads across the radiator length, one can obtain a current distribution other than the sinusoidal, depending on their magnitudes and points of connection. The experimental results show that a radiator with linear or exponential in-phase current distribution exhibits good performance (high matching level, required shape of the vertical directivity pattern) in a wide frequency range. In particular, such distribution is created with help of capacitive loads [19, 22]. These results confirm the known fact that the maximum radiation in a direction, perpendicular to the dipole axis, is attained, if the current is in-phase along the entire length of the antenna. Moreover, a long radiator with an in-phase current has high resistance of radiation, which allows increasing the matching level.

As is known, an ordinary two-wire long line is an equivalent of a metal radiator. An impedance of a two-wire long line is an analogous equivalent of an impedance radiator. But unlike the metallic radiator a tangential component of an electric field on the surface of an impedance radiator is not equal to zero, and that causes an additional voltage decrease on each element of wire length. An impedance long line, equivalent the impedance radiator with a constant surface impedance, is shown in Fig. 1.5b, and the line, which is equivalent the radiator with a piecewise constant impedance, is shown in Fig. 1.5c. Characteristics of these lines are given in Section 1.5. In accordance with the equation for current $I_m(z)$ one can write the expression for the generalized wave propagation constant γ_m at the segment m of a radiator in the form

$$-\gamma_m^2 = k^2 - j2k\chi Z^{(m)}/(a_m Z_0). \tag{3.1}$$

Here χ is a small parameter, a_m is a middle radius. In the general case γ_m is a complex magnitude. In a particular case, when this magnitude is purely imaginary ($\gamma_m = jk_m$), the current distribution at the segment m of the line has sinusoidal character.

If the law for changing the propagation constant, which allows ensuring the required current distribution, is found, expressions (3.1) and (1.48) can be used first for calculating the surface impedance $Z^{(m)}$ and next for calculating the concentrated loads Z_m. The current of the segment m of a stepped line at arbitrary γ_m is

$$J(z_m) = I_m \sin h(\gamma_m z_m + \varphi_m), \ 0 \le z_m \le b, \tag{3.2}$$

where I_m and φ_m are the amplitude and phase of the current along the segment m, respectively, and z_m is the coordinate, measured from the segment end, i.e., $z_m = (N - m + 1)b - z$.

Suppose we want to obtain the given distribution of the current along the long line

$$J(z) = J_A f(z), \ 0 \le z \le L, \tag{3.3}$$

where J_A is the input current of the line (the current of the generator), $f(z)$ is a real and positive distribution function, which corresponds to the in-phase current. If we equate currents $J(z)$ and $J(z_m)$ at the beginning and the end of mth segment, then in the case of a small segment length the current distribution along the line is close to the one that is required. In accordance with (3.2) and (3.3), at $z_m = b$ and $z_m = 0$

$$I_m \sin h(\gamma_m b + \varphi_m) = J_A f[(N - m)b], \ I_m \sin h\varphi_m = J_A f[(N - m + 1)b].$$

If we divide the left- and right-hand parts of the first equation onto the respective parts of the second equation and are confined by the first terms of expansion of hyperbolic functions with small arguments into series (considering that b is a small magnitude), we will get

$$\tanh \varphi_m = \gamma_m b \Big/ \left\{ \frac{f[(N - m)b]}{f[(N - m + 1)b]} - 1 \right\}. \tag{3.4}$$

For the segment $(n + 1)$, similarly to (3.4),

$$\tanh \varphi_{m+1} = \gamma_{m+1} b \Big/ \left\{ \frac{f[(N - m - 1)b]}{f[(N - m)b]} - 1 \right\}. \tag{3.5}$$

The voltage and current are continuous along a stepped line, hence

$$\tanh \varphi_{m+1} = (\gamma_{m+1}/\gamma_m)\tanh(\gamma_m b + \varphi_m). \tag{3.6}$$

Equations (3.4) and (3.5) present a set of equations that allow us to relate γ_m and γ_{m+1} with each other. The solution of this set shows that magnitude γ_m is independent of γ_{m+1}:

$$\gamma_m = \frac{1}{b}\sqrt{1 - \frac{2f[(N-m)b] - f[(N-m-1)b]}{f[(N-m+1)b]}}. \qquad (3.7)$$

As is clear from (3.3), function $f(z)$ defines the law, in accordance with which the amplitude of the current changes along the radiator. In the case of in-phase current distribution along the antenna, its directional pattern in the vertical plane has a form

$$F(\theta) = \sin\theta \int_{-L}^{L} f(z) \exp(jkz\cos\theta)\,dz. \qquad (3.8)$$

Calculations show that in this case in contrast to sinusoidal distribution, the radiation maximum with growing frequency does not deviate from the perpendicular to the radiator axis. Increasing L/λ makes the main lobe narrower and increases the maximal directivity.

At linear distribution of the in-phase current amplitude (see Fig. 3.1b, curve 2),

$$J_2(z) = J_A(1 - z/L),$$

where $z = (N - m + 1)b - z_m$, i.e.,

$$f_2(z) = (L - z)/L = [(m-1)b + z_m]/(Nb), \qquad (3.9)$$

The linear distribution is a particular case of the exponential one (see Fig. 3.1b, curves 3 and 4):

$$J_{3,4}(z) = J_A \frac{\exp(-\alpha z) - \exp(-\alpha L)}{1 - \exp(-\alpha L)},$$

that is

$$f_{3,4}(z) = \frac{\exp(-\alpha z) - \exp(-\alpha L)}{1 - \exp(-\alpha L)} = \frac{\sin h\{(\alpha/2)[(m-1)b + z_m]\}}{\sin h(\alpha Nb/2)}, \qquad (3.10)$$

where α is the logarithmic decrement. If α is positive, the curve of a current is concave, i.e., the current quickly decreases from the maximum value near the generator to zero near the free end of the antenna. If α is negative, the curve of a current is convex, i.e., the current is more uniformly distributed along the dipole. The steepness of a curve depends on the value of α. It is easy to show that if α tends to zero, the expression for $J_{3,4}(z)$ turns into $J_2(z)$.

In the general case of an in-phase current distribution along the antenna, its field E_θ at an arbitrary angle to its axis in accordance with expression (1.55) has the form

$$E_\theta = j\frac{30k\exp(-jkR)\sin\theta}{\varepsilon_r R}\int_{-L}^{L} J(z)\exp(jkz\cos\theta)dz.$$

For an exponential distribution, calculation of the integral gives the expression

$$E_\theta = j\frac{60kJ(0)\exp(-jkR)}{\varepsilon_r R}\frac{\cos\theta}{k^2\cos^2\theta + \alpha^2} *$$
$$\{k\cos\theta - e^{-\alpha L}[k\cos\theta\cos(kL\cos\theta) + \alpha\sin(kL\cos\theta)]\}$$

For a linear distribution we get

$$E_\theta = j\frac{60J(0)}{\varepsilon_r}\frac{\exp(-jkR)}{R}\frac{1-\cos(kL\cos\theta)}{\sin\theta}.$$

The antenna input impedance in the first approximation is equal to the input impedance of a stepped long line:

$$Z_l = -jW_N\coth(\gamma_N b + \varphi_N).$$

Here, as is seen from (1.41), $W_N = \gamma_N W/k$, where W is the wave impedance of a metal monopole of the same dimensions without loads.

If $m = N$, we find for the linear current distribution, using equalities (3.5) and (3.6) and taking into account that $f(0) = f_2(0) = 1$:

$$Z_l = -j(W/kb)[f_2(-b) - 1]. \tag{3.11}$$

As seen from this expression, reducing the reactive component of the input impedance requires a slow variation of function $f(z)$ near the antenna base, so that the difference in square brackets should be a small magnitude—of the order of kb. Otherwise, the reactive component of input impedance will be great.

For the exponential distribution, replacing $f_2(-b)$ with $f_3(-b)$, we obtain from (3.11)

$$Z_{A3} = -j(W/kL)f_x(\alpha L/2), \tag{3.12}$$

where $f_x(x) = x(1 + \coth x)$. The graph of function $f_x(x)$ is given in Fig. 3.2a. In particular for the linear distribution

$$Z_{A2} = -j\frac{W}{kb}\left(\frac{N+1}{N} - 1\right) = -jW/(kL).$$

Figure 3.2b compares the input impedance X_{A1} of a uniform line with sinusoidal current distribution and the input impedance X_{A3} of a non-uniform line with an exponential in-phase current distribution depending on frequency.

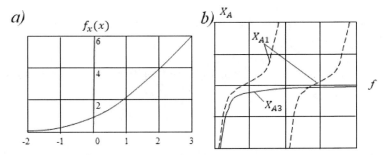

Fig. 3.2. Graph of function $f_x(x)$ (*a*), input impedances of uniform and non-uniform line (*b*).

Here, the magnitude of α is constant. In the first case the input impedance has a shape of a cotangent function, in the second case the curve smoothly approaches to the axis with frequency growth and that allows ensuring good matching in a wide range. As seen from Fig. 3.2*a* and expression (3.12) for Z_{A3}, if α decreases, the input impedance of a long line at a given frequency diminishes. It means that a decrease of α results in a decrease of the reactive component of the antenna input impedance. Simultaneously the effective height grows, since the area, bounded by the curve of the current, increases; hence, the radiation resistance grows too. Thus, at exponential distribution, it is expedient to decrease α, i.e., to reduce the value of α to zero, and then to go to the region of negative values.

According to (1.48) and (3.1),

$$-\gamma_m^2 = k^2 - jk\chi Z_m/(30b). \qquad (3.13)$$

In order to create the in-phase current distribution, magnitude of γ_m should be purely real or purely imaginary along the entire antenna, and magnitude of γ_m^2, correspondingly, only positive or only negative:

$$\text{sign } \gamma_m^2 = \text{const } (m).$$

In order for the in-phase distribution to be realized in a wide frequency range, magnitude of γ_m should be real:

$$\gamma_m^2 > 0. \qquad (3.14)$$

Indeed, if the values of γ_m (and also φ_m) are purely imaginary, the hyperbolic sine in the formula (3.2) would become the trigonometric sine. With frequency growth, an argument of a sine will increase and exceed π, and the sine will change sign. If γ_m is real, then, as seen from (3.4), if function $f(z)$ changes monotonically (decreases with growth of n), the sign of φ_m coincides with sign of γ_m. As it follows from (3.7), in order that γ_m will be real, following condition must be satisfied,

$$f[(N-m)b] \le \frac{1}{2}\{f[(N-m+1)b] + f[(N-m-1)b]\}. \tag{3.15}$$

i.e., function $f(z)$ cannot be convex.

Two variants of carrying-out of condition (3.14) in a wide frequency range follow from (3.13). The first variant takes place if

$$k^2 \ll jk\chi Z_m/(30b). \tag{3.16}$$

i.e.,

$$\gamma_m^2 = jk\chi Z_m/(30b). \tag{3.17}$$

If one takes into account that parameter χ is, strictly speaking, a complex magnitude ($\chi = \chi_1 - j\chi_2$), then the admittance of load in accordance with (3.17) is equal to

$$Y_m = 1/Z_m = j\omega C_m + 1/R_m, \tag{3.18}$$

where

$$C_m = 4\pi\varepsilon\,\chi_1/(b\gamma_m^2), \quad R_m = b\gamma_m^2/(4\pi\varepsilon\omega\chi_2).$$

It follows from (3.18), that if the magnitude of γ_m^2 is positive, each load should be executed as a parallel connection of a resistor and a capacitor (Fig. 3.3a). The resistance of the resistor should vary in inverse proportion to frequency, and the capacitance of the capacitor should remain constant. While creating an actual antenna, it is expedient to choose the value of R_m for the middle frequency of the range.

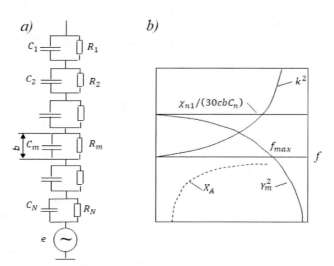

Fig. 3.3. An antenna circuit with capacitors and resistors (*a*), frequency dependence of a propagation constant (*b*).

To obtain required current distribution $f(z)$ along the antenna, the magnitude γ_m should correspond to equality (3.7). Substitution of (3.7) into (3.18) gives

$$C_m = 4\pi\varepsilon\chi_1 b\left\{1 - \frac{2f[(N-m)b] - f[(N-m-1)b]}{f[(N-m+1)b]}\right\}, Rm = \frac{\chi_1}{\chi_2 \omega C_m}. \quad (3.19)$$

By comparing (3.15) and (3.19), one can easily be convinced that, if inequality (3.15) holds, the values of C_m are not negative. For exponential and linear distribution,

$$C_{m3} = \frac{8\pi\varepsilon\chi_1}{\alpha^2 b\{1 + \coth[\alpha(m-1)b/2]\}}, \; C_{m2} = \frac{4\pi\varepsilon}{\alpha}\chi_1(m-1), \; R_m = \frac{\chi_1}{\chi_2 \omega C_m}. \quad (3.20)$$

One can see from (3.20) that in this particular case, if one must obtain a law of current distribution, close to being linear, capacitances of loads should decrease towards the free end of the antenna in proportion to the distance from it:

$$C_{m2} = C_{N2}(m-1)/(N-1). \quad (3.21)$$

where C_{N2} is the capacitance of the capacitor near the antenna base. The resistances of resistors should grow towards the free end of the antenna:

$$R_{m2} = R_{N2}(N-1)/(m-1). \quad (3.22)$$

Thus, to obtain an in-phase current distribution that ensures good electrical characteristics of an antenna in a wide frequency range, each load should represent a parallel connection of a resistor and a capacitor. For the first time the expediency of using a complex load for creating a linear current distribution was demonstrated in [48]. Later on, the main attention was given to antennas with capacitive loads. It follows from the given analysis and the calculation results, that, if resistors are included in parallel with the capacitors, then the linear law of current distribution along the radiator is observed precisely, and the operating frequency range increases. However, use of resistors leads to efficiency reduction, so the question of their application should be decided in each particular case.

Figure 3.3b shows a plot of γ_m^2 versus frequency for an antenna with capacitive loads:

$$\gamma_m^2 = -k^2 + \chi_1 4\pi\varepsilon/(bC_m). \quad (3.23)$$

For the propagation constant to be real at a given frequency f, capacitances of capacitors should not exceed the magnitude

$$C_m \leq \frac{\chi_1}{30 k^2 bc} = \frac{2.54 \cdot 10^5 \chi_1}{f^2 b}. \quad (3.24)$$

Here, c is the speed of light in m/s, capacitance C is expressed in farads, and frequency f is in Hertz's. In the case of a linear distribution, capacitance C_{N2} near the antenna base is greater than others capacitances and should be chosen in accordance with (3.24). Similarly, under other distributions, this expression determines the maximum capacitance.

It follows from (3.23), that, at low frequencies the propagation constant is real along the entire antenna. As the frequency increases, the magnitudes γ_m become purely imaginary (first of all, on segments, adjoining the generator), i.e., the current distribution along these segments of the radiator become sinusoidal, and the main lobe of the directional pattern deviates from the perpendicular to the dipole axis. This effect limits the antenna frequency range from above. From below, the range is limited by frequencies, where the reactive component of input impedance is still great. In order that magnitude γ_m does not become purely imaginary with increasing frequency, the capacitances of capacitors should decrease with growth of frequency (e.g., vary in inverse proportion to square of the frequency).

Regarding the second option of realization of condition (3.24) in a wide frequency range, with its help one can make similar conclusions. This option is feasible, if the second summand of the right-hand part of (3.13) is proportional to k^2:

$$-\gamma_m^2 = k^2[1 - j\chi Z_m/(30kb)], \qquad (3.25)$$

and the magnitude of concentrated load Z_m is equal to the sum of two summands, whose signs are the same when the frequency changes. In this case the load Z_m is

$$Z_m = -j\frac{30bk}{\chi}\left(1 + \frac{\gamma_m^2}{k^2}\right) = -j\omega|\Lambda_m|, \qquad (3.26)$$

i.e., it is a negative inductance $-|\Lambda_m| = -\dfrac{30b(1+\gamma_m^2/k^2)}{(\chi c)}$. At a small ratio, γ_m/k, the inductance depends weakly on the frequency f. In this case, the value $\gamma_m^2 = k^2[\chi|\Lambda_m|c/(30b) - 1]$ is positive, if

$$|\Lambda_m| > 30b/(\chi c). \qquad (3.27)$$

Negative inductance is a circuit element with purely negative reactive impedance, proportional to frequency. This element is equivalent to frequency-dependent capacitance:

$$-j\omega|\Lambda_m| = 1/(j\omega C_m), \qquad (3.28)$$

where

$$C_m = 1/(\omega^2|\Lambda_m|) = C_{m0}f_0^2/f^2. \qquad (3.29)$$

Here C_{m0} is the magnitude of an equivalent capacitance C_m at frequency $f = f_0$. The magnitude C_{m0} does not depend on f.

Thus in order to maintain the in-phase current distribution in a wide frequency range, the capacitances of concentrated loads must change in inverse proportion to the square of the frequency. If it were possible, the capacitor could be made with the capacitance, which satisfies the condition (3.14) and eliminates the constraint (3.16). But, strictly speaking, purely reactive impedances can only grow, and negative inductances do not exist in nature. Therefore, the use of constant capacitors allows creating in-phase current distributions only in limited frequency bands.

The executed analysis allows making a number of practical conclusions. In order that the concentrated loads may efficiently influence the current distribution, the distance between them must be small in comparison with the wavelength. For creating a wide-range radiator, only capacitors should be used as reactive elements, since inclusion of reactive two-poles of a more complex type, whose structure includes inductance coils, results in narrowing of the operating range. Capacitors enable creation of an electromagnetic wave with real propagation constant that is realized in the form of an in-phase current with an exponentially decreasing amplitude (i.e., the current amplitude has the form of the concave curve), along a radiator in a wide frequency range. Obtaining a convex curve of the current with the help of simple concentrated elements (resistors, capacitors, inductance coils) is impossible.

Method of the impedance long line and use of its results for creating wide-range linear radiators were first described in [49].

3.2 Method of a metallic long line with loads. Synthesis of a current distribution

The previous section describes the approximate method, according to which one can build an antenna with in-phase current distribution that allows to provide high electrical characteristics in a wide frequency range. This method is based on the analysis of properties of the impedance long line and on the calculation of load magnitudes that provide the specified current distribution.

Together with the method of impedance long line another approximate method for calculating magnitudes of loads, which provide the required current distribution, has been developed [50]. This method is based on the calculation of the characteristics of an ordinary long line from two metal wires. The method allows us to find the law for changing the equivalent line length and to determine loads, which must be included in each antenna segment to implement this law in accordance with the required distribution.

Use of this method and the method of impedance long line gives analogous results.

Figure 3.4 shows an asymmetrical radiator of height L with N loads, which are uniformly spaced along it at a distance b from each other. A current distribution along the radiator in a first approximation is similar to the current distribution along an open at the end long line, with the loads Z_m included in series. The current distribution along each line segment, located between adjacent loads, has sinusoidal character:

$$J(z_m) = J_m \sin k(z_m + l_{e,m-1}), \ 0 \leq z_m \leq b. \tag{3.30}$$

Here $z_m = (N - m + 1)b - z$ is the coordinate, measured from the end of the segment m, J_m is the amplitude of a current onto the segment m, $l_{e,m-1}$ is the equivalent length of all preceding segments with $(m - 1)$ loads, whose total length is equal to $(m - 1)b$. The magnitudes l_{em} and $l_{e,m-1}$ are mutually related by the expression:

$$-jW\cot kl_{em} = Z_m - jW \cot k(b + l_{e,m-1}),$$

where W is the wave impedance of the line. Note that the length l_{em}, if $m = N$, is equal to the equivalent length L_e of the radiator.

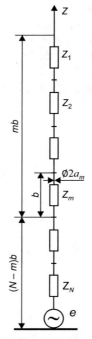

Fig. 3.4. Antenna with a several concentrated loads.

Inverse Problems of Antenna Theory 135

This expression permits to find the magnitude Z_m of mth load:

$$Z_m = -jW[\cot kl_{em} - \cot k(b + l_{e,m-1})].$$

If the distance between loads is small ($kb \ll 1$) then

$$Z_m = -jW \frac{\sin k(b + l_{e,m-1} - l_{em})}{\sin kl_{em} \sin k(b + l_{e,m-1})} \approx -jW \frac{k(b + l_{e,m-1} - l_{em})}{\sin^2 kl_{em}}.$$

By means of the last two expressions one can calculate the magnitudes of loads. For this it is necessary to know the law, in accordance with which the equivalent length grows along the line. The choice of this law depends on the required current distribution along the radiator. In the general case similarly (3.3)

$$J(z_m) = J_A f(z), \quad 0 \le z_m \le b, \quad (N-m)b \le z \le (N-m+1)b, \quad (3.31)$$

where J_A is the current amplitude in the antenna base, and $f(z)$ is the law of current distribution. Henceforth we shall assume that the function $f(z)$ is real and positive, i.e., we shall consider only in-phase distributions.

Suppose we want to obtain a given current distribution $J(z)$ along the antenna. For this we set assume that the current $J(z_m)$ at the beginning and the end of each segment coincides with the current $J(z)$. If the segments' lengths are small, the distribution of the current along the line is close to the required one. In a general case we have in accordance with (3.30) and (3.31)

$$J_{m+1} \sin k(b + l_{em}) = J_A f[(N-m-1)b] \quad \text{at} \quad z_{m+1} = b,$$
$$J_{m+1} \sin kl_{em} = J_A f[(N-m)b] \quad \text{at} \quad z_{m+1} = 0.$$

If we divide the left and right parts of the first equality by the corresponding parts of the second equality and retain only the first terms of the expansion of trigonometric functions in a series, then considering that the magnitude b is small, we will obtain

$$1 + kb \cot kl_{em} = f[(N-m-1)b]/f[(N-m)b],$$

i.e.,

$$l_{em} = \frac{1}{k} \tan^{-1} \frac{kb}{f[(N-m-1)b]/f[(N-m)b] - 1}. \quad (3.32)$$

As is seen from (3.32), the equivalent length l_{em} is frequency dependent. Knowing l_{em} and $l_{e,m-1}$, one may in accordance with the expression for Z_m find the load magnitudes. They also have a frequency-dependent nature:

$$Z_m = -j\frac{W}{kb} \left\{ \frac{f[(N-m-1)b]}{f[(N-m)b]} - 2 + (1 + k^2 b^2) \frac{f[(N-m+1)b]}{f[(N-m)b]} \right\}. \quad (3.33)$$

An input impedance of the long line open at the end is a first approximation to a reactive component of antenna input impedance. In a general case it is equal to

$$Z_A = -jW \cot kL_e = -j\frac{W}{kb}[f(-b)-1]. \quad (3.34)$$

This expression shows that the function $f(z)$ should change slowly near the antenna base and the difference $f(-b) - 1$ should be small, of the order of kb. Otherwise, the reactive component of the antenna input impedance will be great.

We define, for example, the loads' magnitudes providing the exponential distribution (3.10) of the current amplitude along the radiator. The linear distribution (3.9) is its particular case. The equivalent lengths of the long lines for the exponential and linear distribution in accordance with (3.32), (3.10) and (3.9) are equal to

$$l_{em3,4} = \frac{1}{k}\tan^{-1}\frac{kb[1-\exp(-m\alpha b)]}{\exp(\alpha b)-1}, \quad l_{em2} = \tan^{-1}(mkb).$$

As is seen from these expressions, if $\alpha > 0$, the equivalent length of the antenna arm does not or may exceed quarter of the wave length. The input impedance of a long line with the exponential and linear current distribution may be written in the form

$$Z_{A3,4} = -j\frac{W[\exp(\alpha b)-1]}{kb[1-\exp(-\alpha L)]}, \quad Z_{A2} = -jW/(kL). \quad (3.35)$$

Using expressions (3.33) and (3.11), we find the magnitudes of the loads, which provide the exponential law of the current amplitude distribution along the radiator:

$$Z_m = -j\frac{W}{kb}\frac{2(\cosh \alpha b - 1) + k^2 b^2 \exp(-\alpha b)}{1-\exp(-m\alpha b)}\{1-\exp[-(m-1)\alpha b]\}. \quad (3.36)$$

If the product αb is not small, then, neglecting the second term of the numerator, we obtain

$$Z_m = 1/(j\omega C_m), \quad (3.37)$$

where

$$C_m = \frac{b[1-\exp(-m\alpha b)]}{2Wc(\cosh \alpha b - 1)}.$$

As is seen from this formula, the sign of C_m coincides with the sign of α.

Thus, in order to obtain an exponential distribution of current amplitude with a great decrement, one must use capacitive loads. Capacitors allow to create only concave current distributions (with $\alpha > 0$). In order to obtain a convex distribution (with $\alpha < 0$), capacitances must be negative. If $ab \ll 1$, then, by confining the first terms of the functions' expansion into a series, we find from (3.36)

$$Z_m = 1/(j\omega C_m) + j\omega \Lambda_m, \qquad (3.38)$$

where

$$C_m = \frac{m}{cW\alpha}, \quad \Lambda_m = -\frac{Wb(m-1)}{cm}.$$

In order to obtain the exponential distribution with a small decrement α, using capacitors, the negative inductances Λ_m should be included. They can be neglected, if the first term of (3.38) is much larger than the second one, i.e., $\alpha \gg k^2 b(m-1)$. If $\alpha = 0$,

$$Z_m = j\omega \Lambda_m = -jkbW(m-1)/m, \qquad (3.39)$$

i.e., in the case, when the loads are fabricated in the form of negative inductances, proportional to $(m-1)/m$, we obtain a purely linear distribution.

As it follows from the analysis executed in this section, the method of the long line with loads and the method of the impedance long line lead to similar results. Comparison of these results allows us to apply these methods to specific problems, using specific details of the current distribution along the radiators.

3.3 Wide-range V-radiator

As already was mentioned, one of the important tasks of antenna engineering is creating a radiator, which ensures a maximum field in the plane perpendicular to the radiator axis in a wide frequency range. An ordinary linear radiator fails to meet this requirement: if the radiator arm is larger than 0.7λ, the radiation in the plane, perpendicular to the antenna axis, decreases. In this case one can use V-antenna formed by two converging inclined wires. If arm length L is greater than 0.7λ, an ordinary V-antenna has preferential radiation along the bisector of the angular aperture. However, with growing frequency the side lobes of the directional pattern increase, and the main lobe of this pattern in the antenna plane diminishes. If the arm length is greater than about 1.25λ, the main lobe splits, and the radiation along the bisector sharply decreases.

Including capacitive loads in the antenna wire allows expanding the frequency range, in which there are the directed radiation along the bisector of the angular aperture, and to increase the useful signal in this direction.

Consider a symmetric V-dipole with arm length L and arbitrary angular aperture $\alpha = \pi - 2\theta$ (Fig. 3.5). The far field along the bisector of the angular

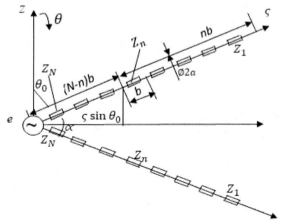

Fig. 3.5. *V*-dipole with loads.

aperture, created by an elementary segment $d\varsigma$ of the upper antenna arm located near the origin of coordinates system is equal to

$$E_\theta(\varsigma)d\varsigma = E_\theta(0)[J(\varsigma)/J(0)]exp(jk\varsigma \sin \theta_0)d\varsigma, \qquad (3.40)$$

where ς is the coordinate measured along the radiator axis, $J(z)$ is the current along the upper arm, $J(0)$ is the current near point 0, and $k\varsigma \sin \theta_0$ is the path-length difference between the points 0 and ς. To ensure that all the far fields of different points coincide in phase, the current distribution along this arm must correspond to the expression

$$J(\varsigma) = J(0)f(\varsigma)exp(-jk\varsigma \sin \theta_0). \qquad (3.41)$$

Here $f(\varsigma)$ is a real and positive function.

Let N loads Z_n be located uniformly along a wire of each antenna arm at a distance b from each other. If the load spacing is small ($kb \ll 1$), then, as in the case of the linear dipole, as well, the replacement of concentrated loads by distributed surface impedance $Z(\varsigma)$ practically does not change the current distribution along the antenna. We assume that the surface impedance of each antenna segment with load Z_n is constant and equal to $Z^{(n)}$ in accordance with (1.48).

As is said in Section 1.5, the current distribution along the antenna with piecewise constant surface impedance coincides in the first approximation with current distribution along an equivalent impedance line, i.e., along the line with a stepped variation of the propagation constant. Here, the wave propagation constant γ_n along the segment n is related to surface impedance $Z^{(n)}$ in accordance with (3.1). If the law of changing propagation constant is known, one can use magnitude γ_n for calculating with the help of (3.13) concentrated loads Z_n, which are needed for the embodiment of this law. The current along the segment n of a stepped line is

$$J(\varsigma_n) = I_n \sinh(\gamma_n \varsigma_n + \varphi_n), \quad 0 \le \varsigma_n \le b, \qquad (3.42)$$

where I_n and φ_n are the amplitude and phase of a current onto a segment n, respectively, and ς_n is the coordinate, measured from the segment end, i.e., $\varsigma_n = (N - n + 1)b - \varsigma$. We equate current $J(\varsigma_n)$ at the beginning and the end of each segment to current $J(\varsigma)$, to ensure the phase coincidence of far fields from all antenna segments. The current inside each segment does not coincide with current $J(\varsigma)$. However, if the segments' lengths are small, the current distribution along the line is close to $J(\varsigma)$.

According to (3.41) and (3.42), at $\varsigma_n = b$ and $\varsigma_n = 0$:

$$I_n \sinh(\gamma_n b + \varphi_n) = J(0)f[(N-n)b]exp[-jk(N-n)b\sin\theta_0],$$
$$I_n \sinh \varphi_n = J(0)f[(N-n+1)b]exp[-jk(N-n+1)b\sin\theta_0].$$

If we divide the left- and right-hand sides of the first equation by the corresponding sides of the second equation, then considering that b is a small in magnitude and retaining only the first terms of the series expansions for trigonometric functions with small arguments, similar to (3.4) for segment n, we get,

$$\tanh \varphi_n = \gamma_n b \bigg/ \left\{ \frac{f[(N-n)b]}{f[(N-n+1)b]}(1 + jkb\sin\theta_0) - 1 \right\}.$$

Analogously, for the segment $(n + 1)$

$$\tanh \varphi_{n+1} = \gamma_{n+1} b \bigg/ \left\{ \frac{f[(N-n-1)b]}{f[(N-n)b]}(1 + jkb\sin\theta_0) - 1 \right\}.$$

Voltage and current are continuous along a stepped line. Therefore, (3.6) is true. It forms a set of equations that allows us to relate γ_n and γ_{n+1}. From the solution of this set of equations it follows that magnitude γ_n is independent of γ_{n+1}:

$$\gamma_n = \frac{1}{b}\left\{1 - \frac{2f[(N-n)b] - f[(N-n-1)b]}{f[(N-n+1)b]} - 2jkb\sin\theta_0 \frac{f[(N-n)b] - f[(N-n-1)b]}{f[(N-n+1)b]}\right\}^{\frac{1}{2}}.$$

(3.43)

This expression generalizes expression (this is also doubtful because it is an expression for current later in the chapter for parallel vertical wires) for a linear dipole and transforms it at $\theta_0 = 0$.

The possibility of implementation of propagation constant γ_n is determined by the possibility of implementation of concentrated loads. According to

(3.13), at low frequencies, when inequalities (3.16) and (3.17) resulting from it, are true, the load value is

$$Z_n = -\frac{j30(\gamma_n b)^2}{(kb\chi)}. \qquad (3.44)$$

By substituting (3.43) into (3.44), we get

$$Z_n = R_n + 1/j\omega C_n, \qquad (3.45)$$

where

$$R_n = \frac{60}{\chi}\sin\theta_0 \frac{f[(N-n-1)b] - f[(N-n)b]}{f[(N-n+1)b]},$$

$$C_n = 4\pi\varepsilon_0 b\chi \bigg/ \left\{1 - \frac{2f[(N-n)b] - f[(N-n-1)b]}{f[(N-n+1)b]}\right\}.$$

As seen from (3.45), each load should be a series connection of a resistor and a capacitor, where the resistance of the resistor is positive, if function $f(\varsigma)$ decreases monotonically with growing ς, and the capacitance of the capacitor is positive, if function $f(\varsigma)$ is concave. The resistance depends on the angular aperture of the antenna and the form of function $f(\varsigma)$, whereas the capacitance depends only on the latter.

For a linear radiator with loads ensuring the maximal radiation in the plane, perpendicular to its axis, each load should, when condition (3.13) holds, represent a capacitor. Capacitors ensure real wave propagation constant γ_n and an in-phase distribution of the current along an antenna. In V-antenna with capacitors, the resistor must be included in series with the capacitor, and that leads to a phase lag of a current wave along an antenna wire. Such phase delay is necessary for a V-dipole, since it compensates the path-length difference from individual segments to an observation point and ensures coincidence of phases for fields, created by different segments in the far zone along the bisector of the angular aperture.

The use of resistors in a transmitting antenna is inexpedient. This means that the loads of a V-dipole should not differ from the loads of a linear radiator, which ensure an in-phase current distribution along an antenna wire.

At high frequencies, when condition (3.16) is not met, to create the in-phase current distribution along a linear radiator, the load must represent a negative inductance (a capacitance, which is inversely proportional to square of frequency). Similarly, the load for a V-dipole should be a series connection of a capacitor with a frequency-dependent capacitance and a resistor. In order for the propagation constant to be real and the current along an antenna to be in-phase, the capacitances should not exceed the value determined by inequality (3.24).

3.4 Method of mathematical programming

Use of mathematical programming methods [51] plays a major role in solving the inverse problems of creating antennas. These methods allow determination of optimal parameters of an antenna, in particular geometric dimensions and magnitudes of concentrated and distributed loads for creating antennas with specified characteristics, or more precisely, with characteristics that are as close as possible to the given ones.

This remark is due to the fact that the variation intervals of radiator parameters are bounded, i.e., not every value of an electrical characteristic can be realized practically. Different characteristics are optimal for different parameters. Moreover, an antenna should have certain properties not at a single fixed frequency, but in the entire operation range. Therefore, the selected parameters are a result of a compromise, reached with the help of the mathematical programming method.

The problem of mathematical programming in the general case is stated as follows: it is necessary to find vector $\vec{x}$ of parameters that minimizes some objective function $\Phi(\vec{x})$ under imposed constraints $\Phi_i(\vec{x}) \geq 0$. Depending on the type of functions $\Phi(\vec{x})$ and $\Phi_i(\vec{x})$, mathematical programming is divided into linear, convex and non-linear one. In the case at hand, the problem is solved by non-linear programming methods since the type of function $\Phi(\vec{x})$ is unknown.

The objective function $\Phi(\vec{x})$ (or general functional) is a sum of several partial functionals $\Phi_j(\vec{x})$ with weighting coefficients p_j and penalty function Φ_{ij}:

$$\Phi(\vec{x}) = \Sigma_{(j)} p_j \Phi_j(\vec{x}) + \Sigma_{(i)} \Phi_{ij}. \tag{3.46}$$

The partial functional is an error function for one of the antenna characteristics. The weighting function allows taking into account the importance of this characteristic and the sensitivity of the corresponding functional to change of vector $\vec{x}$. A penalty function is zero, if the parameters lie within a given interval, and has great magnitude, even if only one of the parameters falls outside the interval limits.

For an antenna with concentrated loads the controlled parameters x are magnitudes of loads, coordinates z_n of their connection points and the wave impedance W of the cable. Under loads understood to be simple elements: capacitors with capacitances C_n, coils with inductance Λ_n and resistors with resistance R_n, values z_n, W, C_n, Λ_n and R_n should be real, positive and frequency-independent, and z_n smaller than antenna length L. These requirements naturally limit the variation interval of parameters.

Different ways of an error function formation are known. Good results are produced by means of quasi-Tchebyscheff criterion:

$$\Phi_j(\vec{x}) = \frac{1}{N_f}\left[\frac{f_{j0}}{f_{jmin}(\vec{x})} - 1\right]\left\{\sum_{(n_f)}\left[\frac{(f_{j0}/f_j\langle\vec{x}\rangle) - 1}{(f_{j0}/f_{jmin}\langle\vec{x}\rangle) - 1}\right]^S\right\}^{1/S}. \quad (3.47)$$

Here N_f is a number of points of the independent argument (e.g., a number of frequencies in given range), n_f is frequency number, $f_j(\vec{x})$ is one of electrical characteristics of an antenna, $f_{jmin}(\vec{x})$ is its minimal magnitude in the considered interval, f_{j0} is a hypothetical value of the characteristic, which must be reached, S is the power, allowing the control of method sensitivity.

Another criterion is called a root-mean-square criterion. It uses another error function:

$$\Phi_j(\vec{x}) = \frac{1}{N_f N_l}\sum_{n_f=1}^{N_f}\sum_{n_l=1}^{N_l}[f_j(\vec{x}) - f_{j0}]^2. \quad (3.48)$$

Here N_f and N_l are numbers of points of an independent argument (e.g., a number of frequencies in a given range and a number of division points on the wire), n_f is the frequency number, n_l is the point number, $f_j(\vec{x})$ is one of the electrical characteristics of the antenna (e.g., a current or a voltage), and f_{j0} is a hypothetical value of the characteristic, which must be reached.

The choice of optimizable characteristics depends on a stated problem. For creation of a wide-range radiator one must use in the capacity of characteristics $f_j(\vec{x})$, a travelling wave ratio (*TWR*) in a cable and a pattern factor (*PF*), which is equal to the average level of radiation at predetermined angle ranges. If resistors with resistances R_n are used as loads, it is necessary to supplement the set of $f_j(\vec{x})$ by the characteristic of antenna efficiency (η_A):

$$TWR = \frac{2a}{a^2 + b^2 + 1 + \sqrt{(a^2 + b^2 + 1)^2 - 4a^2}}, PF = \frac{1}{K}\sum_{k=1}^{K}F(\theta_k), \eta_A = 1 - \frac{1}{J_A^2 R_A}\sum_{n=1}^{N}|J_n|^2 R_n, \quad (3.49)$$

Here $a = R_A/W$, $b = X_A/W$ are ratios of active and reactive components of antenna impedance to the wave impedance of a cable, respectively, K is number of angles θ_k in a vertical plane within the limits of an angular sector from θ_1 to θ_K (for example, from 60° to 90°), k is the angle number within this sector, and $F(\theta_k)$ is a magnitude of normalized directional pattern in the vertical plane for an angle θ_k. N is the number of loads, J_n and J_A are the current in the nth load and in the antenna base, respectively.

If it is necessary to obtain a given current distribution $J(z)$, it is expedient to use characteristics $f_j(\vec{x})$ with both real and imaginary current components:

$$f_1 = \text{Re } J(z, f), \quad f_2 = \text{Im } J(z, f), \quad (3.50)$$

or amplitude and phase of the current:

$$f_3 = |J(z,f)|, \quad f_4 = \tan^{-1}[Im(z,f)/Re(z,f)]. \tag{3.51}$$

In cases, when an analytical expression for an objective function $\Phi(\vec{x})$ is absent, the minimum of function must be found by a numerical method, based on gradient search. Gradient search is an iterative procedure, which becomes step by step from one set of parameters $\vec{x}_m$ to another set $\vec{x}_{m+1}$ in a direction of maximal decrease of the objective function. Therefore, this method is called the method of steepest descent:

$$\vec{x}_{M+1} = \vec{x}_M - \alpha_M grad\Phi(\vec{x}_M). \tag{3.52}$$

Here M is the iteration number, α_M is the scale parameter. Each iteration in essence searches the minimum of a functional (of an objective function) in the direction of anti-gradient. The result of this search is the determination of the coefficient α_M, at which the objective function $\Phi(\vec{x}_M)$, that enters together with α_M into equation (3.52), becomes minimal. Also, the values of parameters $\vec{x}_M$, which correspond to this minimum, are determined.

A modification of the steepest descent method is the method of conjugate gradients. In this case the iteration 1, $(Q-1)$, $(2Q+1)$ and so on are calculated according to the anti-gradient (here Q is number of parameters) and the rest of the steps correspond to the expression

$$\vec{x}_{M+1} = \vec{x}_M - \alpha_M \vec{G}_M, \tag{3.53}$$

where

$$\vec{G}_M = grad\Phi(\vec{x}_M) + \left|\frac{grad\Phi(\vec{x}_M)}{grad\Phi(\vec{x}_{M-1})}\right|^2 \vec{G}_{M-1}.$$

The calculation ends when the decrease of the objective function from iteration to iteration becomes smaller than a preset value, or the number M of iterations exceeds certain limit M_0.

The minimum of an objective function $\Phi(\vec{x}_M)$ and the values of parameters, which correspond to this minimum, are determined in each iteration. In essence, each iteration searches a parameter α_M.

The most rational method consists in a sequential increase (for example, doubling) of magnitude α and further interpolating the function $\Phi(\vec{x}_M)$ in a considered interval by a polynomial of a given power. It is convenient to use cubic interpolation, since the number of interpolation nodes is large enough (four), and the root of the derivative (the value α that turns the derivative into zero) is found analytically. If the first step results in an increase, rather than decrease of the objective function, the step should be reduced by a factor of 10^p, where $p = 1, 2...$, whereupon the linear search goes on again with doubling a step.

The mathematical programming method (synthesis) presupposes multiple computations of the antenna's electrical characteristics at different initial parameters (analysis). Performing such calculations requires incorporation of a special program into the synthesis software. This program allows determination of all electrical characteristics of an antenna, i.e., calculating the functions $f_j(\vec{x})$ for vector $\vec{x}$ of initial parameters at given emfs and loads.

The most laborious of these calculations is computation of self and mutual impedances between antenna segments (between so-called short dipoles). Therefore, in order to speed up the calculations, it is expedient to fixate, for example, points of placing concentrated loads, in order that the coordinates of short dipoles and their mutual impedances do not change from iteration to iteration. If there are many loads, i.e., the distances between them are small in comparison with the wave length, this restriction will have no effect on the synthesis results.

As regards initial magnitudes of concentrated loads, these magnitudes must be found by the approximate physical method, described in Sections 3.1 and 3.2. The results of the calculations show that the computational process in this case speeds up. But the most important result of using these magnitudes consists in reducing the error probability since an arbitrary choice of the initial parameters may cause an optimization process, which will lead to a local, rather than true, minimum of the objective function.

The described iterative procedure is essentially in motion in the vector space $\vec{x}_m$ along the surface $\Phi(\vec{x}_M)$ in the direction of a minimum of a general functional. A particular case of such motion, is when the vector $\vec{x}_m$ consists of two parameters (x and y) is shown in Fig. 3.6. The algorithm of the iterative procedure is presented in Fig. 3.7.

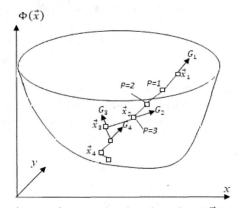

Fig. 3.6. Iterative procedure as motion along the surface $\Phi(\vec{x}_M)$ to an optimum.

Inverse Problems of Antenna Theory   145

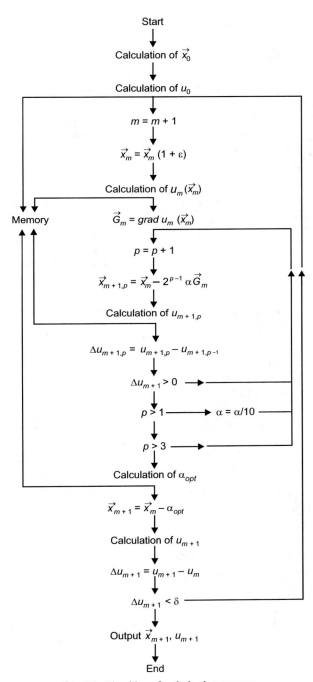

Fig. 3.7. Algorithm of optimization program.

It is necessary to emphasize that the method of mathematical programming makes it possible to realize the synthesis of a wide-range radiator, i.e., chooses the optimum capacitive loads that provide the highest possible level of TWR, PF and efficiency in a given frequency range. The synthesis program, using the mathematical programming method, allows us to lead the problem solution to the desired goal.

Other methods of solving often stop halfway and do not reach the goal. For example, earlier the synthesis of the antennas with given electrical characteristics was broken up into two stages: in the first stage the distribution of current was computed. The parameters of the antenna, providing such distribution, should have been determined at the second stage. The first stage was investigated sufficiently. It covers a wide class of the tasks (one of the possible options is the task of creating a broadband antenna). Much less attention was paid to the second stage of the synthesis.

In principle, if the required current distribution along a wire antenna is known, one can split the wire of an antenna into short dipoles and define currents at the centers of these dipoles. The amplitudes of piecewise sinusoidal basis functions are equal to the currents at the corresponding antenna points. It is easy to calculate the magnitudes of loads, which one must connect at these points to obtain the desired currents.

But the impedances of the loads, calculated by this method, consist of active and reactive components, which are change with frequency. The calculated active component of load impedance may be negative, and this is an evidence of impossibility to create such a distribution with the help of passive elements. As to the reactive component, it is still necessary to solve the problem of its implementation in the given frequency range with the help of a set of simple elements. Therefore, it is necessary to solve the problem of creating an antenna with the chosen type of loads in order to ensure the current distribution that is as close as possible to the desired distribution in the desired range, not the given current distribution. This problem, like the problem of creating a wide-range radiator, may be solved only by the mathematical programming method.

The method of mathematical programming offers a wide scope for the solution of various problems of synthesis. During its application, the results of calculations of antennas with concentrated capacitive and capacitive-resistive loads were analyzed and taken into account. Practical conclusions made as a result of the analysis are given at the end of Section 3.1. The results of applying the described technique to specific tasks are described in the following sections. They allow us to make a general conclusion that the linear antenna with concentrated capacitances decreasing in proportion to the distances from their free ends, allow the creation of in-phase currents with a linear current amplitude distribution and obtains good electrical characteristics in a wide frequency range.

3.5 Application of results to concrete tasks

Figure 3.8 gives *TWR* of three asymmetrical antennas in a cable with wave impedance 75 Ohm. The calculations are performed for an antenna of height 12 m and radius 0.03 m, with ten capacitors located at distance 1.2 m from each other (the upper and lower capacitors are placed at the distance 0.6 m from the antenna ends).

Curve 1 in Fig. 3.8 corresponds to a radiator without loads, i.e., to a whip antenna, curve 2—to a radiator with loads, whose capacitances are frequency independent. Here, the capacitance C_{N0} of the bottom capacitor was chosen equal to 177 pF. In this case the propagation constant is real along the entire antenna up to frequency 10 MHz. Capacitances C_{n0} of others capacitors decrease in the direction of the free end of the antenna in proportion to the distance from it. That allows achievement of the current distribution along the radiator, which is close to the linear law. Curve 3 is given for a radiator with frequency-dependent capacitive loads. Their capacitances are changed in accordance with (3.29), where $f_0 = 20$ MHz.

Table 3.1 shows lower f_1 and upper f_2 frequencies of the operating range of each antenna. At the frequency f_1 *TWR* becomes greater than 0.2, at the frequency f_2 a field strength along a perpendicular to an antenna axis becomes less than 0.7 of the maximum field strength (as a rule, f_2 corresponds to the second maximum on curve of *TWR*). *TWR* of a whip antenna with growth of frequency quickly decreases below level of 0.2, and the frequency,

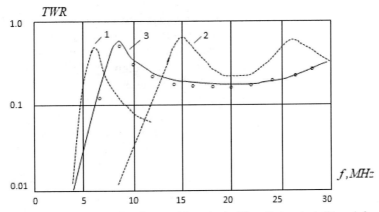

Fig. 3.8. Input characteristics of radiators without loads (1), with constant (2) and frequency-dependent (3) capacitive loads.

Table 3.1. Frequency ratio of radiators.

Version of antenna	Frequency, MHz		Range width Δf, MHz	Frequency ratio k_f
	f_1	f_2		
1	5.2	7.7	2.5	1.5
2	12.3	26.0	13.7	2.1
3	6.3	34.0	27.7	5.4

corresponding to this point, is taken as f_2. Besides, Table 3.1 for each antenna reports a range width $\Delta f = f_2 - f_1$ and frequency ratio $k_f = f_2/f_1$.

As is seen from Fig. 3.8 and Table 3.1, at low frequencies the matching level of the variant 3 approaches to the matching level of the whip antenna, and at high frequencies the upper boundary of the operating range for variant 3 is shifted to the right in comparison with variant 2. In addition, the minimal *TWR* in the middle of the operating range increases.

Figure 3.8 also presents the results of experimental verification of *TWR* for the variant 3. The measurements were performed by means of an antenna model in a scale 1:10. The frequency range was split into short intervals, and the capacitances of loads used in each interval were equal to capacitances calculated for the middle of the interval. The calculated and experimental results are reasonably well consistent with each other.

The most practical and accessible realization of antennas with frequency-dependent capacitances is the use of tunable, in particular simply switched capacitors; for example, on the signal from the console (control panel).

The methods of impedance long line and metallic long line with loads allowed to define potential capabilities of antennas with loads. The results, obtained by means of these methods, can be used for solving the optimization problem of an antenna with loads by the mathematical programming method. Application of synthesis program for selection of the optimal capacitive loads permits to obtain maximal *TWR* and *PF* in the predetermined range of frequencies $f_1 - f_2$. During this work all weighting coefficients p_j as a rule were taken by identical. The calculations show that for a synthesis of a simple antenna four to five iterations are enough. The number of optimized electrical characteristics in this case has little effect on the synthesis time. For example, the time of optimizing *TWR* and *PF* is almost equal to the time of optimizing only *TWR*. The choice of criterion has practically no effect on the results of synthesis of wide-range radiators and the calculation time. In particular, the root-mean-square criterion has no advantage as compared with the quasi-Tchebyscheff criterion and is rarely used. As a hypothetical value of the characteristic f_{j0} it is advisable to choose a maximum magnitude, since its decrease leads to a decrease of the result in the synthesized antenna.

An increase of index S in (3.47) accelerates the process convergence. In the calculations it was assumed that $S = 6$.

Figure 3.9 gives the basic dimensions of an optimizable antenna with loads. It is a monopole of height $L = 12$ m with nine capacitors spaced equidistantly. The capacitance C_0, equal to capacitance of a typical ceramic insulator (15 pF), is located at the base of the antenna in parallel with its input.

The results of the antenna synthesis are presented in Table 3.2, where the basic characteristics of the radiators are given. In Table 3.2 the following

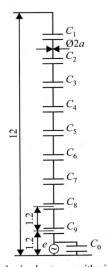

Fig. 3.9. Synthesized antenna with nine capacitors.

Table 3.2. Main characteristics of antennas.

Version	L, m	a, m	N	$f_1 - f_2$, MHz	N_f	M	TWR min	PF min
1	12	0.03	9	7.5–15	16	4	0.123	0.819
2	"	"	"	15–30	"	"	0.273	0.610
3	"	"	"	30–60	"	5	0.360	0.562
4	"	0.15	"	7.5–15	"	4	0.205	0.813
5	"	"	"	15–30	"	5	0.414	0.680
6	"	"	"	30–60	"	4	0.380	0.605
7	"	0.03	"	8.5–13	10	3	0.217	0.870
8	"	"	"	13–22	"	5	0.216	0. 0,790
9	"	"	"	22–60	20	8	0.204	0. 0,437
10	"	0.15	"	8.5–13	10	5	0.314	0.829
11	"	"	"	13–22	"	4	0.278	0. 0,859
12	"	"	"	22–60	20	5	0.322	0.565

designations are used: a is antenna radius, N is number of capacitors, N_f is number of used frequencies, M is a required number of iterations. Frequency ratio for the antenna variants with numbers 1–6 was adopted as equal to two. As seen from table, the increase of antenna radius from 0.03 m to 0.15 m at frequencies up to 30 MHz results in growing minimal *TWR* approximately by a factor 1.5. The variation of radius has a weaker influence on *PF* when it is minimal.

Figure 3.10 shows the electrical characteristics of variants 1–3 (antenna radius is 0.03 m) and of a whip antenna with the same geometrical dimensions and the same capacitance C_0 of the insulator. The characteristics of the antenna with a radius 0.15 m (variants 4–6) are similar. As seen from the calculations, the curve of *TWR* can have two maximums at high frequencies. The curves of *PF* do not decrease monotonically with frequency, but have maximums too. In addition to calculated curves, the pictures demonstrate (by dots and other symbols) the results of experimental verification, carried out on the model in the scale of 1:5. The calculation and experiment coincide well.

As it follows from Table 3.2 (variants 1–6), if the frequency ratio for various ranges are identical, the level of antenna matching with a cable is different in the different ranges. This level substantially rises, if the frequency grows. In order to obtain more uniform and, on the whole, better characteristics over the entire frequency range (at unchanged number of sub ranges), it is expedient to split the total range into parts such that the frequency ratio in different sub ranges increases with increase of frequencies. The results of solving this problem are presented in Table 3.2 as variants 7–12.

The electrical characteristics of variants 10–12 with radius 0.15 m as well as of a monopole without loads with the same radius and with capacitance C_0 at the base are given in Fig. 3.11. Data of Table 3.2 confirm a general increase of *TWR* level in comparison with variants 1–6. In all the sub ranges, increase in the antenna radius causes a rise of minimum *TWR* (approximately by a factor 1.5), together with a rise of minimum *PF* at high frequencies.

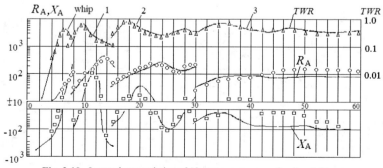

Fig. 3.10. Input characteristics of 12-meter antennas of radius 0.03 m.

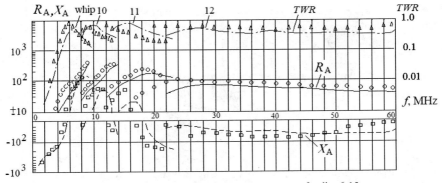

Fig. 3.11. Input characteristics of 12-meter antennas of radius 0.15 m.

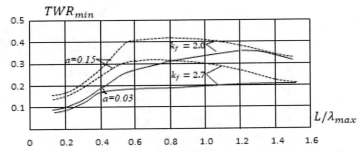

Fig. 3.12. The maximum level of matching for the antenna with constant capacitances.

The results of optimization of 12-meter antennas with capacitances $C_0 = 15\ pF$ in the base of the antenna are used to plot the curves in Fig. 3.12 for the minimal *TWR* depending on relative antenna length L/λ_{max} (λ_{max} is the maximum wavelength) at various frequency ratios k_f and different antenna radii a. These curves determine the maximum attainable characteristics, which can be obtained by means of antennas with constant capacitive loads.

Results of calculations and experimental verification presented in this section show that the described procedure is an effective way to optimize the characteristics of antennas with capacitive loads. Application of this procedure confirms its validity and usefulness.

The calculation results also show that if it is necessary, the antenna range can be expanded in the direction of high frequencies at a sufficiently high *TWR*; for example, one can obtain $TWR \geq 0.2$ in the range with a frequency ratio about 10. But the directional patterns in the additional (high-frequency) range deteriorate substantially. In this connection, the frequency ratio of an antenna with constant capacitive loads does not exceed 3 (at $PF \geq 0.5$ and $TWR \geq 0.2$).

As it became clear later, this circumstance was caused by the fact that under calculations of the objective function in accordance with (3.47) weighting coefficients equal in magnitude were adopted, i.e., the different sensitivity of partial functionals to a change in the vector $\vec{x}$ was not taken into account. Using a multi-tiered structure (see Chapter 5) allows to ensure desired directional pattern in a wide range. Joint application of both principles, i.e., using a multi-tiered structure and capacitive loads in the wires of each tier allows high level of matching and desired directional pattern in a wide range.

Thus, the method of mathematical programming is an efficient method of optimization of antennas with capacitive loads. Its software may be used for optimization of antennas with loads of other kinds. The procedure developed for solving the described problem can be effectively used to solve other problems. We will begin to talk about that from creating the desired current distribution in a certain frequency range.

It should be emphasized that the requirement to create a specific distribution does not mean a strict coincidence of the given and the obtained distribution at all frequencies.. This requirement only means obtaining a distribution, which is close to the required one as far as possible.

Synthesis of antenna with a given current distribution in a certain frequency range was considered as an example in [50]. The calculation was performed for the monopole of height 6 m and radius 6.5×10^{-3} m with ten capacitive loads, which are located along it at a distance 0.6 m from each other. The distance from the upper load to the antenna end and from the bottom load to its base is half as much. The capacitance of the bottom load was taken equal to 18 pF, in which case the propagation constant γ_n is real along the entire antenna up to frequency 40 MHz. The capacitances of the remaining loads were selected in accordance with (3.21).

Initial calculation of load magnitudes was executed by means of approximate method for a metallic long line with loads. Figure 3.13a shows the equivalent lengths l_{em} of this line measured from the free end of the monopole to the points n of capacitors location. The capacitances C_n of these loads are given in Fig. 3.13b. The corresponding curves are indicated by labels '*lin*' and '*exp*'. Equivalent lengths and capacitances were calculated in accordance with equalities (3.38) and (3.43) respectively at the frequency $f = 40$ MHz.

This data was used to design two antennas with in-phase currents: it was necessary to obtain a linear distribution of the current amplitude in one antenna at a frequency $f = 40$ MHz, and an exponential distribution of the current amplitude (with a logarithmic decrement $\alpha = 2$) in the second antenna at the same frequency. The Capacitive loads of the antennas were calculated by approximate methods. The design results allowed a rigorous calculation of the amplitude and phase of the current along both the antennas with loads.

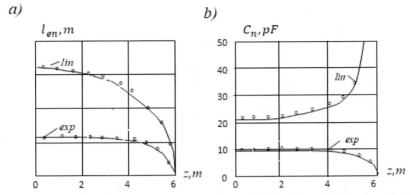

Fig. 3.13. Results of approximate calculation of equivalent lengths (*a*) and the capacitances (*b*).

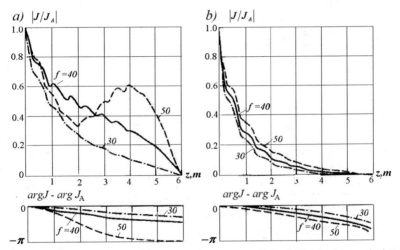

Fig. 3.14. Amplitudes and phases of the currents in antennas with loads (the loads are calculated by approximated method at f = 40 MHz for creating, linear (*a*) and exponential (*b*) distribution of the amplitude).

They are given in Fig. 3.14*a* for a linear distribution and in Fig. 3.14*b* for an exponential distribution. As can be seen from the figures, at $f = 40$ MHz the amplitude distribution is close to the required one, and the phase curves have a slight slope. When the frequency changes, that is, at $f = 30$ and $f = 50$ MHz, the distribution of current amplitude and phase is not preserved.

In order to provide the required current distribution in a range from 40 to 80 MHz, the synthesis of the antenna was performed by the method of mathematical programming. These results are shown in Fig. 3.15 for

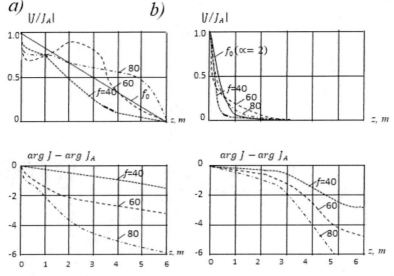

Fig. 3.15. Results of synthesis of linear (*a*) and exponential (*b*) current distribution in a range from 40 to 80 MHz.

the linear (*a*) and exponential (*b*) distributions respectively. The current amplitudes and phases were obtained by means of optimizing antennas' electrical characteristics. The objective function was formed, using the root-mean-square criterion. Parameters calculated by the method of a metallic long line with loads at the middle frequency $f = 60$ MHz, were taken as an initial approximation. The number N_f of used frequencies in a given range is equal to 9 and the number N_l of division points on a wire is equal to 11.

The results were improved significantly. In each figure, four curves for current amplitude are given: the curve, labeled by f_0, corresponds to the required distribution, and curves labeled by $f = 40$, 60 and 80, corresponds to the synthesis result at frequencies 40, 60 and 80 MHz. As is seen from the figures, the obtained distribution is, on the whole, close to the required distribution, but is not identical to it. However, the reason for this difference is limited opportunities rather than inexact methodology. Thus, in addition to the successful solution of a problem, the used methods permit determination of the potential opportunities of antennas.

Antennas with the required electrical characteristics permit to also solve the problem of reducing an influence of nearby metal superstructures.

The loads included in the antenna allow attainment of the concrete electrical characteristics for a selected antenna length. If it possible to manufacture and

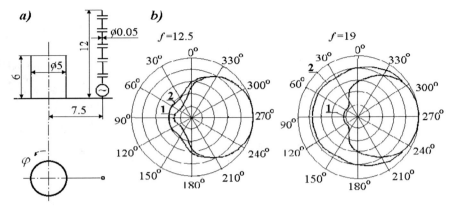

Fig. 3.16. An antenna near a superstructure (*a*) and its horizontal directional pattern (*b*).

install antennas of the required length, then the freedom in choosing this length permits a weakened effect of closely located metal structures, for example, of superstructures on the directional pattern of antenna and antenna array. Figure 3.16 demonstrates results of calculating the directional pattern of a monopole, situated near a metal superstructure in the shape of a circular metal cylinder of finite length. The directional patterns in the horizontal plane are presented at two frequencies of *HF* range.

We examine two variants of antennas: 1 – the monopole without loads of height 6 m and of diameter 0.016 m, 2 – the radiator of height 12 m and of diameter 0.06 m with nine capacitive loads, ensuring optimal electrical characteristics in the frequency range from 8 to 22 MHz. The relative arrangement of the superstructure and the antennas as well as the superstructure dimensions are given in Fig. 3.16*a*. The calculations were performed by means of the program based on the Moments method. The circular cylinder during calculation was replaced with a wire structure of equidistant wires, located along generatrixes of the cylinder and the radii of its upper surface. As is seen in Fig. 3.16*b*, the radiation of an ordinary monopole (curve 1) in the direction of superstructure decreases sharply, and the use of the antenna with loads (curve 2) allows weakening of this effect.

Figure 3.17 demonstrates similar results for a uniform linear array, situated near the superstructure. The same two variants of antennas are adopted as the array elements. The mutual arrangement of the superstructure and antennas as well as the superstructure dimensions are given in the figure and the phase shift between antenna currents is adopted as zero. The calculation results show that in an upper part of the frequency range the superstructure influence

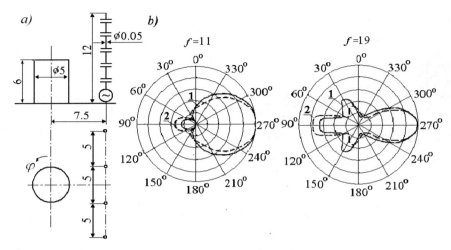

Fig. 3.17. A linear array near a superstructure (*a*) and its directional pattern in the horizontal plane (*b*).

on the directional pattern of array, consisting of the monopoles without loads, is slighter than its influence on the directional pattern of the sole monopole. This is, apparently, concerned with the fact that the superstructure does not hinder the propagation of electromagnetic waves from the side antennas. Nevertheless, the use of radiators with loads in this case also allows reduction in the superstructure influence and increase in the signal in its direction.

As it is shown in Section 3.3, capacitive loads can be used to improve electrical characteristics of a directional V-antenna. As an example of a V-dipole we shall consider the antenna with arm length $L = 1.5$ m and radius 0.025 m. Fifteen capacitors are included in each arm with spacing 0.1 m between loads (the first and last loads are placed at a distant 0.05 m from the antenna end and center). In order that the propagation constant remains real at frequencies up to 100 MHz, the capacitance closest to the generator is chosen to be 33.5 pF. The capacitances of other loads decrease along the antenna according to the linear law. As shown in Section 1.5, using capacitors, one can ensure the current distribution along an antenna close to linear and high level of matching with a cable.

Figure 3.18*a* shows the directivity of *V*-dipole with capacitive loads (curve 1) and without loads (curve 2) along the bisector of angular aperture with width 90°. For the sake of comparison, Fig. 3.18*b* shows similar curves for a linear dipole (curve 3 with loads, curve 4 without loads). The loads magnitudes and the antenna arm length are the same. The calculations were executed in a frequency range from 100 to 500 MHz.

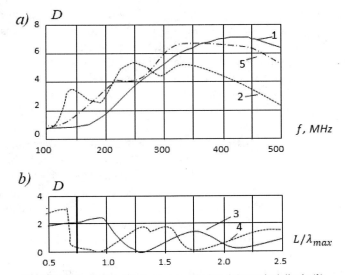

Fig. 3.18. Directivity of V-dipole (a) and of straight vertical dipole (b).

As seen from the figures, the directivity of a straight vertical dipole without loads in the direction, perpendicular to the dipole axis, quickly decreases at $L \approx (0.6$–$0.7)\lambda$. For a straight dipole with loads, this threshold value is found at $L \approx (1$–$1.2)\lambda$. The directivity of V-dipole along the angular aperture bisector is much higher and retains it in a substantially wider frequency range—from 350 ($L = 1.75\lambda$) to 500 MHz ($L = 2.5\lambda$). The loads increase the directivity of V-dipole by a factor between 1.4 and 2.8.

3.6 Calculating directional characteristics of linear and self-complementary radiators

In the previous sections it was shown that antennas with in-phase current have better electrical characteristics than ordinary antennas produced of metal wires with sinusoidal currents. In-phase current is a current whose phase is close to constant or at least does not change sign along the antenna wire. In this case, the amplitude of the current falls along the wire towards the free end of the antenna, in particular according to a linear or exponential law. To obtain in-phase currents, one can, for example, include concentrated capacitive loads along the antenna wire. Their capacitances should not exceed certain magnitudes depending on the signal frequency.

The in-phase current distribution not only allows providing a high level of antenna matching with a cable or a generator, but also leads to a significant change in an antenna pattern. Figure 3.19 demonstrates the normalized radiation patterns in a vertical plane, created by vertical symmetrical metal radiators with in-phase currents distributed in accordance with exponential (with different damping decrement α) and linear ($\alpha = 0$) law. The length of the radiator arm L changes from $3/4\lambda$ up to 4λ. For the sake of clarity, the curves are plotted in a rectangular coordinate system. Here, for comparison, the directional patterns $F_1(\theta)$ of a radiator with a sinusoidal current distribution are also presented.

As is seen from Fig. 3.19, in the case of a linear and exponential in-phase distribution, unlike a sinusoidal one, the radiation maximum with increasing frequency does not immediately deviate from the perpendicular to the radiator axis. The increase of L/λ tapers the main lobe and increases the directivity. The width of the main lobe decreases with decreasing α, including transition of α to negative values. The antenna input impedance in the first approximation is equal, as it is shown in Section 3.1, to the input impedance of the stepped long line.

In the process of analyzing the properties of antennas with in-phase currents the main attention at first has been paid to the problem of matching antenna impedance with the cable or the generator. As a result, the desired character of a directional pattern was obtained in a narrower frequency band than the required level of matching. As a result, to eliminate this drawback, it was necessary to carefully study the directional characteristics of the antennas.

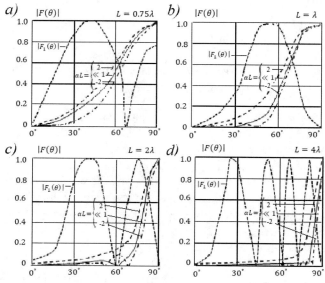

Fig. 3.19. Directional patterns of radiators with the arm length L, equal to $\frac{3}{4}\lambda$ (a), λ (b), 2λ (c), 4λ (d), and different distribution laws of the current.

In order to analyze the directional characteristics of radiators with a given distribution law, a calculation method was proposed that provides high quality comparison of these characteristics with each other. The possibility of antenna operation in a wide frequency range is determined by the properties of the radiator, whose length is larger than the wavelength. Therefore, the characteristics of antennas with a large length are of a substantial interest. The directivity of the antenna along with the level of matching with the source of excitation is the most important electrical characteristic of any antenna. Therefore, the importance of this issue is connected not only with the need to correct errors admitted in the synthesis of the radiators with concentrated capacitive loads.

The main characteristics of the directivity are two: first, this is the maximum directivity D_{max}, i.e., a ratio of maximum radiation intensity to its average value (to a value of isotropic radiation). In accordance with this formulation, the directivity is the coefficient of increasing total radiated power P, if a maximum power is radiated in all directions. In this case, a total power is

$$DP = 4\pi S_{max}, \quad (3.54)$$

where S_{max} is a power flux in the direction of maximum radiation, i.e., total power is equal to a product of a power in the direction of a maximum radiation by the area $S = 4\pi R^2$ of a sphere with a large radius R. The power flux per unit area is determined by the Poynting vector $S = [E, H]$, where E and H are intensities of electric and magnetic fields. In a cylindrical coordinate system, the radiated power of a typical antenna with an axial structure, symmetrical with respect to two planes, passing through coordinates origin, is

$$P = \int_0^{2\pi} d\varphi \int_0^{\pi/2} S_m F^2(\theta,\varphi) d\theta = 4\pi \int_0^{\pi/2} S_{max} F^2(\theta,\varphi) d\theta. \quad (3.55)$$

According to (3.54) and (3.55) the antenna directivity is

$$1/D = P/(4\pi E_m^2) = \int_0^{\pi/2} \frac{S_m F^2(\theta)}{E_m^2} d\theta, \quad (3.56)$$

where E_m is the electric field in the direction of maximum radiation. In order to calculate the directivity magnitude of the antenna in an arbitrary direction, it is necessary to replace the field E_m in (3.56) by a field E in this direction.

The second characteristic is the pattern factor (PF), which is equal to an average radiation level in a given range of angles. In order to increase the distance of radio communication, signals must be radiated along the earth's surface, for example, between angles θ equal to 60° and 90°, and the share of power radiated in this angular sector can serve as the measure of an antenna quality. In accordance with (3.49) the pattern factor for the vertical plane is

$$PF = \frac{1}{k}\sum_{k=1}^{K} F(\theta_k), \qquad (3.57)$$

where $F(\theta_k)$ is the magnitude of normalized vertical directional patterns at an angle θ in an angular sector from θ_1 to θ_K. If the directivity maximum of an antenna is large, but is located outside the necessary angular sector, then its magnitude is useless for increasing the radio communication range.

Expression (3.56) makes it possible to determine the antenna directivity, if the current distribution along the radiator is known. Calculation is based on determination of power fluxes (of Poynting vectors) through each part of a sphere. Replacing the integration by summation, we obtain:

$$1/D = \sum_{m=1}^{N} S_m \frac{F^2(\theta)}{E_m^2} \qquad (3.58)$$

Calculations are accomplished in accordance with (3.58). They use known values of the current at points uniformly located along the antenna arm [53]. Let the axis of an antenna coincide with the z-axis, and an antenna center coincide with the origin of a cylindrical coordinate system (see Fig. 3.20). An antenna arm is divided into N segments of equal length, the field of each segment depends on the magnitude of its current. The angle $\theta = \frac{\pi}{2}$ in the vertical plane is divided into N angles of equal width Δ. The fields of the segments are added together along the lines that are uniformly diverging in a vertical plane under elevation angles from the structure center. The elevation angles (directions of a radiated signal) are equal to $\theta_n = \frac{n\pi}{2N}$, where $n = 1, 2...N$. Power flow S_n passes at the angle θ_n through a section of the

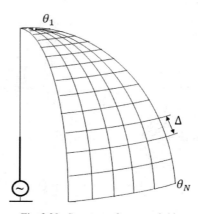

Fig. 3.20. Structure of antenna field.

sphere of a large radius with an angular width Δ (through a strip between adjacent horizontal lines). Here n is a strip number.

The sphere of large radius is divided along azimuth into M equal strips of the same width Δ (between adjacent lines connecting the poles) with the same fields. The width of a vertical strip is equal to $\Delta \sin \theta_n$, and the area of the sector n is equal to $\Delta^2 \sin \theta_n$. The power flow through each sector is $S_n = E^2_n \sin \theta_n$. The total power flow is equal to

$$P = M \sum_{n=1}^{N} E_n^2 \Delta^2 \sin \theta_n. \qquad (3.59)$$

Accordingly, the power flow in the horizontal direction is $ME_N^2\Delta^2$. Thus, the directivity in the horizontal direction is equal to

$$D = E_N^2 \Big/ \sum_{n=1}^{N} E_n^2 \sin \theta_n. \qquad (3.60)$$

In the performed calculations, the number N was adopted to be 20, i.e., the magnitude Δ in radians is equal to $\Delta = \dfrac{\pi}{2N} = 0.0785$.

Variants of symmetrical radiators with sinusoidal current and in-phase current were considered in [27], including straight vertical dipole, V-dipoles with different angles between the arms and V-dipoles with the arms from several segments located at different angles to each other. Here we will consider straight radiators and self-complementary antennas. Symmetrical straight radiators are presented in Fig. 3.21: a – with sinusoidal current, b – with in-phase current.

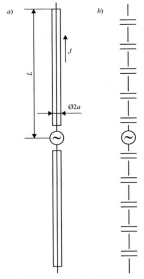

Fig. 3.21. Symmetrical straight radiators: a – with sinusoidal current, b – with in-phase current.

An electrical field of the straight dipole in a far zone is determined by a magnitude of a far field of Hertz dipole (of an elementary linear radiator) located along the z axis. The field in the far zone of the sinusoidal current $J(z) = J(0) \sin k(L - |z|)$ at the distance R from the source and is equal to

$$E_\theta = j \frac{60 I_0}{\sin kL} \cdot \frac{\cos(kL \cos \theta) - \cos kL}{\sin \theta} \cdot \frac{\exp(-jkR)}{R}. \qquad (3.61)$$

Here $\dfrac{I_0}{\sin kL} = J(0)$ is the generator current, k is the propagation constant of the wave in air, L is the arm length. The second multiplier of the expression (3.61) is the directional pattern of the straight metallic dipole.

Inclusion of concentrated capacitive loads in the antenna wire allows creating the in-phase current with linear or exponential distribution and to expand the frequency range with a high level of matching. Let the loads decrease the current towards the free end of the antenna in proportion to the distance from it. The electrical field of the upper arm of the straight dipole with in-phase current is equal to

$$E_{\theta 1} = j 30 J(0) \sin \theta \cdot \frac{\exp(-jkR)}{R} \int_0^L (L-z) e^{jkz \cos \theta} dz =$$

$$j \frac{30 J(0) L \sin \theta}{k^2 \cos^2 \theta} \cdot \frac{\exp(-jkR)}{R} [1 - \cos(kL \cos \theta) + jkL \cos \theta - j \sin(kL \cos \theta)].$$

Since the structure of radiators is symmetrical with respect to the plane passing through the origin of coordinates and the direction of the currents in both arms is the same, the field of the bottom arm is obtained in this expression by replacing a sign of $\sin(kz \cos \theta)$ with the opposite sign. Therefore, the total field of the straight dipole with in-phase current is equal to

$$E_\theta = j \frac{60 J(0) L \sin \theta}{k^2 \cos^2 \theta} \cdot \frac{\exp(-jkR)}{R} [1 - \cos(kL \cos \theta)]. \qquad (3.62)$$

The magnitudes D and PF of a straight dipole with sinusoidal (1) and in-phase current (2) depending on the arm length are given in Fig. 3.22. As can be seen from this figure, these magnitudes in the case of in-phase currents steadily increase with increasing arm length. We must, of course, bear in mind that these characteristics are valid only if the in-phase nature of the current is kept.

A similar result is obtained for an exponential distribution of the in-phase current. The electric field of the upper arm of the radiator in the case of such a distribution is equal to

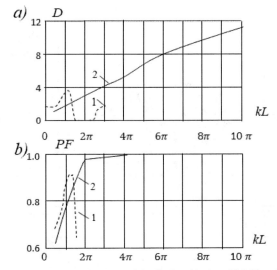

Fig. 3.22. Magnitudes D (a) and PF (b) of a straight dipole with sinusoidal (1) and linear (2) in-phase current.

$$E_\theta = j\frac{60J(0)L\sin\theta}{k(1+\cos^2\theta)} \cdot \frac{\exp(-jkR)}{R}\{1 - e^{-kL}\cos(kL\cos\theta)e^{-kL}\cos\theta\sin(kL\cos\theta) +$$

$$j[\cos\theta - e^{-kL}\cos\theta\cos(kL\cos\theta) - e^{-kL}\sin kL]\}.$$

Taking into account the field of the bottom arm of the antenna, we find

$$E_\theta = j\frac{60J(0)L\sin\theta}{k(1+\cos^2\theta)} \cdot \frac{\exp(-jkR)}{R}[1 - e^{-kL}\cos(kL\cos\theta) + e^{-kL}\cos\theta\sin(kL\cos\theta)]. \tag{3.63}$$

The magnitudes D and PF of the straight dipole with an exponential distribution of in-phase current at different values of α, depending on arm lengths are given in Fig. 3.23.

The obtained results allow us to make a number of essential conclusions. First, it is necessary to emphasize the usefulness of parameter PF. It detects frequencies at which the signal along the earth's surface is small (maximal directivity is great, but at a large angle to the horizon). Comparison of directional properties of radiators with different current distribution shows that radiators with in-phase currents allow obtaining a higher level of directivity, which is preserved in a wide range of frequencies. Straight vertical and V-dipoles maintain a smooth variation of directivity in a range with frequency ratio of at least 10. The directivity of V-dipoles, as a rule, is more than the directivity of straight radiators.

164   ANTENNAS: Rigorous Methods of Analysis and Synthesis

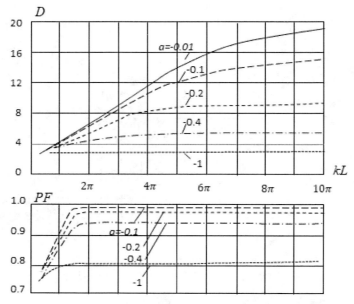

Fig. 3.23. Magnitudes of D and PF of a straight dipole with exponential distribution of in-phase current.

A similar method of extending the frequency range and increasing PF can be used in self-complementary antennas including capacitive loads in metal radiators. As is shown in Fig. 3.24, for this purpose it is necessary to include capacitive loads in the form of slots in the metal radiator and to include inductive loads in the form of metal strips of the such same shape and dimensions in the slot radiator, i.e., one must create an in-phase electric current in the metal radiator and an in-phase magnetic current in the slot. In accordance with this approach, two variants of asymmetric flat antennas are presented in the figure in order to compare their characteristics—in the shape of a semicircle (a) and in the shape of a rectangle (b). The electric radiator can be made in the form of a wide thin plate or a set of metallic wires connected to each other, as shown in the figure.

Equivalent schemes of both metal (electric) radiators are shown in Fig. 3.25. Each radiator is replaced by five monopoles, excited at a common point and located at an angle $\pi/8$ to each other. The lengths of the central monopoles are the same. The lengths of the side monopoles of the first antenna are equal to the length of the central one, and the lengths of the side monopoles of the second antenna were increased in accordance with its shape.

The field calculation of the radiators with in-phase current is performed in accordance with (3.62). For a radiator located at an angle α to the vertical, the result should be multiplied by $\cos \alpha$. Directional characteristics of radiators

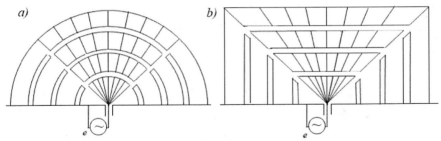

Fig. 3.24. Self-complementary antenna in the shape of a semicircle (*a*) and a rectangle (*b*) with in-phase currents in metal and slot radiators.

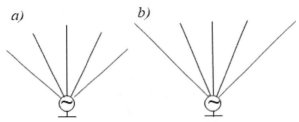

Fig. 3.25. Equivalent circuits of antennas—in the shape of a semicircle (*a*) and of a rectangle (*b*).

were determined in accordance with the above described method. The calculation results for a gain D and a pattern factor PF of a vertical radiator with an in-phase current depending on the electrical length kL are shown in Fig. 3.26. The inclination of the radiator changes the magnitude of the field, since this magnitude is proportional to a vertical projection $L \cos \alpha$ of the radiator. At the same time the PF and maximal D do not change. The magnitudes D and PF depend on the electrical length of the radiator. This circumstance significantly affects the characteristics of a rectangular antenna. In particular, for $kL = \pi$ magnitudes D of three different radiators of this antenna are 3.2, 3.24 and 3.66, and magnitudes PF are 0.83, 0.845 and 0.892. As a result, the general parameters of the antenna change. These parameters can be easily calculated by adding up the fields of the radiators. As a result, we obtain $D = 3.4$, and $PF = 0.86$. It is easy to verify that these results obtained for the rectangular antenna with in-phase currents are close to the average values and are not inferior to the parameters of the antenna in the shape of a semicircle.

At the same time the antenna shape much stronger affects the parameters of radiators with sinusoidal currents. As can be seen from Fig. 3.22, the characteristics of the antenna with sinusoidal currents are not inferior to the characteristics of an antenna with in-phase currents only in the narrow frequency range. With increasing electrical length, they become much worse. But even here, in this

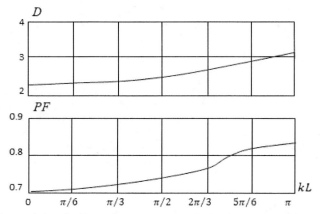

Fig. 3.26. Characteristics of self-complementary antenna in the shape of a semicircle with in-phase current.

small range at $kL = \pi$, the different lengths of the radiators reduce the overall gain D to the magnitude 1.67 and the overall PF to 0.47.

The results of applying the described method to flat self-complementary antennas for calculating the directional characteristics confirms the high efficiency of this method in analyzing the properties of the radiators. The calculation helps to obtain new useful information about antennas, the characteristics of which were considered quite well known.

3.7 Method of electrostatic analogy

The initial sections of this book are devoted to improving electrical characteristics and expanding the frequency range of simple radiators consisting of one or two linear elements, in which are included or not included loads. Increasing the number of linear elements in the antenna complicates its analysis and synthesis problem. These are in particular director-type, log-periodic and self-complementary antennas. Solving problems concerning these antennas requires new approaches and new mathematical methods.

While solving mathematical problems, an analogous form of equation often lends essential assistance. For example, the first two Maxwell's equations that became a basis for classical electromagnetic theory, allowed substantiating the principle of duality [39, 54]. Let's rewrite equations (1.1) in the form:

$$\mathrm{curl}\vec{H} = \vec{J} + \varepsilon_0 \frac{\partial \vec{E}}{\partial t}, \quad \mathrm{curl}\vec{E} = -\mu_0 \frac{\partial \vec{H}}{\partial t}, \qquad (3.64)$$

where $\vec{E}$ and $\vec{H}$ are strengths of electric and magnetic field respectively, $\vec{J}$ is a conduction current, ε_0 and μ_0 are permittivity and permeability of a surrounding medium. By replacing variables in accordance with the equalities $\vec{E}_1 = \sqrt{\varepsilon_0}\vec{E}$, $\vec{H}_1 = \sqrt{-\mu_0}\vec{H}$ and introducing a magnetic conduction current conventionally, we obtain

$$curl\vec{H}_1 = \vec{J}_1 + \sqrt{-\varepsilon_0\mu_0}\frac{\partial \vec{E}_1}{\partial t}, \; curl\vec{E}_1 = \vec{J}_{m1} + \sqrt{-\varepsilon_0\mu_0}\frac{\partial \vec{H}_1}{\partial t}, \qquad (3.65)$$

where $\vec{E}_1$ and $\vec{H}_1$ are also strength vectors, and $\vec{J}_1$ is a conduction current, but in other units.

The obtained equations are completely symmetric with respect to $\vec{E}_1$ and $\vec{H}_1$. An introduction of a magnetic conduction current makes them symmetric with respect to electric and magnetic conduction currents, i.e., makes it possible to treat (3.65) not only as equations for an electric radiator, but also as equations for a magnetic radiator with a magnetic conduction current $\vec{J}_{m1}$. Expressions for fields and characteristics of a magnetic radiator can be recorded by means of the expressions analogous to expressions for an electric radiator. For example, an input impedance of a magnetic radiator Z_m, whose shape and dimensions coincide with the shape and dimensions of an electric radiator, is equal to

$$Z_m = (120\pi)^2/Z_e. \qquad (3.66)$$

Here Z_e is an input impedance of an electric radiator. A real embodiment of a magnetic radiator is a slot.

A similar method of solving problems is used in many other cases. For example, there is the known analogy between a picture of electrostatic field of charged conducting bodies located in a homogeneous and isotropic dielectric and a picture of constant currents in a homogeneous, weakly-conducting medium. In this case, bodies placed in the medium must have high conductivity and their shape and geometric dimensions must coincide with the shape and dimensions of the conducting bodies located in a dielectric.

If a picture of an electric field of linear charges is known, then, using the correspondence principle, one can construct a picture of a magnetic field of constant linear currents, provided the currents and charges are distributed in space identically. The difference between these images is only that lines of equal magnetic potential are located at places of lines of electric field strength, and lines of magnetic field strength are located at places of lines of equal electric potential [55].

Generalizing the principle of correspondence, it is advisable to compare electromagnetic fields created by high-frequency currents of linear radiators with electrostatic fields of charges, which are placed on linear conductors. Both fields are directly proportional to the magnitude of the current or the magnitude of the charge, and in the far zone, they are inversely proportional to the distance from the source.

The offered method is based on an analogy between two structures consisting of high-frequency currents and constant charges. It is assumed that shapes and dimensions of radiators coincide with shapes and dimensions of conductors. In the case of several radiators a ratio of emf in their centers is equal to a ratio of charges placed on the conductors. The positive charge, equal to Q_0, is located on the conductor 0, which corresponds to the active radiator. The negative charges (their number is equal to N) are located on the conductors i, corresponding to passive radiators. They are equal to $-Q_i$ and their sum is $\Sigma_{i=1}^{N}(-Q_i) = -Q_0$, i.e., the sum of all charges is zero and the conductors form an electrically neutral system. In this system

$$Q_i/Q_0 = C_{0i}/\Sigma_{(i)}C_{0i}, \qquad (3.67)$$

where C_{0i} is the partial capacitance between conductors 0 and i. It follows from (3.67), the charges of the conductors i are directly proportional to the partial capacitances C_{0i} between these conductors and the conductor 0 (see, for example, [30]).

Equivalent replacement of a complex structure of high-frequency radiators by a structure with constant charges placed on conductors sharply simplifies the problem, reducing it to an electrostatic problem. In accordance with what has been said, it is natural to call the proposed method "the method of electrostatic analogy".

The considered method allows analyzing the problem in a general view, for example, to study and to compare different laws of current distribution along the individual radiators. This is an undoubted advantage of the method. Characteristics of complex antennas are usually calculated using complex programs based on the Moments method. For discussed problems, such a method is in essence a trial-and-error method. The Moments method does not permit comparison of antennas with different distribution of currents in a common view. Therefore, this method is not applicable here. The approximate method does not give exact results. If this method is correct, i.e., corresponds to the physical essence of the problem, its accuracy is the same for different distributions of the current, i.e., the method allows us to choose the best option.

As stated in the Introduction, a reasonable sequence of solving each synthesis problem requires that at the first stage, an approximate method of solving is developed, whose results are used as initial values for the numerical solution of the problem by the method of mathematical programming. The method of electrostatic analogy can be used as an approximate method of this type.

It is expedient to consider the procedure of applying the electrostatic analogy method on a concrete example. As an example, the director-type antenna (Yagi-Uda antenna) described in [56] was adopted. It should be recalled that the problem of optimizing the antenna characteristics by choosing dimensions of its elements and using mathematical programming methods was first solved with respect to the director-type antenna, and this decision confirmed the correctness of the chosen approach. Work by [56] was one of the first profound efforts devoted to the optimization of the director-type antenna.

The antenna circuit is given in Fig. 3.27. The antenna consists of four metal radiators (active radiator, reflector and two directors). In Fig. 3.27, the antenna dimensions are shown in metres. They were determined by solving the optimization problem in a strict formulation. Let's start with the capacitance between the conductors. If the radii of the conductors i and 0 are the same and equal to $a = 0.001$ m and the lengths l_0 and l_i of these wires are slightly different from each other, then the partial capacitance C_{0i} between these wires in the first approximation is equal to

$$C_{0i} = \pi\varepsilon_0 l_i / ln(b_i/a), \qquad (3.68)$$

where ε_0 is the medium permittivity and b_i is the distance between the wires.

If we divide the wire 0 of the antenna into three wires and to denote these wires by indices $0i$, then the antenna circuit is divided into three circuits. Each circuit consists of two conductors: of wire i and wire $0i$ (Fig. 3.28). The generator is also divided into three generators located in the centers of the wires $0i$. The values of their emfs are defined as

$$e_i = eQ_i/Q_0, \qquad (3.69)$$

where e is the emf of the active radiator 0.

As shown in the theory of folded antennas, the circuit of two identical parallel vertical wires located at a distance b_i can be divided into a dipole and

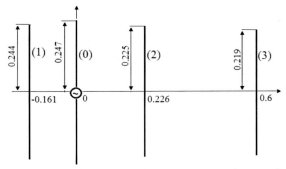

Fig. 3.27. Dimensions of the optimized director-type antenna from metal radiators.

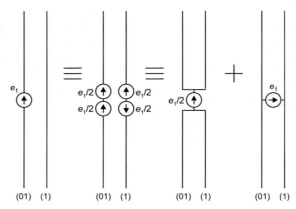

Fig. 3.28. Circuit from two parallel vertical wires.

a long line, open at both ends. The current at the center of each dipole wire is equal to

$$J_{id} = e_i/(4Z_{id}). \qquad (3.70)$$

The reactive component of its input impedance is $X_{id} = -120 \ln(2L_0/a_{ei}) \cot kL_0$, where $L_0 = l_0/2$ is the arm length of the active radiator, $a_{ei} = \sqrt{ab_i}$ is its equivalent radius, k is the propagation constant. The current at the center of each wire of the long line is equal to

$$J_{il} = e_i/(2jX_{il}), \qquad (3.71)$$

where $jX_{il} = -j120 \ln(b_i/a) \cot kL_i$ is the reactive input impedance of the long line with length $L_i = l/2$. The currents of the active and passive radiators are equal to the sum and difference of the current J_{id} of dipole wire and the current J_{il} of each long line, i.e., the total current J_0 of the active radiator and the total current J_i of each passive radiator are equal to

$$J_0 = \sum_{i=1}^{3}(J_{id} + J_{il}), \; J_i = J_{id} - J_{il}. \qquad (3.72)$$

The amplitude and phase of the fields created by each radiator depend on its structure. The radiator structure determines the law of current distribution along it. If the radiator arm is a straight metallic conductor, its current is distributed according to the sinusoidal law $J(z) = J(0) \sin k(L - |z|)$. In this case, as is shown in Section 1.6, the far field of the radiator is equal to

$$E_\theta = \frac{AJ(0)}{\sin\theta} \cdot \frac{\exp(-jkR)}{R}[\cos(kL\cos\theta) - \cos kL], \qquad (3.73)$$

where $A = j60$, where R is the distance to the observation point. If the concentrated capacitive loads realize the in-phase current, distributed along the radiator axis in accordance with a linear law $J(z) = J(0)(L - |z|)$, the far field of the radiator, as is shown in Section 3.1, is equal to

$$E_\theta = j \frac{60 J(0)}{\varepsilon_r} \frac{\exp(-jkR)}{R} \frac{1 - \cos(kL \cos\theta)}{\sin\theta}. \tag{3.74}$$

From Fig. 3.27 it is clear that the maximum radiation of the antenna considered is directed to the right, that is, towards the radiator 3. Since the radiator 1 is located on the left of the active radiator 0 at a distance b_1 from it, its field lags behind the field of the active radiator, first, per phase corresponding to a time of signal propagation from radiator 0 to the radiator 1 and secondly, per phase corresponding to the propagation time of the signal in the opposite direction—from the radiator 1 to the radiator 0 (the signal of radiator 1 must come to the radiator 0 an angle θ, i.e., the path length between the wires is $b_1/\sin\theta$). The total phase difference is equal to $\psi_1 = -kb_1 (1 + \sin\theta)/\sin\theta$. Similarly, in the case of radiators 2 and 3, this phase difference is equal to $\psi_2 = kb_2 (\sin\theta - 1)/\sin\theta$ and to $\psi_3 = kb_3 (\sin\theta - 1)/\sin\theta$, respectively.

The described procedure permits determination of the total field of the director-type antenna shown in Fig. 3.27. In accordance with (3.69), emfs of the different radiators are equal to $e_1 = 0.388e$, $e_2 = 0.335e$, $e_3 = 0.277e$. The total field of this antenna with radiators in the form of straight metal wires is

$$E_\theta = \frac{AJ(0)}{\sin\theta} \sum_{i=1}^{3} e_i \cdot \frac{\exp(-jkR)}{R} \left\{ \left(\frac{1}{4Z_{id}} + \frac{1}{2Z_{il}} \right) [\cos(kL_0 \cos\theta) - \cos kL_0] + \right.$$

$$\left. \left(\frac{1}{4Z_{id}} - \frac{1}{2Z_{il}} \right) \exp(j\psi_i)[\cos(kL_i \cos\theta) - \cos kL_i] \right\} \tag{3.75}$$

The emfs of the radiators do not change.

The inclusion of concentrated capacitive loads along the linear radiators, whose magnitudes vary in accordance with the linear or exponential law, permits the creation of radiators with in-phase current. While retaining the dimensions of the radiators and the distances between them, we get the director-type antenna shown in Fig. 3.29. The total field of such an antenna is calculated by the formula

$$E_\theta = \frac{AJ(0)}{\sin\theta} \sum_{i=1}^{3} e_i \cdot \frac{\exp(-jkR)}{R} \left\{ \left(\frac{1}{4Z_{id}} + \frac{1}{2Z_{il}} \right) [1 - \cos(kL_0 \cos\theta)] + \right.$$

$$\left. \left(\frac{1}{4Z_{id}} - \frac{1}{2Z_{il}} \right) \exp(j\psi_i)[1 - \cos(kL_i \cos\theta)] \right\}. \tag{3.76}$$

172   ANTENNAS: Rigorous Methods of Analysis and Synthesis

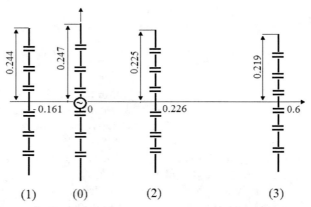

Fig. 3.29. Director-type antenna with straight in-phase radiators.

The results of calculating the directivity and the pattern factor of a director-type antenna with straight metal wires are given in Fig. 3.30 (curves 1). Since here an approximate calculation procedure was used, these results are not identical to the results presented in [56], but are similar to them. Practically the antenna operates at the same frequency. The results of calculating these characteristics for the antenna with in-phase currents are also given in Fig. 3.30 (curves 2). They speak for themselves. This antenna operates over a wide frequency range and its directivity steadily and smoothly increases with increasing frequency, that is, the quality factor of this antenna is small. Of course, we must bear in mind that these characteristics are valid only if the current is in-phase. But the frequency ratio of antennas with capacitive loads with a high level of matching is a magnitude of the order of 10.

As has already been said, the method of electrostatic analogy is based on the resemblance of a structure consisting of high-frequency currents and a structure consisting of constant charges. Comparison of electromagnetic fields created by high-frequency alternating currents of linear radiators with electrostatic fields of charges placed on linear conductors of electrically neutral systems shows similarity of the mathematical structures of both fields.

This method allows us to propose a simple and effective procedure for calculating the directional characteristics of the director-type antennas consisting of linear radiators. The procedure uses knowledge about the basic antenna dimensions and current distributions along the radiators. It does not require detailed information on the types and magnitudes of concentrated loads. As calculations have shown, the director-type antennas consisting of linear radiators with in-phase currents provide a high directivity and smooth variation of characteristics in a wide frequency range. The results obtained with help of this method can be used to solve the problem of optimizing various director-type antennas by mathematical programming methods.

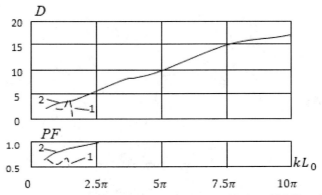

Fig. 3.30. Directional characteristics of director-type antennas with sinusoidal (1) and in-phase (2) currents.

Similar results can be obtained for V-radiators. In this case, the operating range decreases, but *TWR* increases at concrete frequencies. The circuit of such director-type antenna with V-radiators is given in Fig. 3.31. Calculations were performed for antennas with the same lengths of radiators arms and the same distances between them as in the antenna shown in Fig. 3.28. The angles between the arms of the radiators and the vertical in each antenna are the same and are equal to 10°, 20°, and 30°, respectively. In this case, the total antenna field is equal to

$$E_\theta = \frac{AJ(0)}{\sin\theta} \sum_{i=1}^{3} e_i \cdot \frac{\exp(-jkR)}{R} \left\{ \left(\frac{1}{4Z_{id}} + \frac{1}{2Z_{il}} \right) [1 - \cos(kL_i \cos\langle\theta - \theta_0\rangle)] + \left(\frac{1}{4Z_{id}} - \frac{1}{2Z_{il}} \right) \exp(j\psi_i)[1 - \cos(kL_i \cos\langle\theta - \theta_0\rangle)] \right\}. \quad (3.77)$$

The directional characteristics of the director-type antennas with V-radiators are shown in Fig. 3.32. The magnitude of *PF* does not drop below 0.85. The results show that these antennas, in comparison with the antennas consisting of straight in-phase radiators allow increasing directivity but with a narrower frequency range. In this case the directivity and frequency range depend on the inclination of the antenna arm: if this angle increases, the directivity at low frequencies also increases but the frequency range becomes narrower.

As indicated in the Introduction, a reasonable sequence of solving each problem is to propose and develop an approximate method of analysis in the first stage and then use its results as initial values for the numerical solution of the problem by methods of mathematical programming. The method of

174   ANTENNAS: Rigorous Methods of Analysis and Synthesis

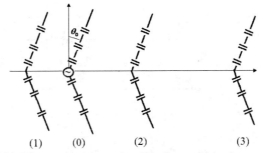

Fig. 3.31. Director-type antenna from V-radiators with in-phase currents.

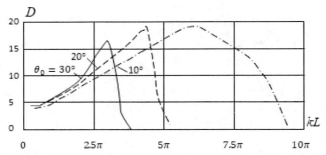

Fig. 3.32. Directivity of director-type antennas from V-radiators with in-phase currents.

electrostatic analogy allows taking the first step. The results obtained with its help can be used to solve the problem of optimizing various director-type antennas by mathematical programming methods.

The electrostatic analogy method was proposed in 2017 [57]. The need for such a method was understood long ago. A.F. Chaplin with his pupils stubbornly paved the way for its creation. Also A.F. Yakovlev persistently tried to improve the electrical characteristics of a log-periodic antenna. His results will be discussed in the next section.

3.8 Application of the method of electrostatic analogy to log-periodic antennas

As is known, log-periodic antennas belong to the class of frequency-independent antennas. They are based on the principle of complementarity and have the ability to automatically "cut-off" current. In accordance with the principle of electrodynamic similarity any radiator has the same electrical characteristics at different frequencies, if its geometric dimensions vary with frequency in proportion to the wavelength (in the first approximation the requirement of corresponding change of the material conductivity may be neglected). Not only the tunable antennas, but also the antennas, whose shape is completely determined by the angular dimensions, conform to the principle of electrodynamic similarity. In this case changing of the distance scale does not change the antenna, i.e., a radiator shape and dimensions in wavelengths are the same at different frequencies.

The property of the automatic "cut-off" of currents means that the field at each frequency is radiated by a current along a small antenna segment, which is named the active area, and that the electric current outside the boundaries of this area is quickly attenuated. Here, coordinates and dimensions of radiated segments are rigidly related with the magnitude of a wavelength. If the frequency changes, the antenna segment radiating the field shifts along the antenna. Both longitudinal and cross electrical dimensions of this radiating area remain constant and ensure the invariability of the characteristics. Thus, the log-periodic antenna has a constant input impedance and invariable directivity characteristics in a wide band.

If the antenna has finite dimensions, its frequency range is finite, but in this finite range the antenna has the properties of an infinite antenna. The maximal wavelength depends on the maximal cross dimension of the antenna (on its width), and the minimal wavelength depends mostly on the accuracy of observing required dimensions near the excitation point.

Log-periodic dipole antenna (LPDA) shown in Fig. 3.33, are a collection of dipoles, with dimensions forming a geometric progression with denominator $1/\tau$:

$$R_{n+1}/R_n = l_{n+1}/l_n = 1/\tau. \tag{3.78}$$

Here R_n is the distance from the vertex of the angle α to dipole n, l_n is the arm length of dipole n, α is the angle between the antenna axis and the line passing through the dipole ends. Accordingly, the antenna electrical characteristics are repeated at frequencies forming the geometric progression with the same denominator. It means that directivity characteristics and input impedance of the antenna are periodic functions of the logarithm of frequency f, i.e., if the electrical characteristics are drawn as a function of $\ln f$, their

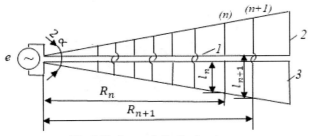

Fig. 3.33. Log-periodic dipole antenna.
1 – longitudinal wire, 2 – cross wire, 3 – interval between the cross wires.

176 ANTENNAS: Rigorous Methods of Analysis and Synthesis

values are repeated with period equal to $\ln\tau$. This peculiarity defined the antenna name.

Weak variation of antenna characteristics within the periodic interval is an indispensable condition for their weak dependence on frequency. In order to meet this condition, the periodic interval must be small. But this is an insufficient condition.

LPDA shown in Fig. 3.33 consists of two structures situated in one plane. Each structure is shaped as a straight wire, with linear conductors attached to it at right angles alternately from the left and from the right. Their lengths increase with growing distance from the excitation point in accordance with the law of geometric progression. Such an antenna is a simplified and modified variant of a flat log-periodic structure shown in Fig. 3.34, which is a self-complementary structure, i.e., it consists of metal plates and slots coinciding with each other in shape and dimensions. The input impedance of a flat infinite self-complementary structure is purely active, independent of the frequency, and is equal to 60π Ohm. Designing log-periodic antenna in the form of a self-complementary or similar structure ensures a small variation of electrical characteristics of the antenna within each interval.

Each of the two structures, forming LPDA, differs from the structure, forming an arm of a flat log-periodic antenna. The metal sector 1' is replaced with a longitudinal wire 1. The metal strip 2', situated along the arc of a circumference, is replaced with a lateral wire 2, tangential to the arc. The slot 3' is replaced with an interval 3 between the lateral wires. Such a construction is essentially simpler for implementation and, at the same time, is close to the original one in the electrical properties.

The active area of LPDA consists of dipoles with arm length close to $\lambda/4$. In their input impedance an active component is predominant, and the reactive component is small. In actual practice the number of dipoles, forming the

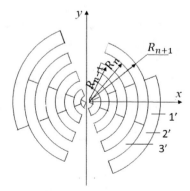

Fig. 3.34. Flat self-complementary log-periodic antenna.
1' – metal sector, 2' – metal strip, 3' – slot strip.

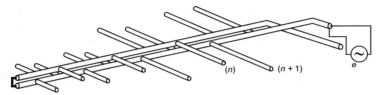

Fig. 3.35. Design of antenna LPDA.

active area, usually is equal to five. For the sake of simplicity, we will assume that the active area consists of only three dipoles, with the arm length of the central dipole being $\lambda/4$.

As is seen from Fig. 3.33, the upper arms of dipoles are connected alternately to one or more conductors of the distribution line. It is equivalent to a mutual crossing of conductors on the segments between the dipoles. With allowance for this crossing, the phase of the electrical current in the longer dipole exceeds the phase of the current in the resonance radiator, and the phase of the current in the shorter dipole is less than the phase of the current in the resonance radiator, i.e., the longer dipole acts as a reflector, and the shorter dipole acts as a director. As a result, the fields of individual radiators are summed in the direction toward the excitation point (to the shorter dipoles) and cancel each other in the opposite direction.

The waves in the distribution line reflected from dipoles of the active area cancel each other to a large degree, since the reactive components of the input impedances of short and long dipoles are opposite in sign. This explains a high level of matching of the active area of the antenna and the distribution line. In addition, the electrical length of the line from the excitation point to the active area remains unchanged when the frequency changes. Therefore, an impedance of active area transformed to the antenna input is the same at different frequencies.

The dipoles located outside the active area are excited weakly due to the great reactive impedance. The short dipoles at the beginning of the structure practically fail to radiate, since the fields created by them are summed up almost in anti-phase because of crossing wires and the proximity of dipoles to each other (as compared with the wavelength). As a result, the EM wave along this segment of the line does not weaken, i.e., the distribution of currents and voltages at the line segment between the excitation point and the active area is close to the traveling wave. The short dipoles act as capacitances shunting the distribution line and thereby decreasing its wave impedance slightly. The long dipoles situated behind the active area radiate weakly too, since, first, their input impedances are great and, second, the power of the EM wave at that segment of the line drops substantially as a result of attenuation in the active area.

The method of LPDA calculation [58] is based on antenna presentation in the form of a parallel connection of two multipoles (Fig. 3.36). The first multipole consists of dipoles and is defined by matrix Z_A of mutual impedances. The second multipole is a distribution line, which is defined by matrix Y_l of admittances. For each cross-section n of the structure, consisting of two multipoles the following equation is true:

$$J_{nA} Z_{nA} = J_{nl}/Y_{nl}, \quad J = J_{nA} + J_{nl}, \qquad (3.79)$$

i.e., $J_{nl} = J_{nA} Z_{nA} Y_{nl}$. Here J_{nA} is the current at the dipole input, Z_{nA} is the input impedance of the dipole (with allowance for coupling with neighboring dipoles), J_{nl} is the current of the distribution line, Y_{nl} is the admittance of the line in the cross-section n, and J is the extraneous current at given point. It should be noted that in calculating J_{nA} the mutual coupling with neighboring dipoles is accounted for, and in calculating Y_{nl} we consider that the distribution line is shorted at the terminals of neighboring dipoles (while calculating the current of the source the other sources must be shorted in accordance with Kirchhoff's law).

A first equation from (3.79) is written for a voltage along a closed circuit, and a second equation is written for a current at the input point. From here,

$$J = (1 + Z_{nA} Y_{nl}) J_{nA}. \qquad (3.80)$$

Accordingly, a matrix equation for the column-vector $[J_A]$ of the dipole input currents is written in the form

$$[J] = ([E] + [Z_A][Y_l])[J_A], \qquad (3.81)$$

where $[E]$ is an identity matrix, $[J]$ is the column-vector of currents exciting the lines connecting the multipoles with each other. Since the source of excitation is only the generator located at the input of the distribution line, we put that the current of this generator equal to $|J_0|$, then $[J] = \begin{vmatrix} J_0 \\ 0 \\ \ldots \\ 0 \end{vmatrix}$. Solving equation (3.81), we find the column-vector $[J_A]$, and then matrix $[V_A] = [Z_A][J_A]$ of the voltages at the dipoles' inputs. The first element of the matrix at

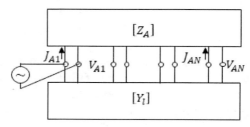

Fig. 3.36. Equivalent circuit of LPDA.

the input of the shortest dipole is the input voltage. If the exciting current J_0 is equal to 1, this first element is equal to the input impedance of the antenna.

When designing LPDA, it is important to choose the geometric dimensions in such a way that the electrical characteristics changed weakly in a range from f to τf. The magnitude of τ and all antenna characteristics depend essentially on the parameter σ, which is equal to the distance between the half-wave dipole and the neighboring shorter dipole (in wavelengths):

$$\sigma = 0.25(1 - \tau)\lambda \cot \alpha. \qquad (3.82)$$

In substance, this expression is dependence of the parameter σ on the angle α in view of τ. As it is shown in [58, 59], the characteristics change weakly, if $\tau > 0.8$ and $0.05 \leq \sigma/\lambda \leq 0.22$. Under these conditions, the currents of the dipoles, located near the resonant (half-wave) radiator, reach a maximum and the wave along the distribution line is so attenuated in the active area that the following dipoles practically do not radiate.

In [60] by means of generalization of data available in the literature the optimum relationship of the basic parameters is defined in the form:

$$\sigma/\tau = 0.191. \qquad (3.83)$$

This ratio does not depend on the values of α, l_n/a_n and Z_0. Here a_n is the radius of dipole n, $Z_0 = 60 \cosh^{-1}[(D^2 - 2a^2)/(2a^2)]$ is the wave impedance of the distribution line, a is the radius of the distribution line's wires, and D is the distance between the axes of these wires. Substituting (3.83) into (3.82), authors of [60] obtained simple expressions connecting the optimal parameters τ and σ with the antenna dimensions, as under:

$$\tau = 1/(1 + 0.765 \tan \alpha) = L/[L + 0.765(l_1 - l_N)], \quad \sigma = 1/(4 \tan \alpha + 5.23). \quad (3.84)$$

The value L in these expressions is the distance between the first and the last (Nth) dipole.

The antenna with $\sigma/\tau = 0.191$ has a narrow directional pattern and high front-to-back ratio. SWR of this antenna in a properly designed LPDA is typically smaller than 1.5. But since log-periodic antennas have rather large overall dimensions, the task of reducing the dimensions of a log-periodic antenna has always been the main task of the developers. In order to decrease the transverse dimensions, it is expedient to shorten the longest dipoles using loads of a different kind or structures with slowing-down properties, i.e., by means of mechanisms used usually for reducing the monopole's and dipole's length. The options of slowed-down structures are numerous. It should be noted that the slowing factor is always less than the length of the wire.

Slowing-down allows shorting the monopole, i.e., to reduce the length of the monopole m times for the first resonance frequency or to decrease the resonance frequency m times for the given length of the monopole. But the resistance of radiation at the resonance frequency as a consequence of length reduction decreases m^2 times and the antenna wave impedance is increased m times. Both results impair matching of each element of LPDA and an antenna on the whole with a cable.

Attempts to decrease longitudinal dimensions of an antenna by using slowing-down in the distribution line or at the expense of additional connecting dipoles failed, since violation of geometric progression relationships and increasing the number of dipoles causes, as a rule, sharp deterioration of electrical characteristics and gives an insignificant decrease in overall dimensions at the expense of connecting additional dipoles.

In [60] it is described that the variant of log-periodic antennas, which operate in two adjacent frequency bands allow making the antenna shorter than the antenna designed for operation in the total range. Basically, the authors' proposal reduces to the use of linear-helical dipoles, i.e., radiators, each of which consists of straight and helical dipoles. They arranged coaxially and have the common excitation point. The dipoles' lengths are the same, but the helical wire length is twice as much as the straight rod's length. A linear-helical dipole in contrast to straight and helical dipoles has two serial resonances, and the ratio of the resonant frequencies of two dipoles of equal length is equal to the slowing factor of the helical dipole.

As said before, resonant dipole and its nearest neighbors create an active area, and passing through it the electromagnetic wave actively radiates energy. LPDA with linear-helical dipoles has two active areas, and they provide a signal radiation in two bands of the frequency range. The experimental check of log-periodic antenna with linear-helical dipoles confirms that this proposal is promising. With comparable electrical characteristics the length of log-periodic antenna, in which only straight dipoles are used, is 1.8 times as large as the length of the antenna with linear-helical dipoles. If only helical dipoles are used, the length of an antenna with linear-helical dipoles is 1.3 times less than the length of an antenna with linear dipoles.

The length of a log-periodic antenna can be reduced by increasing the angle α between the antenna axis and the line passing through the dipoles ends. This option seems the most simple and natural. But, as it is seen from (3.82), an increase in α, if τ is constant, leads to a decrease in the distance between the dipoles and to an increase in their mutual influence, and consequently to decreasing both the directivity and the active component of input impedance. As a result, frequency-independent characteristics deteriorate.

One can increase the angle α in another manner. The LPDA consists (see Fig. 3.33) of two asymmetric structures located in the same plane and excited in opposite phases. If these structures are located at an angle, $\psi > \alpha$

to each other, the resulting three-dimensional structure will incorporate two planar structures, distant from each other. The monopoles are attached to the conductor of the distribution line alternately from the left and right. The distance between the monopoles, situated on the one side of the conductor, is almost twice as large as in a planar LPDA. This reduces their mutual influence and allows for increasing the angle α. However, this antenna occupies a great volume, and that makes its installation difficult and changes its characteristics, for example, it increases input resistance and creates additional problems in the utilization of the antenna.

Asymmetrical coaxial log-periodic antennas, described in [60], do not have these disadvantages. The two-wire distribution line in this antenna is replaced by a coaxial line, and dipoles are replaced by monopoles. The antenna assembly is shown in Fig. 3.37. The antenna consists of two structures, whose circuits are given in Fig. 3.38. The first one (Fig. 3.38a) is a straight conductor. The wire segments of required length located in one plane are connected to defined points of this conductor at the right angle alternately from left and right. This conductor is the central wire of the coaxial distribution line and the wire segments are monopoles, which are excited by means of the conductor.

The second structure (Fig. 3.38b) is designed as a long cylindrical metal tube with short metal tubes embedded in it, which open inside and outside a long tube. The long tube is the outer envelope of a coaxial distribution line, the short tubes are the outer coaxial shells, inside which monopoles connected to the central conductor of the coaxial distribution line are placed. As can be seen from Fig. 3.37, the monopoles are inserted into the coaxial shells, so that their axes coincide. The monopoles are the radiators where excitation points shifted from the base.

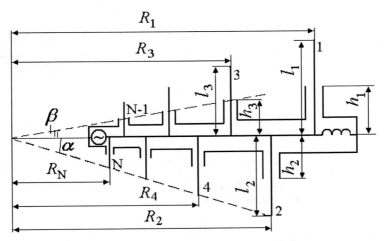

Fig. 3.37. The circuit of asymmetrical coaxial log-periodic antenna.

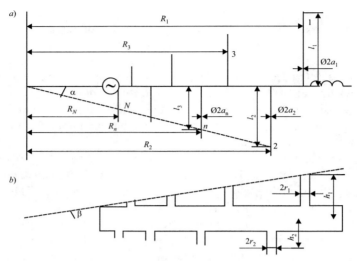

Fig. 3.38. Internal (*a*) and external (*b*) structures, forming an asymmetrical coaxial log-periodic antenna.

In accordance with the usual practice of designing log-periodic antenna, their dimensions must correspond to the geometric progression with ratio $1/\tau$:

$$R_{n+1}/R_n = l_{n+1}/l_n = h_{n+1}/h_n = 1/\tau. \tag{3.85}$$

Here h_n is the distance from the axis of the distribution line to an excitation point of radiator *n*. Other dimensions have been defined earlier. In addition, it is necessary for the wave impedance of each short tube to coincide with the first series resonance frequency and the resistance located in this tube monopole. This condition determines the choice of the ratio of the tube diameter to the radiator diameter.

From above it follows that the two-wire distribution line is replaced in the proposed antenna by a coaxial distribution cable, and the dipoles are replaced by monopoles connected to the internal conductor of this cable. The outer envelope of the cable is used as a ground. This envelope in turn serves as a ground for monopoles, which are excited in anti-phase with it, and that substantially distinguishes the given ground from large metal sheet. This means that the proposed structure realizes an asymmetrical version of the usual log-periodic antenna (symmetrical version of such an antenna is implemented as the LPDA antenna). Accordingly, it is possible to significantly increase the angle α and to shorten the antenna without fearing decrease of directivity and increase of losses, i.e., deterioration of frequency-independent characteristics. Since the radiating elements of the antenna are the monopoles, then by analogy with the LPDA, it is expedient to name this antenna by LPMA.

The operating principle of a symmetrical antenna was reviewed earlier by the analysis of processes in its active area. The processes in the active area of an asymmetrical antenna practically do not differ from the processes in a symmetrical antenna, since the waves in a coaxial distribution line are similar to the waves in the two-wire line and are dependant on the monopoles' influence which is similar to the influence of dipoles on the waves in a symmetrical structure. In the surrounding space the equally excited dipoles and monopoles produce similar fields.

The mock-up of an antenna has been manufactured and tested by the authors. It was designed for use in the range of 200–800 MHz. Antenna characteristics were measured for the two variations of mountings on the metal mast: (1) installation with help of a cantilever, and the radiators are mounted vertically, (2) installation on the mast top, and the radiators are mounted horizontally, and the gravity center coincides with the mast axis. The distribution line was manufactured in the shape of a truncated pyramid with a square cross-section and the internal conductor was made in the shape of a horizontal plate of variable width.

The history of the creation of LPMA, described in detail, makes it possible to get acquainted with the difficulties that arise when trying to reduce the dimensions and to achieve a minimal improvement in the properties of log-periodic antennas. These difficulties are caused by the complex structure of the antenna itself. This structure does not withstand rough intervention, and, as it was mentioned earlier, attempts to decrease longitudinal dimensions of an antenna by violating geometric progression relationships causing a sharp deterioration of electrical characteristics and gives an insignificant decrease in overall dimensions.

In this chapter it was shown that placement of concentrated capacitive loads along a linear radiator allows getting high directional characteristics in a wide frequency range. It was shown that in-phase currents in director-type antennas provide higher directivity in a wide frequency range. In this case, it is expedient to use the method of electrostatic analogy as an approximate method of analysis, whose results can be used later as initial values for solving the problem by the method of mathematical programming. As was shown in Section 3.7, this method is based on an analogy between two structures consisting of high-frequency currents and constant charges. This method allows us to use a new approach for solving the problem of reducing dimensions of the log-periodic antenna.

In accordance with a well-known method of calculating log-periodic antenna [58], we consider an active region of this antenna consisting of three radiators (Fig. 3.39) and determine fields of these radiators when emf e is located in the middle radiator, in its center. We assume that the arm length

184   ANTENNAS: Rigorous Methods of Analysis and Synthesis

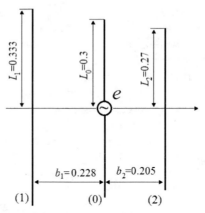

Fig. 3.39. The antenna of three radiators.

of the middle radiator with active area is $L_0 = \lambda/4$ (λ is the wavelength), the arm length of the left (longer) radiator is equal to $L_1 = \lambda/(4\tau)$, and the arm length of the right (shorter) radiator is $L_2 = \lambda\tau/4$, where τ is a denominator of a geometric progression, according to which the radiators' dimensions are changed. A magnitude of another parameter (σ) is given in (3.82). As stated in [60], generalization of information in the literature leads to the conclusion that the minimum changes in the electrical characteristics of the log-periodic antenna with metal dipoles within the frequency interval from f to $f\tau$ occurs when $\sigma\tau = 0.19$. We assume that this relation is also valid for the considered antenna. For definiteness we assume that τ is equal to 0.9. Then $L_1 = L_0/0.9 = 0.278\lambda$, $L_2 = L_0 \cdot 0.9 = 0.225\lambda$, $\alpha = 0.146$, $b_1 = 0.19\lambda$, $b_2 = 0.171\lambda$.

In order to use the theory of folded antennas farther we divide the active radiator (conductor 0) into two parallel conductors (with numbers 01 and 02), and the obtained group of four conductors between two circuits: with numbers 01 and 1 in the first circuit, and with numbers 02 and 2 in the second circuit. In accordance with the method of the electrostatic analogy of two structures, i.e., according to the physical content of the problem, one must assume that the ratio of the emf in the centers of the wires 01 and 02 is equal to the ratio of the partial capacitances C_{01} and C_{02}, i.e., $e_1 = 0.52e$, $e_2 = 0.48e$.

As shown in the theory of folded antennas, the circuit of two parallel vertical wires located at a distance b_i can be divided into a dipole and a long line, open on both ends. The current in the center of each dipole is equal to $J_{id} = e_i/(4Z_{id})$. The current in the center of each conductor of a long line is equal to $J_{il} = e_i/(2X_{il})$. Accordingly, the current in the center of the active radiator is the sum of the currents of the dipole and the long line, and the current in the center of the passive radiator is the difference between these values, i.e., $J_0 = \sum_{i=1}^{2}(J_{id} + J_{il})$, $J_i = J_{id} - J_{il}$. In the case of metal wires in the first approximation $Z_{id} = R_d + jX_{id}$, $R_d = 80 \tan^2(kL_0/2)$, $X_{id} = 120 ln(2L_0/a_{ei}) \cot kL_0$, $jX_{il}(L_i, b_i) = -j120\ ln(b_i/a) \cot kL_i$, where $a_{ei} = \sqrt{ab_i}$ is the equivalent radius of the dipole.

If each element of a log-periodic antenna is the radiator with concentrated capacitive loads, and the loads are changed along a radiator axis in accordance with linear law, i.e., a current along each radiator is the in-phase current, then (see, for example, Chapter 1) $R_d = 20k^2L_0^2$, $X_{id} = -120ln(2L_0/a_{ei})/kL_0$, $jX_{il} = -j120\ ln(b_i/a)/kL_i$.

The field amplitude and the phase of each radiator depends on its structure and location in an antenna. If the radiator arm is manufactured in the form of a straight metallic wire, the current along it is distributed by the sinusoidal law $J(z) = J(0) \sin k(L - |z|)$. In this case the far field of the radiator is equal to $E_\theta = \dfrac{AJ(0)}{\sin\theta}[\cos(kL\cos\theta) - \cos kL]$, where $A = j60$, where, k is the propagation constant, R is the distance to the observation point. If concentrated capacitive loads located along the axis of the radiator allow the formation of an in-phase current, distributed along this axis by the linear law $J(z) = J(0)(L - |z|)$, the far field of the radiator is,

$$E_\theta = \dfrac{AJ(0)\sin\theta}{\cos^2\theta}[1 - \cos(kL\cos\theta)].$$

Consider an influence of the radiator location on the far field of the antenna by a specific example of the director-type antenna, shown in Fig. 3.39. The arm length of the middle (active) radiator is equal to 0.3 m, i.e., the wave length of the its first (series) resonance is equal to 1.2 m. Radii of all conductors are the same and are equal to 0.001 m. The magnitude of τ is 0.9. It is obvious that the maximal radiation of the antenna should be directed to the right, toward the radiator 2. Since the radiator 1 is located to the left of the active radiator 0, at a distance b_1 from it, its field lags behind the active radiator field, first by kb_1 in phase, i.e., in accordance with the propagation time of the signal from the active radiator to the passive radiator 1, and, second, by $kb_1/\sin\theta$ in phase, i.e., in accordance with the time of signal propagation in the opposite direction, from the radiator 1 to the active radiator (signal of the wire 1, radiated at an angle θ, must come to the active radiator at the same angle θ, i.e., it travels the distance $b_1/\sin\theta$ and not the distance b_1. The total change in phase is equal to $\psi_1 = -kb_1 \cdot \dfrac{1 + \sin\theta}{\sin\theta}$. Similarly, in the case of radiator 2, this phase change is equal to $\psi_2 = -kb_2 \cdot \dfrac{\sin\theta - 1}{\sin\theta}$.

On the basis of the aforesaid the total field at angle θ of the antenna structure with the in-phase current distribution, shown in Fig. 3.39, may be written in the form

$$E_\theta = \frac{AJ(0)\sin\theta}{\cos^2\theta} \sum_{i=1}^{2} e_i \left\{ \left(\frac{1}{4Z_{id}} + \frac{1}{2Z_{il}} \right) [1 - \cos(kL_0 \cos\langle\theta - \theta_0\rangle)] + \right.$$
$$\left. + [1/(4Z_{id}) - 1/(2Z_{il})] \exp(j\psi_i)[1 - \cos(kL_i \cos\theta)] \right\} \quad (3.86)$$

The directivity magnitude is determined by the expression

$$D = |E_\theta(\pi/2)|^2 / \sum_{n=1}^{N} [|E_\theta(\theta_n)|^2 \Delta \sin\theta_n], \quad (3.87)$$

where Δ is the interval between neighboring values θ_n, N is the number of these intervals between $\theta = 0$ and $\theta = \pi/2$.

The results of calculating directivity of the structure, presented in Fig. 3.39, depending on an electrical length kL_0 of the active radiator arm are given in Fig. 3.40 by a curve 1. This curve shows the directivity of the structure with in-phase currents in each element. The directivity of the structure with sinusoidal currents is given for comparison by means of a curve 2. It is known that radiators with concentrated capacitances distributed in accordance with linear or exponential law along each arm allow us to ensure a high level of matching with a cable in the admissible range with a frequency ratio of the order of 10. Therefore, graphics are made so that to show the directivity of the structure in the range from $kL_0 = \pi$ to $kL_0 = 10\pi$.

Radiation of neighboring elements of log-periodic antenna with the arm length 0.27 m and 0.333 m respectively requires calculating fields in the structures presented in Fig. 3.41. Results of these calculations are given in Fig. 3.40: curve 3 corresponds to Fig. 3.41a, curve 4—to Fig. 3.41b.

The directivity magnitudes in Fig. 3.40 for specific values of kL correspond to the same frequencies, i.e., the same values of kL correspond to the elements of equal length in the all three schemes (for example, to the elements with the arm length 0.3 m). The figure shows that the directivity magnitudes at the same kL values are close to each other. This is natural, since the directivity in each scheme increases slowly with increasing frequency. Small increments in

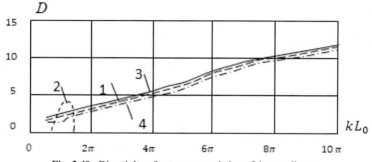

Fig. 3.40. Directivity of antennas consisting of three radiators.

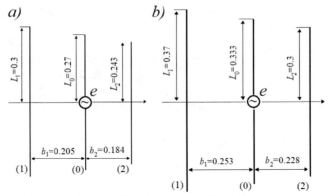

Fig. 3.41. Structures of smaller (*a*) and greater (*b*) neighboring radiators.

the active radiator length cause a small increase in the resonant wavelength at the same kL value, i.e., a small decrease in the resonance frequency and the directivity.

The obtained results show that each active radiator with in-phase current, included in the structure of the log-periodic antenna, provides high radiation directivity in a wide frequency range. Neighboring radiators have similar directivity magnitudes. The direction of the radiation is the same. A signal propagates along the distribution line from short elements to the long dipoles (in Figs. 3.39 and 3.41 to the left), and the radiated signal propagates in the opposite direction. The total path difference is quite large. For example, for the signal radiated by a half-wave neighboring radiator is equal to 0.38λ, i.e., it is close to a half wavelength. Crossing wires of the distribution line in an interval between the elements permits a dramatic reduction in the path difference.

Summarize the results of an analysis of the different variants of log-periodic antennas. The known log-periodic antenna with sinusoidal current distribution along the radiators have a property of automatic current "cut-offs", i.e., a separate antenna segment (active region) radiates a signal in a narrow frequency band. Outside this band and outside the borders of the active region the signal decays rapidly. Wide frequency range is provided by a large antenna length, which is equal to the sum of the lengths of the active regions. Attempts to reduce the antenna length by disturbing the geometric progression and increasing the number of radiators leads to a small decrease of dimensions and a sharp deterioration in electrical characteristics.

The more effective methods are include a two-fold use of each active region by an application of linear-spiral radiators (this method allows decreasing the antenna length by 25–30%) and an employment of an asymmetrical log-periodic antenna with coaxial distribution line.

Replacement of the straight metal radiators by the radiators with concentrated capacitive loads provides reusable active area and allows us to obtain a high directivity in a wide frequency range, using a simple structure with three radiators. Results obtained with the help of the method of electrostatic analogy may be used for solving optimization problem by methods of mathematical programming. Increasing the number of radiators in the structure may allow a dramatic increase in its directivity.

The obtained results open up a promising prospect of improving electrical characteristics and reducing dimensions of log-periodic antennas and require serious work to realize this perspective.

Chapter 4
New Methods of Analysis

4.1 Reduction of three-dimensional problems to a plane task

We know that, the problem of calculating electric fields of charged bodies is substantially simplified, if all geometrical dimensions depend only on two coordinates (such a field is called the plane-parallel field). A three-dimensional problem is solved only in a few particular cases, whereas a two-dimensional problem is considered more frequently—for a different number of metal bodies and wires and various cross-sectional shapes and various options for their connection and mutual locations. In this connection it is of interest to attempt the use of the solution of two-dimensional problems for field calculations in three-dimensional problems, when a mutual location of metal bodies reminds the two-dimensional variant.

The field calculation of two infinitely long charged filaments converging to one point (Fig. 4.1a) is an example of such problem. Let the linear charge densities of both filaments be the same in magnitude but opposite in sign:

$$\tau_1 = -\tau_2 = \tau. \tag{4.1}$$

Analogue of this three-dimensional problem is a two-dimensional problem for two parallel filaments (Fig. 4.1b). The scalar potential of two such filaments is equal to

$$U(x,y) = \frac{\tau}{2\pi\varepsilon} \ln(\rho_2/\rho_1) \tag{4.2}$$

where ε is the dielectric permittivity of the medium, ρ_1 and ρ_2 are the distances from the observation point M to the axes of the filaments, and

$$\rho_2^2 = (b-x)^2 + y^2, \quad \rho_1^2 = (b+x)^2 + y^2.$$

Here b is the half of the distance between the filaments' axes.

As seen from Fig. 4.1b, the lines of equal potential $U = $ const in the plane field of two charged filaments are the circumferences with the centers

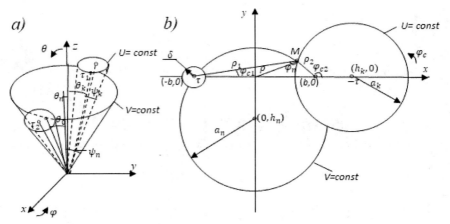

Fig. 4.1. Three-dimensional (*a*) and two-dimensional (*b*) problems for two infinitely long charged filaments.

on the abscissa axis. From here in particular it follows that the field of two parallel not coaxial metallic cylinders has similar character, since one can always dispose the axes of the equivalent filaments so that in their field two surfaces of equal potential coincide with the surfaces of the metallic cylinders (Fig. 4.2*a*). Lines of field $V = const$ are the circumferences with the centers on the ordinate axis.

In accordance with the uniqueness theorem the solution of an electrostatic problem must satisfy Laplace's equation, and the surfaces of conductive bodies must coincide with the equipotential surfaces. The three-dimensional problem for two convergent charged lines (see Fig. 4.1*a*) is a particular case of a conical problem, in which conductive bodies have a shape of a cone with the top in the origin (Fig. 4.2*b*). The conical and cylindrical problems are compared with each other in [61], where it is shown that Laplace's equation remains true in going from one problem to another problem, if the replaceable variables are related by equalities:

$$\rho = tan(\theta/2), \quad \varphi_c = \varphi. \tag{4.3}$$

Here ρ and φ_c are cylindrical coordinates, θ and φ are spherical coordinates.

The result of such a transformation of variables is transforming (mapping) a spherical surface with an arbitrary radius R into the plane (ρ, φ_c). Here the line of intersection of the spherical surface with any circular cone (with a vertex in the coordinates origin) transforms it into the circumference. Therefore, the three-dimensional conical problem may be reduced to the two-dimensional one. Cylindrical coordinates of the conductive bodies in the two-dimensional problem are related with the spherical coordinates of these bodies in the three-dimensional problem by equalities (4.3).

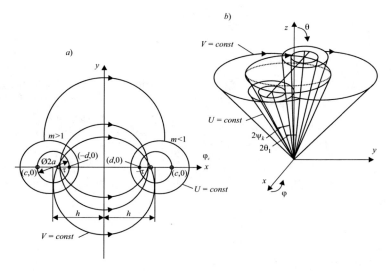

Fig. 4.2. The two-dimensional problem for metal cylinders (*a*), and the three-dimensional problem for metal cones (*b*).

In accordance with an analogy between the conical and the cylindrical problem the case of the two convergent charged filaments located at an angle of $2\theta_0$ to each other in the plane xOz (see Fig. 4.1a) corresponds to two parallel filaments spaced at the distance $2b = 2tan(\theta_0/2)$ from each other (see Fig. 4.1b). The case of the two metal cones with the angle 2ψ at the vertex of each cone and with the angle $2\theta_1$ between the cones' axes (see Fig. 4.2b) corresponds to two metallic cylinders of the radius $a = (c - d)/2$. The distance between their axes (see Fig. 4.2a) is equal to $2h = c + d$. Since according to (4.3)

$$c = tan[(\theta_1 + \psi)/2], \quad d = tan[(\theta_1 - \psi)/2], \tag{4.4}$$

it is not difficult to verify that

$$a = \frac{sin\psi}{cos\theta_1 + cos\psi}, \quad h = \frac{sin\theta_1}{cos\theta_1 + cos\psi}. \tag{4.5}$$

In particular it follows from (4.4) that

$$\theta_1 = tan^{-1}c + tan^{-1}d = tan^{-1}\frac{2h}{1-(h^2-a^2)},$$

$$\text{and,} \quad \psi = tan^{-1}c - tan^{-1}d = tan^{-1}\frac{2a}{1-(h^2-a^2)}. \tag{4.6}$$

It is necessary to emphasize that upon transition from the cone to the cylinder the cone axis doesn't coincide with the cylinder axis. The surfaces of equal potential $U = const$ of the electric field created by the two convergent charged filaments are the circular cones, whose axial lines lie in the plane xOz. The surfaces of field strength $V = const$, in which lines of field are located, are also circular cones.

The scalar potential $U(\theta, \psi)$ of the electric field created by the two convergent charged filaments is calculated in accordance with (4.2), where

$$\rho_2^2 = [\tan(\theta_0/2) - \cos\varphi\tan(\theta/2)]^2 + \sin^2\varphi\tan^2(\theta/2),$$
$$\rho_1^2 = [\tan(\theta_0/2) - \cos\varphi\tan(\theta/2)]^2 + \sin^2\varphi\tan^2(\theta/2).$$

In the given case the surfaces of equal potential $U = const$ are the circular cones, the axial lines of which lie in the plane xOz. Each surface satisfies an equation:

$$\rho_2/\rho_1 = m = const.$$

For the plane problem the line of equal potential is the circumference with the center in the point $h_m = (1 + m^2)b/(1 - m^2)$ and with the radius $a_m = 2mb/|1 - m^2|$. Using these magnitudes and the equality (4.6), we find the angle θ_m between the z axis and the axis of an equipotential circular cone in the conical problem,

$$\theta_m = \tan^{-1}\frac{2h_m}{1-(h_m^2 - a_m^2)} = \tan^{-1}\left(\frac{1+m^2}{1-m^2}\tan\theta_0\right),$$

and the angle ψ_m between the cone generatrix and its axis

$$\psi_m = \tan^{-1}\frac{2a_m}{1-(h_m^2 - a_m^2)} = \tan^{-1}\frac{2m\tan\theta_0}{|1-m^2|}.$$

The surfaces of field strength $V = const$, on which lines of the field are located, are also circular cones. Actually, in the plane problem the flux function appears as

$$V = -\frac{\tau}{2\pi\varepsilon}(\varphi_{c2} - \varphi_{c1}). \tag{4.7}$$

The sense of the angles φ_{c2} and φ_{c1} is clear from Fig. 4.1b. The equation for the line of field

$$\varphi_{c2} - \varphi_{c1} = \varphi_n = const$$

is the equation of a circumference, whose radius is equal to $a_n = b/\sin\varphi_n$, and the axis is at a distance $h_n = b\cot\varphi_n$ from axis z. In the conical problem the axial lines of the circular cones $V = const$ also lie in the plane yOz forming with axis z the angle

$$\theta_n = \tan^{-1}(\cot\varphi_n \tan\theta_0).$$

The angle ψ_n between the cone axis and generatrix is equal to

$$\psi_m = \tan^{-1}(\tan\theta_0/\sin\varphi_n).$$

If to reduce a conical problem to a cylindrical, this permits to calculate the capacitance per unit length and the wave impedance of the long line consisting of two convergent filaments or cones. It is known, for example, that the capacitance per unit length and the wave impedance of the line consisting of two conductors with radius a located at the distance $2h$ from each other are equal to

$$C_l = \frac{\pi\varepsilon}{\ln\left[h/a + \sqrt{(h/a)^2 - 1}\right]} = \frac{\pi\varepsilon}{ch^{-1}(h/a)}, \quad 1/(cC_l) = 120 ch^{-1}(h/a). \quad (4.8)$$

Here c is the velocity of light.

From (4.5) and (4.8), we obtain, for two convergent cones.

$$C_l = \frac{\pi\varepsilon}{ch^{-1}(\sin\theta_1/\sin\psi)}, \quad W_1 = 120 ch^{-1}\frac{\sin\theta_1}{\sin\psi}. \quad (4.9)$$

In particular for a dipole with an angle 2α between axes of conic arms (see Fig. 4.3), $\theta_1 = \alpha$, i.e.,

$$W_2 = 120 ch^{-1}\left(\frac{L\sin\alpha}{a}\right). \quad (4.10)$$

The case of two convergent charged shells, located along the surface of a circular cone with the angle $2\theta_0$ at vertex (Fig. 4.4a) is of specific interest. Let the arc length in a cross-section of a charged shell be equal to 2α. The two-dimensional problem in the form of two coaxial cylindrical shells of radius $a = tan(\theta_0/2)$ with the same arc length in a cross-section (see Fig. 4.4b) corresponds to this three-dimensional problem.

One can obtain a concept about the character of an electrostatic field of two cylindrical shells, if we sum fields of pairs consisting of symmetrically located parallel filaments 1-1', 2-2', 3-3' and so on with charges equal in magnitude and opposite in sign (see Fig. 4.4b). The lines of equal potential for each pair are the circumferences with their centers on the curve passing through the filaments. The curve going around filaments along circumferences with the same value of constant m is the line of equal potential for the field of the envelopes. It is a curve of a complicated shape, extended along both sides of each shell and smoothly bent around its ends. The symmetry axis of structure, i.e., y-axis, is also one of the lines of equal potential.

The lines of field strength for each pair of filaments are circumferences with centers on the symmetry axis y, which pass through the filaments. Two lines coincide with the circumference, on which the shells are situated,

194 ANTENNAS: Rigorous Methods of Analysis and Synthesis

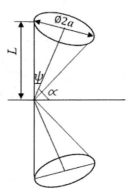

Fig. 4.3. An inclined dipole.

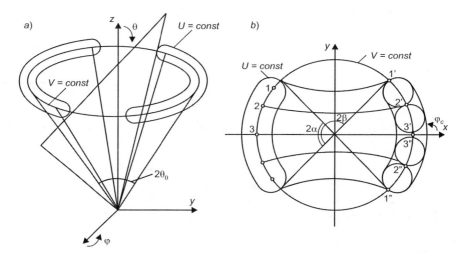

Fig. 4.4. Three-dimensional (*a*) and two-dimensional (*b*) problems for non-closed coaxial shells.

i.e., these lines close two gaps between the shells. Field lines inside this circumference connect the symmetrically placed filaments with each other and cross the lines of equal potential at right angles. The field structure in the case of two convergent conical shells has a similar character (see Fig. 4.4*a*), the only difference being that the surfaces of equal potentials, i.e., $U = $ const and the surfaces of field strength, i.e., $V = $ const coincide with the conical surfaces, and not with the cylindrical ones.

The capacitance C_l per unit length and the wave impedance W of the long line formed by two cylindrical shells [30, 62] are equal to

$$C_l = \varepsilon K(\sqrt{1-k^2})/K(k), \quad W = 120\pi K(k)/K(\sqrt{1-k^2}), \tag{4.11}$$

where $K(k)$ is the complete elliptic integral of the first kind from an argument

$$k = \tan^2(\beta/2). \tag{4.12}$$

Here 2β is the angular width of the slot, $2\alpha = \pi - 2\beta$ is the angular width of the metal shell, i.e., C_l and W depend only on angular slot width 2β and hence on angular width 2α of the cross-section of the metal shell.

Magnitudes C_l and W are independent of the cylinder radius a. This means that both expressions are constant along the conical line with the same cross-section, i.e., two-wire line consisting of the convergent shells is a uniform line. It is known, that an input impedance of a uniform two-wire line, tends to its wave impedance, when the line length increases infinitely. Therefore, the input impedance of an infinitely long line with angular width 2α excited by a generator situated near the cone vertex is

$$Z_l(k) = \frac{120\pi K(k)}{K(\sqrt{1-k^2})}, \tag{4.13}$$

i.e., such a long line has a high level of matching with the generator in an unlimited frequency range.

It could be considered that the structure under study presented in Fig. 4.4 on the one hand is a two-wire line and on the other hand is an antenna. The antenna is a symmetrical V-radiator, with the arms shaped as two convergent metal shells located along the surface of a circular cone. Finally, we can consider this structure as a slot antenna in a conical screen. In fact, the slot impedance is the input impedance of the metallic radiator situated next to it. Z_E is the input impedance of the metallic radiator, which is identical to the slot in shape and dimensions.

It is useful to compare the input impedances of metallic and slot radiators with the same width. If, for example, the metal shell width is $2\alpha = 2\pi/3$, then $\beta/2 = \pi/12$, $k^2 = 0.00515$ and $K(\sqrt{1-k^2})/K(k) = 2.56$, i.e., $Z_E(2\pi/3) = 120\pi/2.56$. If the slot width is $2\beta = 2\pi/3$, then $\beta/2 = \pi/6$, $k^2 = 0.111$, $K(\sqrt{1-k^2})/K(k) = 1.56$. Accordingly, $Z_S(2\pi/3) = 120\pi/1.56$. Therefore, the impedances of metallic and slot radiators of the same width $2\pi/3$ are related to each other by the expression $Z_E(2\pi/3) \cdot Z_S(2\pi/3) = \dfrac{(120\pi)^2}{(2.56 \cdot 1.56)}$, from whence it follows that

$$Z_S(2\pi/3) = (60\pi)^2/Z_E(2\pi/3).$$

In the general case, if the structure length is great, it is easily verified that

$$Z_S(2\alpha) = (60\pi)^2/Z_E(2\alpha), \tag{4.14}$$

when $Z_E(2\alpha)$ is the input impedance of a metallic (electric) radiator with angular arm width 2α, and $Z_S(2\alpha)$ is the input impedance of a slot antenna with width 2α.

Radiators with the same width of the metal shell and the slot are of particular interest. Setting $2\alpha = 2\beta = \pi/2$, we obtain: $k^2 = 0.0294$, $K(\sqrt{1-k^2})/K(k) = 2.0$, i.e.,

$$Z_E(\pi/2) = Z_E(\pi/2) = 60\pi. \tag{4.15}$$

As can be seen from (4.15), the infinitely long slot radiator with angular arm width $2\alpha = \pi/2$ mounted on a cone has a constant and purely resistive input impedance, and hence it has a high level of matching with a source (cable or generator) in an unlimited frequency range. If the radiator has finite dimensions, the frequency range is limited, but remains sufficiently wide. Such a slot in a metal cone is shown in Fig. 4.5. In this option slot edges coincide with the cone generatrixes.

The infinitely long metal and slot radiators identical in shape and dimensions, which fill the whole plane, are called self-complementary. In [38] it was shown first that the structure consisting of electric and magnetic radiators of the same shape and dimensions may not be flat. The volumetric self-complementary structure has properties similar to that of the plane structure. It has a constant and purely resistive input impedance, which ensures a high level of matching with a cable in a wide frequency range. Equal power is radiated into the inner and outer corner of the cone, regardless of the angle at the apex of the cone and the width of the metal and slot radiators.

This consideration shows that a choice of surface for placement of the self-complementary radiation structure is not accidental. It is the surface of rotation. As it is shown earlier, using equalities (4.3), relating the replaceable variables of the spherical and cylindrical systems of coordinates, one can reduce a three-dimensional conical problem to a two-dimensional problem. This result had played a big role in the development of electromagnetic theory and in its use in antenna engineering. A similar result can be obtained, if we

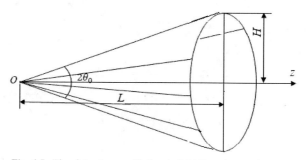

Fig. 4.5. The slot antenna with the straight-line edges at the cone.

reduce the parabolic problem to the plane problem. The generatrix of a circular cone is a straight wire. The curve line also may be used as a generatrix. If, e.g., a generatrix has a parabolic shape, the metallic surface assumes the shape of a paraboloid of rotation. Calculating the field of two infinitely long charged filaments of parabolic shape located on the paraboloid surface and converging to its vertex is a matter of unconditional interest.

Use of parabolic coordinates (σ, τ, φ) facilitates an analysis of such a structure. This is a system of orthogonal curvilinear coordinates [63]. The coordinate surfaces of this system are first confocal paraboloids of rotation (σ = const, τ = const), whose focal point coincides with the origin of the coordinate system, and second half-planes (φ = const), passing through the axis of rotation (see Fig. 4.6).

Rectangular coordinates are related to parabolic coordinates by the equalities:

$$x = \sigma\tau \cos\varphi, \quad y = \sigma\tau \sin\varphi, \quad z = (\tau^2 - \sigma^2)/2. \qquad (4.16)$$

Parabolic lines are located along the surface of a paraboloid, or more exactly they are lines of intersection of this surface with the half-planes, passing through the axis of rotation. As in the case of convergent straight wires, it is expedient to reduce the calculation problem of an electrostatic field of two charged parabolic filaments to the calculation of the field of two parallel filaments (see Fig. 4.1b). To this end the Laplace's equation in accordance with the uniqueness theorem must remain true at the transition from one problem to another, and the surfaces of conductive bodies must coincide with the surfaces of equal potential.

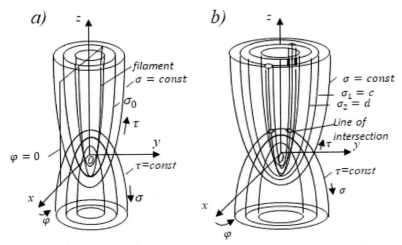

Fig. 4.6. Three-dimensional problem for two infinitely long parabolic filaments (a), and solid geometrical figures (b).

In the cylindrical coordinate system (ρ, φ_c, z) the Laplace's equation for a potential U has the form

$$\frac{\partial}{\partial \rho}\left(\rho \frac{\partial U}{\partial \rho}\right) + \frac{1}{\rho}\frac{\partial^2 U}{\partial \varphi_c^2} = 0. \tag{4.17}$$

Here, $\partial U/\partial z = 0$, is taken into account, i.e., the lines parallel to z-axis have a constant potential (the field is plane-parallel). Laplace's equation in the system of parabolic coordinates has the form

$$\left[\frac{1}{\sigma}\frac{\partial}{\partial \sigma}\left(\sigma \frac{\partial U}{\partial \sigma}\right) + \frac{1}{\tau}\frac{\partial}{\partial \tau}\left(\tau \frac{\partial U}{\partial \tau}\right) + \left(\frac{1}{\sigma^2} + \frac{1}{\tau^2}\right)\frac{\partial^2 U}{\partial \varphi^2}\right] = 0. \tag{4.18}$$

As seen from (4.18), this equation is symmetrical with respect to σ and τ, that is, the equation for each unknown quantity is true irrespective of the other equation. In particular, for σ we obtain

$$\frac{1}{\sigma^2}\left[\frac{1}{\sigma}\frac{\partial}{\partial \sigma}\left(\sigma \frac{\partial U}{\partial \sigma}\right) + \frac{1}{\sigma^2}\frac{\partial^2 U}{\partial \varphi^2}\right] = 0. \tag{4.19}$$

Here $U(\sigma) = U(\tau)$, if other coordinates are the same. If, for example, $\partial U/\partial z = 0$, then in accordance with (4.16) $\sigma = \sqrt{\tau^2 - 2z}$, $\tau = \sqrt{\sigma^2 + 2z}$, i.e., $\partial \sigma/\partial z = -1/\sigma$, $\partial \tau/\partial z = 1/\tau$, and

$$\frac{\partial U}{\partial z} = \frac{\partial U}{\partial \sigma}\frac{\partial \sigma}{\partial z} + \frac{\partial U}{\partial \tau}\frac{\partial \tau}{\partial z} = 0,$$

whence $(1/\tau) \partial U/\partial \tau = (1/\sigma) \partial U/\partial \sigma$.

Comparison of (4.17) and (4.19) shows that these equations coincide, if the replaceable variables are related by equations:

$$\rho = \sigma, \quad \varphi_c = \varphi. \tag{4.20}$$

Here ρ and φ_c are the cylindrical coordinates, and σ and φ are the parabolic coordinates. Hence, the Laplace's equation holds true in the transition from the parabolic problem to the cylindrical, if expressions (4.20) are true. These expressions were first presented in [22].

The substitution of variables in accordance with (4.20) results in the transforming (mapping) of parabolic surface $\tau = $ const into the plane (ρ, φ_c) for an arbitrary τ. The line of this surface intersection with any paraboloid $\sigma = \tau$ is transformed into a circumference. Two parabolic lines situated along surface $\sigma = \sigma_0$ in the plane xOz (see Fig. 4.6a) are transformed into two parallel lines spaced at distance $2b = 2\sigma_0$ in the cylindrical coordinates system (see Fig. 4.1b). Two continuous geometrical figures located between parabolic surfaces $\sigma_1 = c$ and $\sigma_2 = d$ (see Fig. 4.6b) are transformed into two solid cylinders of radius $a = (c - d)/2$, whose axes are located at the distance $2h = c + d$ (see Fig. 4.2a).

The scalar potential of the electric field for two parabolic charged filaments situated along surface $\sigma = \sigma_0$ (with linear charge density equal to $\pm q_0$) is analogous to (4.2). The surfaces of equal potential $U = const$ in the given case are the volumetric geometrical figures, and their planes of symmetry coincide with the plane xOz. The surfaces of field strength $V = const$, where force lines are located, are paraboloids of rotation. Their axial lines lie in the plane yOz on the parabolic surfaces. Similarly from (4.8) we find:

$$C_l = \frac{\pi\varepsilon}{ch^{-1}[(\sigma_1 + \sigma_2)/(\sigma_1 - \sigma_2)]}, \quad W = 120ch^{-1}[(\sigma_1 + \sigma_2)/(\sigma_1 - \sigma_2)]. \quad (4.21)$$

Use of the equalities (4.21) allows us to reduce the parabolic problem to the cylindrical problem and to calculate capacitance C_l per unit length and wave impedance W of a long line consisting of two volumetric figures with the parabolic axes. The input impedance of a uniform two-wire line, when the line length increases, tends to its wave impedance $Z_{AB} = 120ch^{-1}\frac{\sigma_1 + \sigma_2}{\sigma_1 - \sigma_2}$.

As seen from (4.11), capacitance C_l per unit length and wave impedance W of an equivalent line are dependent only on arc length 2β of the slot antenna in the cross-section and accordingly on arc length $2\alpha = \pi - 2\beta$ of the metallic shell. For this reason, expressions (4.11) are true for both the parabolic and conical shells. Magnitudes C_l and W are constant along the line, i.e., the two-wires line is uniform. The results of the calculation show that on the whole decreasing an inclination angle of the antenna arm with respect to ground (and corresponding increase of the arm length without height increase) improves the electrical characteristics of the antenna. If the inclination angle of the radiator located on the surface of a cone and a paraboloid is the same, then the radiator on the paraboloid surface has the greater gain particularly in the lower part of the frequency range. The directional patterns of the radiators in a vertical plane (pattern factor) are approximately the same. The directivity of both antennas is greater than the directivity of the flat vertical antenna.

Coordinate systems, considered earlier, have a circular structure (a circular cross-section). In order to analyze electric fields of charged bodies elongated in the horizontal direction, we need to use corresponding structures with a horizontal section, for example, in the shape of an ellipse. Vertical generatrix of such a structure can have the shape of a parabola (Fig. 4.7), or the shape of a broken line. In the first case the structure has the form of elliptical paraboloid. In the second case a vertical cross-section has the shape of a triangle, rectangle or trapezoid. In both the cases the possibilities of

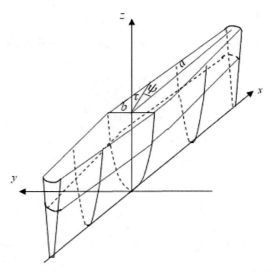

Fig. 4.7. Three-dimensional body, elongated in the horizontal direction.

calculating electric fields created by charged three-dimensional structures are significantly expanded. For example, one can use these structures to calculate the field in a phantom, to determine the fields of flat radiators, and so on.

As a coordinate system for calculating the fields of an elliptical paraboloid, one can use a system of orthogonal curvilinear coordinates (τ, ψ, σ) [63, 64]. Coordinate surfaces of this system are created by two mutually orthogonal families of horizontal ellipses and vertical parabolas. The coordinate τ in this system is the radius of the ellipse, ψ is the angle between this radius and the axis x (see Fig. 4.7). In a rectangular coordinate system, the elliptic equation has the form $x^2/a^2 + y^2/b^2 = 1$ (here a and b are the major and minor semi-axes of the ellipse). Correspondingly, the rectangular and curvilinear coordinates are related by relations $x = \tau \cos \psi$, $y = \tau \sin \psi$. The vertical coordinate σ is equal to cz^2 or cz.

Since it was not possible to find detailed information about the proposed coordinate system, the problem is considered in a general form [64]. The Laplace's equation for a curvilinear orthogonal coordinate system in the general case has the form

$$\Delta V = \frac{1}{e_1 e_2 e_3} \left[\frac{\partial}{\partial u} \left(\frac{e_2 e_3}{e_1} \frac{\partial V}{\partial u} \right) + \frac{\partial}{\partial v} \left(\frac{e_3 e_1}{e_2} \frac{\partial V}{\partial v} \right) + \frac{\partial}{\partial w} \left(\frac{e_1 e_2}{e_3} \frac{\partial V}{\partial w} \right) \right] = 0. \quad (4.22)$$

Here

$$u = \tau, \; v = \psi, \; w = \sigma, \; e_1^2 = \left(\frac{\partial x}{\partial u}\right)^2 + \left(\frac{\partial y}{\partial u}\right)^2 + \left(\frac{\partial z}{\partial u}\right)^2, \; e_2^2 = \left(\frac{\partial x}{\partial v}\right)^2 + \left(\frac{\partial y}{\partial v}\right)^2 + \left(\frac{\partial z}{\partial v}\right)^2,$$

$$e_3^2 = \left(\frac{\partial x}{\partial w}\right)^2 + \left(\frac{\partial y}{\partial w}\right)^2 + \left(\frac{\partial z}{\partial w}\right)^2, \text{ i.e., } e_1 = 1, \; e_2 = 1, \; e_3 = 1/(2\sqrt{c\sigma}), \text{ or } 1/c, \; e_1 e_2 e_3 = \tau e_3.$$

The substitution of the said magnitudes in (4.22) yields

$$\Delta V = \frac{1}{\tau}\left[\frac{\partial}{\partial \tau}\left(\tau \frac{\partial V}{\partial \tau}\right) + \frac{1}{\tau}\frac{\partial^2 V}{\partial \psi^2} + \frac{\tau}{e_3^2}\frac{\partial^2 V}{\partial \sigma^2}\right] = 0. \tag{4.23}$$

As was mentioned, in accordance with the uniqueness theorem, when the three-dimensional problem is reduced to a planar task, the Laplace's equation must remain valid and the metallic surfaces must coincide with lines of equal potential. Let us compare the resulting equation (4.23) with the Laplace's equation (4.17) in a cylindrical coordinate system. In (4.17) $\partial U/\partial z = 0$, is taken into account, i.e., the potential does not depend on z. If we assume that

$$\tau = \rho, \; \psi = \varphi_c, \tag{4.24}$$

then equation (4.23) coincides with (4.17). The essential difference between expressions (4.17) and (4.23) is that the value ρ in (4.17) does not depend on the angle φ_c, and the value τ in (4.23) depends on ψ. This fact changes the field's structure, drastically.

We will consider a particular case of such a structure with a cross-section that does not depend on the coordinate z. As shown in [65], the complex potential of a solitary wire of circular cross-section with charge q per unit length located in a homogeneous medium with dielectric permittivity ε is equal to

$$\varsigma(z) = V(x,y) + jU(x,y) = -\frac{q}{2\pi\varepsilon} j \ln z + C, \tag{4.25}$$

where $V(x, y)$ is the flux of the electric field strength, $U(x, y)$ is the potential of the total sum of charges. Equation $V(x, y) = \text{const}$ is the equation of the line of the field strength, and $U(x, y) = \text{const}$ is the equation of the line of equal potential in the plane of the cross-section, x and y are rectangular coordinates in this plane. Introducing the notation $z = \rho e^{j\varphi_c}$, where ρ and φ_c are polar coordinates in the indicated plane, we obtain

$$V(x, y) = -A\varphi_c + C_1, \; U(x, y) = A \ln \rho + C_2. \tag{4.26}$$

Here A, C_1, C_2 are constants. Equations for the line of field strength and the line of equal potential assume the shape

$$\varphi_c = const, \quad \rho = const. \tag{4.27}$$

Go over to the complex potential of a solitary wire with elliptic cross-section, we find in accordance (4.26) that the field strength along the corresponding line is equal to $V(x, y) = -A\varphi_c + C_1$ and the potential along the line of equal potential is equal to $U(x, y) = A \ln \tau + C_2$.

The corresponding field is presented in Fig. 4.8a.

The complex potential of a line consisting of two conductors located at a distance $2b$ from each other is equal to $\varsigma(z) = Aj\ln\dfrac{z+b}{z-b} + C$. Denoting $z + b = \rho_1 e^{j\varphi_1}$, $z - b = \rho_2 e^{j\varphi_2}$, where ρ_1 and ρ_2 are the lengths of the segments between the axes of the conductors and the observation point M, φ_1 and φ_2 are the angles between these segments and the axis x, and assuming that $C = 0$, we find:

$$V = A(\varphi_2 - \varphi_1), \quad U = A \ln \rho_1/\rho_2. \tag{4.28}$$

For the line consisting of two conductors with elliptic cross-section in accordance with (4.24)

$$V = A(\psi_2 - \psi_1), \quad U = A \ln \tau_1/\tau_2. \tag{4.29}$$

In this field the line of zero potential is the ordinate axis, the initial line of the field strength is the section of the abscissa axis between the line conductors and infinity. Lines of equal potential are the ellipses with focal points on the axis x, and the lines of field strength are the ellipses with focal points on the axis y passing through the wires' axes.

The capacitance per unit height of two identical conductors with an elliptical cross-section, if the small axes of the ellipses are located on the same straight line (Fig. 4.9a), is equal to (see [30])

$$C_l \approx \dfrac{\pi\varepsilon}{sh^{-1}d/\sqrt{a^2-b^2} - sh^{-1}b/\sqrt{a^2-b^2}}.$$

If the great axes of the ellipses are located on the same straight line (Fig. 4.9b), this capacitance is equal to

$$C_l \approx \dfrac{\pi\varepsilon}{sh^{-1}d/\sqrt{a^2-b^2} - sh^{-1}a/\sqrt{a^2-b^2}}.$$

The new coordinate systems, in particular coordinates of elliptic paraboloid allow us to substantially simplify the solution for separate tasks. One of these tasks is a problem of phantoms.

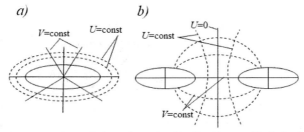

Fig. 4.8. Field of one (*a*) and two (*b*) three-dimensional bodies, elongated in the horizontal direction.

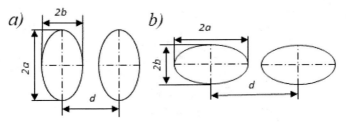

Fig. 4.9. Two identical wires with an elliptical cross-section.

The so-called phantoms that allow to measure the specific rate of absorption (SAR) of energy in the human' body were developed to determine the energy that a person receives when irradiated. Designers tried to create such a device, the shape of which coincides with the shape of the human body. As a result, different devices essentially differ from each other by shape and dimensions. But on the one hand people are not the same, i.e., excessive accuracy does not make sense. On the other hand, since the developed phantoms have a complex shape and unknown characteristics, measurements on different phantoms give different results, and it is difficult to compare these results with each other, and users doubt the results of measurements.

An attempt to compare the measurement results in different phantoms depending on their shape and dimensions was previously made by using a system of parabolic rotation coordinates [66]. The coordinate system for an elliptical paraboloid significantly permits the simplification of the solution to this problem.

The human body in the phantom is modeled by a vessel with a liquid. The dielectric constant and conductivity of the liquid must be equal to the average values of these magnitudes for tissues of the human body at operating frequencies. A probe immersed in the liquid records the field values at the points inside the phantom. The field is created by the transmitting antenna (by a horizontal dipole) located under the vessel with the liquid. The field magnitude at the measurement point depends on the medium between the radiating antenna and the point of observation, and in fact it depends on the capacitance of the indicated interval. This interval consists of three parts: a free space between the transmitting antenna and the vessel, the dielectric envelope of the vessel and the space between the envelope and the probe. The envelope serves as a boundary between different media, so its surface is a surface of equal potential, and other such surfaces pass inside the vessel parallel to the vessel envelope.

The dielectric constant of the fiberglass envelope is equal to $\varepsilon_1 = 2.7$, the permeability of the liquid is $\varepsilon_2 = 41.5$. The envelope capacitance is equal to $C_1 = \varepsilon_1 S_1/d_1$, where $S_1 = Ll$ is the area of the vessel side surface from the lower boundary to the level of the vessel's liquid surface (see Fig. 4.7), d_1 is the envelope thickness, l is the length of parabola between the lower boundary and the liquid level, and L is the vessel's perimeter (the ellipse length) along the midline between these boundaries. A similar formula permits the calculation of the capacitance of the liquid's layer between the envelope and the probe. The air with permeability 1 practically does not affect the capacitance value.

Calculating capacitances of the side phantom used by Holon Technological Institute gave the following results: the envelope capacitance is equal to $C_1 = 0.027 \mu \Phi$, the capacitance of the liquid's layer with a thickness of 2 cm is $C_2 = 0.021 \mu \Phi$, and the total capacitance is $C_1 = 0.012 \mu \Phi$. The previous results presented in [66] ($0.0086 \mu \Phi$; $0.010 \mu F$; $0.0047 \mu F$) were substantially corrected. The field created by the external antenna is successively reduced after passing each layer.

The effective method of solving three-dimensional electromagnetic problems by means of results of two-dimensional problems is based on such a replacement of variables, at which the Laplace's equation remains valid. The beginning of this method development was the work of Carroll, which allowed solving spherical problems in accordance with results obtained for the cylindrical structures. Later on, the region of method employment was significantly expanded by means of comparing different systems of orthogonal curvilinear coordinates. New works in this direction permit to get new useful results.

4.2 Distribution of currents over the cross section of a long line

As stated at the beginning of this chapter, the problem of calculating electric fields of charged bodies is substantially simplified, if the all geometrical dimensions depend only on two coordinates (such a field is called the plane-parallel field). Nevertheless, there are two-dimensional problems that are solved quite simply, but do not arouse interest in their strict solution, because the result is considered known. A known result was obtained long ago and is shrouded in outdated prejudice. The problem of the distribution of the energy flux over a section of a long line (in other words, the problem of the distribution of displacement currents along the cross-section of a transmission line) is one such problem.

The method of electrostatic analogy and the corresponding principle make it possible to use the solution of a problem of an electrostatic field of charged conducting bodies to calculate constant currents in a weakly conducting medium and to construct a picture of the field of linear high-frequency currents. In particular, based on them, a distribution of an energy flux over the cross-section of a long line can be analyzed.

The cross-section of the simplest long line of two round thin wires is shown in Fig. 4.10a. The currents in the wires create a plane-parallel field around them, similar in structure to the field of wires with charges of the same magnitude and opposite in sign. In the case of plane-parallel electrostatic field, flux function $V(x, y)$ defines the flux of vector $\vec{E}$ through a cross-section of a cylinder, which is located along z-axis between the given and zero surface (see Fig. 4.1b). For two parallel infinitely long charged filaments the flux function is defined by expression (4.7). It is proportional to the difference of angles $\Delta\varphi = \varphi_{c2} - \varphi_{c1}$, where φ_{ci} is the coordinate of the metal filament i, i.e., it is the angle between a given direction and the x axis. For any point of the cylindrical surface, located around the left charged filament located in the point with coordinates $x = -b$, $y = 0$ at the distance $\delta \ll b$ along z-axis (this surface is shown in Fig. 4.1b by the continuous circumference) the angle φ_{c2} is constant and equal to π, i.e., the magnitude of the flux inside the cylindrical surface depends only on the angle φ_{c1}, more precisely it is proportional to this angle. It means that the share of the flux is equal to a fraction of the circumference with radius δ. As is seen from Fig. 4.1b, the arc length of this fraction is equal to a half of the circumference length. Part of the flux falls into the volume bounded by the surface, passing through both charged filaments. This volume is a cylinder, and its cross-section has the form of a circle with the center at the origin of the coordinate system. That means, part of the flux, which falls into this volume, is equal to half of the total flux.

A similar result is obtained in the case of convergent filaments (see Fig. 4.1a). Half of the flux of two convergent infinitely charged filaments is directed inside a circular cone passing through both filaments. The reason

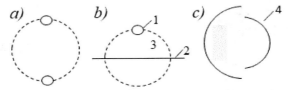

Fig. 4.10. Cross-section of the line in the form of two wires (*a*), a wire over ground (*b*) and unclosed tube (*c*).

for this fact is based on the equality of the angles at the transition from the cylindrical problem to a conical one: in accordance with (4.3) $\varphi_c = \varphi$. Also, it is true for two parabolic infinitely charged filaments (see Fig. 4.6a), since according to (4.20) the angular variables of cylindrical and parabolic coordinates systems are related by the equality $\varphi_c = \varphi$.

The same is also valid for the constant current in a weakly conductive homogeneous medium, in which two wires with high conductivity are located in a bounded volume having the shape of a circular cylinder, cone or paraboloid. If a voltage is included between the wires, then the fraction of the constant current inside this volume is equal to half of the total current.

Also, the magnetic field of linear currents coincides with the electric field of linear charges in accordance with the conformity principle, if the currents and the charges are identically distributed in space. This means, in particular, that a half of the energy flux propagating along a transmission line, consisting of two parallel (divergent) wires, is concentrated inside the circular cylinder (cone, paraboloid) passing through these wires. Similar postulates are true for the system of two wires with finite radius (see Fig. 4.2), if we consider the flux inside the cylindrical (conical, parabolical) surface passing through equivalent filaments creating the field, to coincide with the field of the wires.

If the self-complementary antenna structure is located on the circular cone (or paraboloid of rotation), then the line of two non-closed coaxial shells (see Fig. 4.4a) may be imagined as the sum of pairs of filaments 1-1", 2-2", 3-3" and so on, which are placed opposite to each other and have opposite signed charges (see Fig. 4.4b). The flux of each pair inside the structure is also equal to half of the total flux. That means that if symmetrical slotted antennas are located on a circular metal cone (or paraboloid), then the same power is radiated into input and output space independently at an angle to the cone vertex. This conclusion coincides with the inference in Section 2.9. The field symmetry inside a cylinder, cone and paraboloid is the reason for such equality of powers.

One must also remember that it is true for an infinitely long line. This fact is confirmed for metallic and slotted V-antennas.

From above it follows that in a direction along the lines a shaped in the form two round wires (Fig. 4.10a) only the half of the energy flux goes through the cylinder, whose generatrixes coincide with the wires. The second half of the energy in spite of a various opinions goes through a surrounding space. The same is obtained in the asymmetrical option, when the wire 1 is located over ground 2 (Fig. 4.10b): only half of an energy flux goes between a wire 1 and a ground within the marked cross-section 3.

If a cross-section of a cylinder (or cone, or paraboloid) has unclosed form, and, for example consists of two arcs 4 (Fig. 4.10c) of different radii, then the flux fraction passing through the structure is not equal to half.

4.3 Microstrip antennas

Surprisingly, the closest relatives of self-complementary antennas located on the cone and paraboloid are microstrip antennas (patches). These antennas are widely used on airplanes and spacecrafts, where not only the size and weight of the structures used are very important, but also the simplicity of their installation and the aerodynamic profile, which allows them to be installed on flat surfaces.

Microstrip antenna made as a very thin ($t \ll \lambda$) metallic strip of a small length ($L \ll \lambda$) placed on an object surface is shown in Fig. 4.11. The strip may have any configuration, including the simplest (rectangular). The strip and the ground plane are separated by a dielectric sheet (a substrate). The antenna is excited by a feed line, which is executed also as a conducting strip, but usually of a smaller width. The radiating elements and the feed lines are usually photoetched on the substrate.

In order to analyze the electrical characteristics of microstrip antennas, several models have been proposed. The greatest attention was paid to models in the form of a two-wire long line (transmission-line model) and in the form of a resonator. The first model [67] considers the antenna as a line section terminated at both ends by a radiation admittance Y_T. The second model takes into account higher-order modes, but this does not lead to a significant improvement in results. As is convincingly stated in [68], the transmission-line model "has the advantage of yielding very simple expressions for both the real and imaginary part of Y_T, but it has three important shortcomings:

a) The expressions used for Y_T are inaccurate for narrow patches (i.e., for $W \leq \lambda_0$; λ_0 – free-space wavelength).

b) The mutual coupling between the main radiating slots is neglected.

c) The influence of the side slots on the radiation conductance is neglected."

Here the slots of length W are called the main radiating slots and the slots with length L are called the side slots. Let's give the authors of [68] the right to such terminology, but we consider it necessary to change the order of the

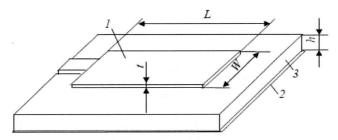

Fig. 4.11. Microstrip antenna.

listed deficiencies, since the main deficiency is the lack of interest in the side slots.

Let us return to the topic of a similarity of microstrip antennas and antennas located on a cone, a paraboloid or a pyramid. The antenna on a cone is shown in Fig. 4.5, the antenna on a paraboloid is shown in Fig. 2.13, and the antenna on a pyramid is shown in Fig. 4.12. All of the listed antennas consist of two symmetrical radiators: metal and slotted—with the arms located at a small angle to each other. They are excited at the point located on the top of the cone, paraboloid or pyramid. As can be seen from Fig. 4.12, the pyramidial structure is closest to the microstrip antenna. Its cross-section has a rectangular shape. The upper and lower faces of the pyramid are made metallic, forming an electric dipole; the two side faces are slots, forming a magnetic dipole. Such a structure is easier to realize, than for example conical, and on medium waves the lower arm of the antenna is created as a reflection of the upper arm in the ground. The antenna, located on a cone and consisting of metal and slot radiators of the same width, is a self-complementary radiator of finite size and has good electrical characteristics. The antenna on the pyramid, strictly speaking, cannot be called either complementary nor self-complementary, because its surface is not a smooth surface of a rotation with axial symmetry. But its characteristics are similar to those of self-complementary antennas.

When we compare the antenna on the pyramid with microstrip antenna, it is easy to be convinced of their structure similarity. The differences are minor. In the case of the microstrip antenna, the triangular or trapezoid metal face of the pyramid is replaced by a rectangular strip, and the metal strips of the antenna and the feed line are joined with each other. The slot radiator is also rectangular. Its width is substantially less than the width of the metal radiator. But basically, the structure of the microstrip antenna is no different from the structure of the antenna located on the pyramid. Such a look at the structure of the microstrip antenna clearly shows that the all elements of the so-called transmission line, located along its length L, do not simply transfer energy from the excitation point to the opening at the opposite end, but radiate the signal. In this regard, the transmission line is no different from the usual two-wire long line, which radiates along the entire length, if the distance between the wires is not infinitely small.

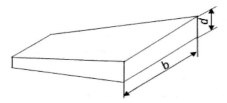

Fig. 4.12. Antenna on the pyramid.

Each of the described radiators can be simultaneously viewed as a long line and as an antenna (see, for example, [27]). This approach is no different from the usual comparison of a linear antenna with an equivalent long line. This greatly simplifies the calculation of the wave impedance and other characteristics. The structure of the microstrip antenna and the dimensions of its cross-section are constant along its length L, i.e., the equivalent long line is uniform, and its characteristic impedance is constant.

A cross-section of the microstrip antenna is shown in Fig. 4.13. The dotted line in the figure shows a mirror image in the metal plane. The capacitance of the symmetric structure consisting of a microstrip antenna and its mirror image is two times less than the capacitance of the asymmetric antenna. Considering the microstrip antenna as a combination of a metal and a slot radiator, we determine an input impedance of each. The capacitance of a metal radiator per unit length can be calculated as the capacitance between two identical plates with a common plane of symmetry [30]. A strict expression for its calculation contains complete and incomplete elliptic integrals of the first and second kind and requires the solution of an equation consisting of such functions. For approximate calculations, approximate functions can be used. If the width of the metal plate is equal to b, and the distance between the plates is d, then at $b/d > 27$ the capacitance between the plates per unit length is $C_l \approx \varepsilon \frac{b}{d}$. When $1 < b/d \leq 27$

$$C_l \approx \varepsilon \frac{b}{d}\left\{1 + \frac{1}{\pi}\frac{d}{b}[1 + ln(2\pi b/d)]\right\}. \quad (4.30)$$

If, for example, $h = 0.7874 \cdot 10^{-3}$ (0.7874 mm), $W = 33.15 \cdot 10^{-3}$ (33.15 mm), then for the capacitance between the metal plates (according to the approximate first formula and the more accurate second one), we obtain, assuming, $b = W$, $d = h$, capacitance,

$$C_{l1} = 42.1\varepsilon, \quad C_{l2} = 44.2\varepsilon.$$

This means that the input impedance of an asymmetrical metal radiator is equal to

$$120\pi \frac{\varepsilon}{C_l} \approx 120\pi \frac{d}{b} \approx 2.8\pi \text{ or } 2.7\pi. \quad (4.31)$$

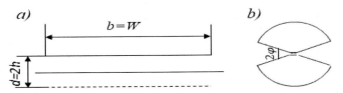

Fig. 4.13. Cross-section of microstrip (*a*) and cylindrical (*b*) antenna.

The input impedance of a symmetric metal radiator (taking into account the mirror image) is twice as large.

When $d/b \leq 1$, the capacitance between the plates is $C_l \approx \dfrac{\pi\varepsilon}{ln(4d/b)}$. The width of the side slotted radiator is equal to h, the distance between the sides is equal to W. This means that the capacitance between the plates is $C_l \approx \dfrac{\pi\varepsilon}{ln(4W/h)} = 0.613\varepsilon$, i.e., the input impedance of an asymmetric metal radiator, identical to the slot in shape and dimensions, is $Z_S = 195.8\pi$. The product of asymmetric impedances is $Z_E Z_S \cong (23\pi)^2$. For a symmetrical antenna this product is 4 times larger, i.e., it is equal to $(46\pi)^2$. As is known, the product of the input impedances of self-complementary antennas is $(60\pi)^2$. In the given case, this magnitude is significantly smaller, since the surface on which they are located cannot be called smooth.

The symmetric structure consisting of a microstrip antenna and its mirror image is similar to the structure of a complementary antenna of two coaxial unclosed shells of the same radius (Fig. 4.13b) separated by angles 2φ. The capacity between them per unit length of shells is equal to (see [30])

$$C_l = \varepsilon \frac{K(\sqrt{1-k^2})}{K(k)}, \qquad (4.32)$$

where $K(k)$ is the complete elliptic integral of the first kind with the argument $k = \tan^2\dfrac{\varphi}{2}$. Let's compare this structure with the structure of the microstrip antenna. For the reduced dimensions of the microstrip antenna, assuming that the structure radius is equal to $R = W/2$ and comparing the angle 2φ with the angle πR, we find: $\dfrac{2\varphi}{\pi} = \dfrac{2h}{\pi W/2}$, i.e., $\dfrac{\varphi}{2} = \dfrac{h}{W} = 0.024$. In accordance with (4.32) we get: $k = 0.00056$, $\sqrt{1-k^2} = 0.9999998$, $K(k) = 1.5708$, $K(\sqrt{1-k^2}) = 8.87$, i.e., the capacitance between shells is $C_{l1} = 5.7\varepsilon$. Replacing the angular width 2φ of the slot with the angular width of the shell (it is equal to $2\alpha = \pi - 2\varphi$), we get: $\dfrac{\alpha}{2} = 0.7616$, $k = 0.909$, $K(k) = 2.3217$, $\sqrt{1-k^2} = 0.416$, $K(\sqrt{1-k^2}) = 1.6464$, i.e., capacitance between the strips-slots is $C_{l2} = 0.71\varepsilon$. Input impedances are equal respectively to $Z_E = 21.05\pi$, $Z_S = 169\pi$. The product of the input impedances is $Z_E Z_S = (60\pi)^2$—in full agreement with the properties of self-complementary structures. The difference between the calculation results for the structures shown in Fig. 4.14a and Fig. 4.14b is, of course, connected primarily with the difference in the distances between the arms of the metal radiator.

A few words must be said about the resistance of a microstrip antenna. Such an antenna has a small length and a sinusoidal current. Its resistance

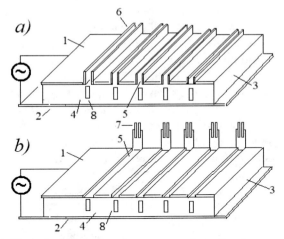

Fig. 4.14. The design of the microstrip antenna: a – first option, b – second option.
1 – metallic strip, 2 – ground plane, 3 – substrate, 4 – side of a substrate, 5 – transverse slot, 6 – metal selvage, 7 – capacitor, 8 – transverse strip.

increases at the expense of a difference from the unit dielectric constant of an adjacent substrate located one hand away from the radiator and the correspond growth of a propagation constant. But this increase is relatively small.

Presented results allow us to analyze the properties of microstrip antennas and to suggest methods for improving these characteristics. The main disadvantage of a microstrip antenna is a low efficiency and a narrow frequency range. A small thickness of the substrate leads to reflections from its boundaries and impairs matching with a signal source (generator or cable). Losses in the substrate reduce an antenna's efficiency.

Despite separate structural elements that distinguish microstrip antennas from other radiators, and peculiarities of the electrical characteristics associated with these elements, a sober look at the operation of these antennas shows that in principle they do not differ from ordinary radiators. And therefore, to improve their performance, we should use conventional methods permitting to improve performances of the antennas. The current along such an antenna is distributed along a sinusoid, as along a conventional linear radiator, and the increase of the propagation constant caused by the substrate does not change the distribution law. (It is known that the structure of a field created by a wire located at the boundary of two media with different dielectric constants coincides with the field structure created by the wire in a homogeneous medium, whose dielectric constant is equal to the arithmetic average of the dielectric constants of both media.)

The operating range of a linear antenna is bounded above by the frequency, at which the length of the segment where the current flows in antiphase is comparable with the length of the main part of the antenna with a current close to the in-phase one. At higher frequencies, the level of matching of the antenna with the signal source drops. Other electrical characteristics (a directional pattern and a gain) also deteriorate sharply. A linear antenna with a sinusoidal current distribution is equivalent to a resonant circuit operating in a narrow frequency range. To expand the frequency range of such a structure, it is necessary to lower its Q (a quality factor), i.e., to increase radiation and loss resistance. In the case of a microstrip antenna, this method can be implemented using a two-story structure, in which the upper antenna (a second floor) connected to the main floor (first one), providing additional radiation, allows to partially extend the operating range (see, for example, [69]). But a significant expansion of the operating range is possible only by means of creating an in-phase current distribution in the linear radiator with the help of capacitive loads included along its length, as is described in detail in Chapter 3. This method is also applicable to the flat and volume antennas.

Section 3.5 compares the input characteristics of thin metal monopoles with sinusoidal and in-phase currents. The amplitudes of in-phase currents fall off toward the ends of the antenna according to a linear law. This type of current distribution is created by the capacitive loads included along the antenna axis. Constant loads allow increasing the operating range several times with the required level of matching. Tunable loads significantly improve this result.

An example of using this method with reference to flat antennas is given in Section 3.6. In Fig. 3.24 two options of a flat self-complementary antenna are presented. Capacitive loads in these antennas are made in the form of transverse slots in flat metal radiators. Inductive loads in the slot antennas in accordance with the principle of duality are made in the form of transverse metal strips.

A similar method can be applied to other antennas, microstrip antennas, in particular. Two options of the microstrip antennas with capacitive and inductive loads are shown in Fig. 4.14. They differ in the fact that the capacitive loads are installed on a metal plate 1 along the axis of the antenna in the form of transverse slots 5, and the inductive loads are installed on the sides 4 of substrates in the form of metal strips 8. The capacitances of loads can be increased by means of metal selvages 6 (see Fig. 4.14a) or concentrated capacitors 7 (see Fig. 4.14b). The inductive loads 8 are included in slotted antennas in accordance with the principle of duality. Capacitances and inductances of loads are calculated using certain expressions, which allow the creation e of an in-phase electric current in a metal plate, and an in-phase magnetic current along the side faces of the substrate.

It follows from Fig. 4.14, that this radiating structure consists of two radiators: electrical and magnetic. Characteristics of such structures are close

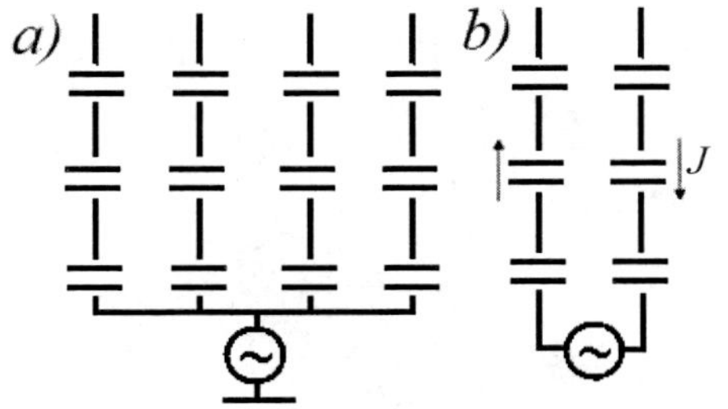

Fig. 4.15. Equivalent scheme of a microstrip antenna with capacitors: a – front view, b – side view.

to the characteristics of complementary antennas. Either radiator has two arms. In particular the first arm of the electric radiator in particular is the microstrip antenna with loads. An equivalent circuit of the first arm is shown in Fig. 4.15: a – front view, b – side view. As can be seen from the figure, the metal plate with the width W is replaced by several equivalent radiators, whose length is equal to the length of the plate L. The second arm of the electric radiator is its mirror image in the ground. The second arm is located in parallel to the first one, and currents along the radiators' arms have opposite directions.

In accordance with the results of calculating characteristics of flat self-complementary antennas with in-phase currents (this antenna is shown in Fig. 3.24a) one can consider that an inclination of the radiator in the vertical plane of the antenna does not affect its characteristics D and PF. Figure 3.26, drawn according to these results, demonstrates the characteristics of a vertical radiator depending on its height. In a wider frequency range, these characteristics, as well as the characteristics of a vertical radiator with a sinusoidal current, are shown in Fig. 3.22. The number of such radiators (a multiplier of the equivalent array) depends on the width of a flat antenna. If this width is small, phases of all equivalent radiators are the same, and the total field in a direction perpendicular to the metal plane on which the antenna is installed is proportional to their number.

The calculation results given in Section 3.6 demonstrate the absolute superiority of antenna structures with capacitors and in-phase currents.

Increasing the width of the microstrip antenna and extending its operating range makes it necessary to increase the level of matching of the antenna with the feed line. To do this, we need to create a smooth conical transition from a narrow strip line to a wider antenna, which simultaneously provides a smooth transition between the wave impedances of antenna and line. As shown in [61], the wave impedance of a triangular plate with an angle α at the apex connecting the antenna to the central conductor of a cable (see, Fig. 4.16) is equal to

$$W = 60\pi K(n)K(\sqrt{1-n^2}), \qquad (4.33)$$

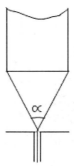

Fig. 4.16. Conical transition between antenna and cable.

Table 4.1. Wave impedance of a metal triangle.

$a°$	80	70	60	50	45	40	30	20	10
W, Ohm	47	59	74	87	94	102	120	146	188

where $K(n)$ and $K(\sqrt{1-n^2})$ are the complete elliptic integrals of the first kind from the arguments n and $\sqrt{1-n^2}$ where, $n = tan^2\left(\dfrac{\pi}{4} - \dfrac{\alpha}{2}\right)$. This wave impedance for different values of angle α (in degrees) is given in Table 4.1. It must be equal to the wave impedance of the cable.

Let us return to the directivity of the microstrip antenna, whose equivalent circuit is shown in Fig. 4.15. It is a line array of radiators, which at a large distance creates a system of fields of the same magnitude with the phases depending on pathlength difference in an arbitrary direction. The total electromagnetic field is a wave with a spherical front, whose center coincides with the field of a middle radiator. The ratio of the total field to the field of an isolated radiator, as is known, is named by a multiplier of the array and is equal to

$$F_N = \sin\left(\dfrac{Nkd}{2}\sin\psi\right) \Big/ \sin\left(\dfrac{kd}{2}\sin\psi\right), \tag{4.34}$$

where N is the number of radiators, d is the distance between adjacent radiators, k is the propagation constant, ψ is the angle between the direction to an observation point and the normal to the array. It is easy to verify that for a small angle ψ the magnitude of F_N is close to N. When $N \gg 1$, the magnitude of first side lobe of the directional pattern is equal to the pattern maximum multiplied by $k_1 \approx \dfrac{2}{3\pi}$, and the magnitude of a lobe p is equal to the pattern maximum multiplied by $k_p \approx \dfrac{2}{(2p+1)\pi}$.

The multiplier of equivalent array shown in Fig. 4.15 . The directivity of an isolated radiator in the frequency range from $kL = \pi$ to $kL = 10\pi$ increases from 1 to 12 (see Fig. 3.22). This means that the directivity of the antenna as a whole grows from 4 to 50. That allows to speak of a high level of obtained characteristics.

4.4 Reflector arrays of microstrip antennas

Lately, reflector arrays have become widespread in the capacity of a flat equivalent of a parabolic reflector. Microstrip antennas can act as re-radiators in such an array. A simplified model of a microstrip antenna in the receive mode is shown in Fig. 4.17a, and an equivalent circuit of the antenna is given in Fig. 4.17b.

An input impedance of a planar electric radiator excited at its center is equal to

$$Z_A = R_\Sigma - jW \cos\frac{k_1 L}{2}, \qquad (4.35)$$

where R_Σ is the resistance of radiation, W is the wave impedance of the planar radiator. Z_L in Fig. 4.17 is the load impedance.

The reflect array, in which microstrip antennas are used as re-radiators, creates a reflected field with a phase depending on the circuit of the reradiating antenna. The phase increment of the reflected field relative to the phase of the incident field is determined in accordance with the reciprocity theorem. This theorem argues that the current in the receiving antenna and all characteristics of the antenna can be found, if the characteristics of the antenna used in transmission mode are known. Applicability of this theorem for the reflected arrays was first substantiated in [22].

Suppose that any two antennas are remote from each other by a distance such that their mutual impedance is zero (Fig. 4.18). This condition is accepted

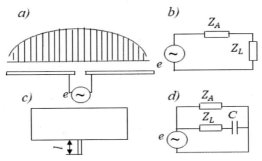

Fig. 4.17. Microstrip antenna in the receive mode (*a*) and its equivalent circuits with different loads (*b, c, d*).

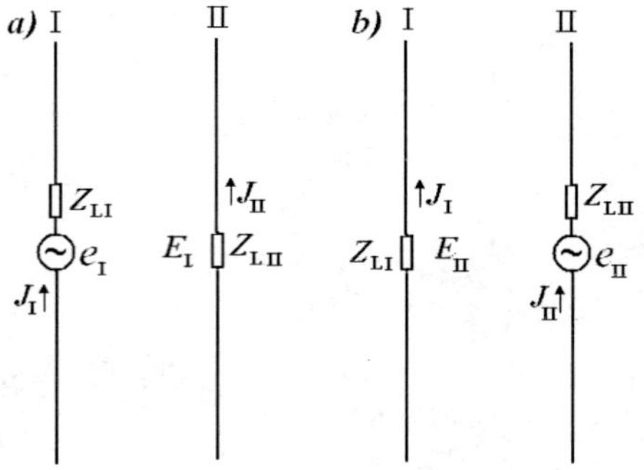

Fig. 4.18. The reciprocity theorem: a – antenna I radiates, b – antenna II radiates.

for the sake of simplicity of proof and has no principle significance. The load Z_L is connected in the middle of each antenna. If emf e_I creates a current $J_I = \dfrac{e_I}{Z_{AI} + Z_{LI}}$ at the input of the first antenna, then the field $E_I = -j\dfrac{30 J_I}{\varepsilon_r} F_I$ arises near the antenna II and creates the current J_{II} in this antenna (see Fig. 4.18a). In a similar case, when emf e_{II} is connected in the second antenna it creates the current J_{II} in it and the current J_I is produced in the first antenna (see Fig. 4.18b).

In accordance with the electrodynamic principle of reciprocity

$$e_I / J_{II} = e_{II} / J_I. \qquad (4.36)$$

From these expressions, which are given for the first antenna, it follows that

$$\frac{e_I}{J_{II}} = \frac{J_I (Z_{AI} + Z_{LI})}{J_{II}} = j \frac{\varepsilon_r E_{II}(Z_{AI} + Z_{LI})}{30 F_I J_{II}}$$

Similarly, for the second antenna

$$\frac{e_{II}}{J_I} = j \frac{\varepsilon_r E_I (Z_{AII} + Z_{LII})}{30 F_{II} J_I},$$

i.e.,

$$\frac{E_{II}(Z_{AI} + Z_{LI})}{F_I J_{II}} = \frac{E_I (Z_{AII} + Z_{LII})}{F_{II} J_I}.$$

If to transfer the magnitudes relating to each antenna into one part of the equation and to assume that each part of the equation is a constant, independent of the properties of the antenna, we get:

$$\frac{J_i (Z_{Ai} + Z_{Li})}{E_i F_i} = \text{const.} \qquad (4.37)$$

If we assume that this expression refers to the receiving antenna, then J_i is the current in the middle of this antenna, E_i is the field strength near it, Z_{Ai} and Z_{Li} are its input impedance and load, F_i is a function characterizing the effective length and the shape of the directional pattern of the antenna in the transmit mode.

The constant on the right side of expression (4.37) can be determined by considering the simple electrical dipole (Hertz' dipole) as an antenna. If the axis of receiving dipole lies in the plane of a an incident wave, this constant is equal to 1. In a general case, it is necessary to take into account the polarization of the field and the azimuth, if the horizontal directional pattern of the antenna differs from a circular one. As a result, we obtain in accordance with (4.37) the current in the middle of the receiving antenna equal to,

$$J_i = \frac{E_i F_i(\varphi)}{Z_{Ai} + Z_{Li}} = \frac{e_i}{Z_{Ai} + Z_{Li}}. \qquad (4.38)$$

The reciprocity theorem shows that the main characteristics of the antenna (input impedance, effective length, directional pattern) coincide for transmit and receive modes.

This theorem allows also to analyze a relationship between the field incident onto an array element (onto one radiator) and the field reflected from that element. It is known that when a wave is incident on a flat perfectly conducting metallic surface it is reflected at an angle equal to an angle of incidence. Amplitudes of the incident and reflected fields are identical, and the wave phase changes in a stepwise fashion by π. However, when a metal surface is replaced with a system of radiators, e.g., when it is replaced with a linear array, direction and phase of the reflected field could be essentially be changed, because they depend on parameters and electric characteristics of an isolated radiator.

An example of such structure is the in-phase reflector array (Fig. 4.19a). It is a flat equivalent of a parabolic reflector. The structure consists of primary exciter 1 of antenna array (e.g., a horn) and an equally spaced array of secondary microstrip radiators 2, situated in one plane along surface 3. In order to sum the signals of secondary radiators in a direction, perpendicular to the array plane, their phases should be identical. Since distances r_i (i is the radiator number) between the primary exciter and an arbitrary reradiator are not identical, this results in a phase path difference, which should be compensated with a phase shift in the reradiating signal.

The method of calculating the phase step of the reflected field in comparison with the incident field phase in the reradiating signal can be constructed on the basis of the reciprocity theorem. The reciprocity theorem for two antennas is described above in the form of (4.36). In [16] the theorem is formulated as

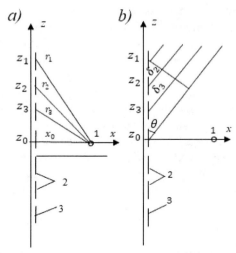

Fig. 4.19. Reflector array with the radiation direction perpendicular to the array plane (*a*) and in any desired direction (*b*).
1 – primary exciter, 2 – secondary microstrip radiators, 3 – surface of radiators' placement.

follows: if emf e_I applied to the terminals of antenna I establishes current J_{II} at the input of an antenna II, then an equal emf e_{II} applied to the terminals of an antenna II will create at the input of the antenna I the same current J_I, i.e.,

$$J_{II}/e_I = Y_{I,II} = Y_{II,I} = J_I/e_{II}, \qquad (4.39)$$

where $Y_{I,II} = |Y_{I,II}| \exp(j\varphi_{I,II})$ and $Y_{ba} = |Y_{II,I}| \exp(j\varphi_{II,I})$ are the mutual admittances between antennas. From this it follows that

$$\varphi_{I,II} = \varphi_{II,I}, \qquad (4.40)$$

i.e., the difference of phases between the exciting emf and the current excited in an adjacent antenna is the same in both cases.

In our example the field E_I acts as a signal source instead of the antenna I with emf e_I. Let's introduce into the circuit in the capacity of a source of field the antenna I, which is excited by the generator e_I with infinitely high output resistance, and as usual take into account that a linear antenna is the aggregate of elementary dipoles with appropriate currents. In this case, because the phase of the radiated field outstrips the phases of current by $\pi/2$, the phase difference between the current J_{II} and emf e_I is (see Fig. 4.18a)

$$\varphi_{I,II} = \varphi_{11} + \varphi_{12} + \varphi_{13} + \varphi_{14}, \qquad (4.41)$$

where φ_{11} is the phase shift between the current in the antenna I and emf e_I (in this case it is absent), φ_{12} is the phase shift between the radiated field E_I and the current of antenna I (it is $\pi/2$), φ_{13} is the phase shift due to the distance between antennas, φ_{14} is the phase difference between current J_{II} and field E_I, that is

$$\varphi_{14} = \varphi_{I,II} - \pi/2 - \varphi_{13}. \tag{4.42}$$

The other source of a signal is the current in antenna II (rather than emf e_{II}), which creates the reflected field E_{II}. The distribution of a current along the receiving antenna differs from a distribution along a transmitting antenna. The antenna II is an aggregate of elementary dipoles, each of which is excited by its generator. The currents of the dipoles create in-phase fields. Let the current in the middle dipole (at the antenna center) be excited by the generator e_{II}; it is equal to the product of emf e_{II} and the dipole admittance. Since the dipole impedance is capacitive, the current phase outstrips the emf phase by $\pi/2$. This assertion holds true for other dipoles of the antenna II. Accordingly, the phase difference between current J_I and emf e_{II} is equal (see Fig. 4.18b)

$$\varphi_{II,I} = \varphi_{21} + \varphi_{22} + \varphi_{23} + \varphi_{24}, \tag{4.43}$$

where φ_{21} is the phase shift between the current in antenna II and emf e_{II} (in this case it is $\pi/2$), φ_{22} is the phase difference between field E_{II} and the current in antenna II, φ_{23} is the phase shift due to the distance between antennas (it is φ_{13}), φ_{24} is the phase shift between the reflected field E_{II} and current J_I (it is zero, since the input impedance of antenna I is infinitely large), that is

$$\varphi_{22} = \varphi_{II,I} - \pi/2 - \varphi_{13}. \tag{4.44}$$

In accordance with the equation (4.42) this implies that the increment of the phase of the receiving antenna current compared to the phase of the incident field and the increment of the phase of the reflected field compared with the phase of the receiving antenna current are the same. As for the amplitude of incident field $|E_1|$ and the amplitude of reflected field $|E_2|$, at the reflection point, then,

$$|E_2| = |E_1| \cos \gamma \cos \delta, \tag{4.45}$$

because the total tangential component of both fields on a perfectly conducting metal surface is zero.

Here γ is the angle of incidence of the ray incident on the antenna, and δ is the angle of reflection.

Let us apply the reciprocity theorem to the reflector array, consisting of microstrip antennas. The amplitude and phase of the current created in a re-radiator depends on the impedance of load. When the load is absent, the current in the antenna is equal to

$$I_A = e/Z_A = e/|Z_A| \exp\{j \cot^{-1}[(W/R_\Sigma) \cot(kL_1/2)]\}. \tag{4.46}$$

Magnitude of $\cot^{-1}[(W_A/R_\Sigma)\cot(kL/2)]$ is the phase increment of the current in the antenna relative to the phase of the incident field. In accordance with the reciprocity theorem, this increment is equal to the phase increment of the reflected field relative to the phase of the antenna current. Hence, the phase step at reradiation is

$$\varphi_1 = 2\tan^{-1}[(W/R_\Sigma)\cot(kL_1/2)]. \tag{4.47}$$

The value of the step is zero for a tuned antenna, negative for an elongated antenna, and positive for a shortened antenna. Increasing the radius a of the dipole lowers its wave impedance W and decreases the phase step (Fig. 4.20a).

The approximate method for calculating the phase step of the reradiated field, based on the reciprocity theorem, is simple and efficient. After only a slight alteration, this method can be used to analyze antenna with a load. For the capacitive load C,

$$I_A = \frac{e}{Z_A - (j/\omega C)} = e\Big/\sqrt{|Z_A| + (1/\omega C)^2}\exp\left\{j\tan^{-1}\left[\frac{1}{R_\Sigma}\left(\frac{1}{\omega C} + W\cot\frac{kL_1}{2}\right)\right]\right\},$$

i.e., the phase step at the reradiating stage is

$$\varphi_2 = 2\tan^{-1}\left[\frac{1}{R_\Sigma}\left(\frac{1}{\omega C} + W\cot\frac{kL_1}{2}\right)\right]. \tag{4.48}$$

Here ω is the circular frequency of the signal. In the tuned antenna, if the load is fabricated in the form of a dissipative stub (Fig. 4.17c),

$$I_A = e(Z_A + Z_1)/Z_A Z_1 = e\sqrt{|Z_1|^2 + R_\Sigma^2}/(R_\Sigma|Z_1|)\exp\{j\tan^{-1}[R_\Sigma/W\tan k_1 L]\}.$$

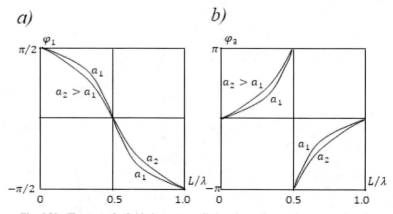

Fig. 4.20. The step of a field phase at reradiating depending on the antenna length.

the phase step at reradiating is

$$\varphi_3 = 2\tan^{-1}[R_\Sigma/(W \tan k_1 L] \text{ (see Fig. 4.20}b).$$

Similarly, for Fig. 4.17d

$$\varphi_4 = 2\tan^{-1}\left[R_\Sigma/W\left(\frac{1}{\omega C} + W \cot k_1 L\right)\right].$$

One can apply the above results to the calculation of the necessary phase increment of a microstrip antenna at the reradiating stage. If the signals' phases of secondary radiators are equal, then the signals are added in the direction perpendicular to the array plane.

Let the coordinates of the primary source be x_0, y_0 and z_0 (see Fig. 4.19a). Then the phase increment in the reradiator i, needs to compensate the phase difference in the direction of the x-axis, and must be equal to $\xi_i = k[\sqrt{x_0^2 + (y_i - y_0)^2 + (z_i - z_0)^2} - x_0]$. The choice of geometric dimensions of the reradiator i allows us to obtain phase step $\varphi_{1i} = \xi_i$.

In Fig. 4.21 the field phase increment φ_1 created at 60 GHz by a microstrip antenna situated on a substrate with a thickness 0.254 mm and a relative permittivity of 2.22 is given as a function of the antenna length. The results were obtained by means of the proposed technique. Curves 1 and 2 correspond to antenna widths of 0.3 and 2.3 mm. The rigorous method for calculating the step magnitude was described in [70]. It relies on the analysis of a plane wave

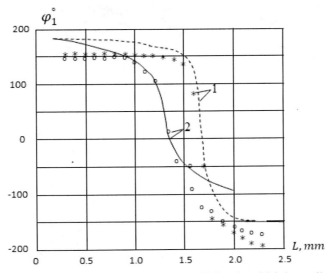

Fig. 4.21. Phase increment φ_1 of microstrip antennas with lengths of 0.3 (curve 1) and 2.3 mm (curve 2).

incident to an infinite periodic array from identical elements, i.e., on solving the analysis problem in the spectral domain and on the Floquet's theorem. The open and closed circles in Fig. 4.21 correspond to the results presented in [70]. As seen from the figure, the correspondence is satisfactory.

Along with simple microstrip antennas, two-story microstrip antennas, permitting to expand the frequency range of reflection arrays, can be used. The field phase increment in a simple antenna at reradiating stage is less than 360° (in the frequency range the smooth phase increment in the case of a sufficiently thick substrate on changing strip length in a dependence from a frequency does not exceed 300°). The maximal phase step, attainable, e.g., in a two-story antenna, is 540°.

In a two-story microstrip antenna (Fig. 4.22) two rectangular metal plates are separated from each other and from the metal plane by a dielectric substrate. The upper plate is smaller than the lower one. Characteristics of a multiple-story antenna can be calculated by the method described above. The phase increment of the field created at 12.5 GHz by a two-story microstrip antenna containing substrates of a thickness of 3 mm with a relative permittivity of 1.03 is shown in Fig. 4.22 as a function of antenna length. The antenna is a square. The calculating curves were obtained with the use of the proposed technique. The open and closed circles in the figure demonstrate the results presented in [71].

In order to get the maximal signal of an antenna array in the prescribed direction, the phases of the radiators' fields in this direction must be equal. Therefore, if this angle varies in the plane xOz, the phases must vary linearly in the dependence of the coordinate z_i: $\psi_i = k(z_1 - z_i) \cos \theta$ (see Fig. 4.19b).

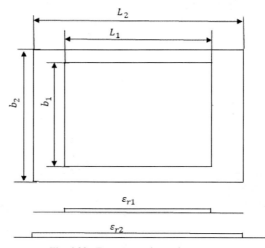

Fig. 4.22. Two-story microstrip antenna.

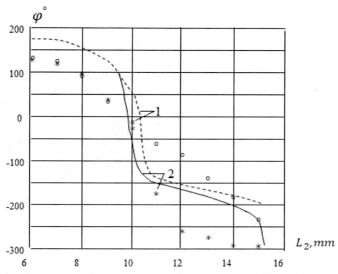

Fig. 4.23. Phase increment φ of a two-story microstrip antennas with $L_1/L_2 = 0.6$ (curve and points 1) and 0.8 (curve and points 2).

To simplify the control procedure, the field phases should be controlled by an electric signal.

The circuits of microstrip antennas with loads are best suitable for continuous control of a phase of a reradiated field. In order to include the load in series into the circuit of a receiving antenna, a slot can be cut in the central part of the planar radiator across its axis. Such a slot filled with a dielectric creates the simplest load in the form of a capacitor with a capacitance C. The capacitance of this load can be controlled easily by placing a special material between the capacitor plates whose permittivity depends on the voltage. In this case the phase increment at reradiation is determined by expression (4.48). The tangent of angle φ_2 depends on two summands. The first summand, $\alpha = 1/(R_\Sigma \omega C)$ is caused by the presence of the capacitor in the radiator center while the second summand $\beta = (W/R_\Sigma)\cot(k_1 L/2)$ is obtained because of the different distances between the separate radiators and the point of observation. The second term can be used to compensate the phase difference caused by the difference in distances r_i between primary radiator 1 and reradiators 2 of the array (see Fig. 4.19a), while the first term can be used for turning the radiation pattern. The total phase increment in the reradiator i should be equal to $\varphi_{2i} = \xi_i + \psi_i$. Here, if reactance $1/(j\omega C_i)$ of capacitors C_i, corresponding to $\varepsilon_r = 1$, varies linearly with coordinate z_i, then, applying equal voltages to the capacitors filled with the same dielectric, we find that the reactance of these capacitors retain a linear dependence on this coordinate. This statement is equally true for the first term in establishing the angle of turning.

Note, however, that the second term takes different values for different re-radiators because they depend on distance r_i. Therefore, phase φ_2 will vary nonlinearly if the same voltage is applied to all capacitors. If the second term is absent, then a linear dependence of phase ψ_i on the coordinate z_i is required for the angular turning the radiation pattern. The value $\tan \psi_i = 1/(R_\Sigma \omega C_i)$ and, accordingly, capacitance C_i does not possess this property. This means that in the general case accordingly, capacitance C_i does not possess this property. Therefore, in the general case, the angular displacement of the radiation pattern requires application of an individual voltage to each capacitor.

The case when angle θ of the maximum radiation of the antenna array is close to $\pi/2$, i.e., $\alpha \ll \beta$, requires a special analysis. In this case, by expanding the function $\varphi_2 = \tan^{-1}(\alpha + \beta)$ into the Taylor series, we obtain

$$\varphi_2(\alpha + \beta) = \tan^{-1}\beta + \alpha/(1 + \beta^2). \tag{4.49}$$

Here, the member $\dfrac{\alpha}{1+\beta^2} = \left[R_\Sigma \omega C \left(1 + \dfrac{1}{R_\Sigma^2} W^2 \cot^2 \dfrac{k_1 L}{2} \right) \right]^{-1}$ is proportional to reactance $1/(j\omega C)$ of the capacitor C. Hence, phases of the fields created by secondary radiators will vary linearly with the coordinate upon application of equal voltages to the capacitors filled with the same dielectric. If the direction of the maximum radiation differs substantially from the perpendicular to the array plane, different voltages can be applied to several groups of antennas in order to bring the law of variation of phase φ_2 along the antenna closer to being linear. The number of voltages can be substantially less than the total number of radiators.

On performing the calculation of the capacitance formed by a slot cut in the plate of a microstrip antenna, it should be noted that this capacitance consists of two terms: capacitance between thin planar plates and capacitance of the planar capacitor between the plate edges.

From the above it follows that the use of capacitive loads in microstrip antennas can play an important role in the control of flat reflected arrays. No lesser important role, can be played by capacitive loads located across the axis of the microstrip antenna in order to expand its frequency range, as shown in Section 4.2.

4.5 Calculating directivity on the basis of main directional patterns

In this section, it is useful to consider the issue of calculating directivity of antennas and antenna arrays. The calculating method is based on measurement results of main directional patterns. These calculations in the case of intricate directional patterns often cause serious difficulties.

As is well known, directivity D is one of the basic electrical performances of any antenna. An antenna gain G is equal to

$$G = D\eta, \qquad (4.50)$$

where η is an efficiency. A knowledge of these magnitudes allows planning an improvement of antenna characteristics. The value of G can be defined by direct measurements. As regards magnitudes of D and η, it is very difficult to measure them or to interpret the measurement results. For example, for evaluating D it is necessary to know the three-dimensional directional patterns of an antenna. But as a rule, these patterns are measured only in two main planes: horizontal and vertical. The calculation difficulties are increased with a decreasing cross-section of the main lobe, i.e., with increasing directivity, caused for example by increased numbers of radiators in the antenna array.

Method of the directivity calculation of the narrow-beam antenna is described in [72]. It is based on the method proposed in [73] for calculating intermediate values in the measured directional pattern.

Usually it is regarded that the magnitude of the directional antenna pattern in an arbitrary direction is equal to

$$F(\theta, \varphi) = F_1(\theta) F_2(\varphi), \qquad (4.51)$$

where $F_1(\theta)$ is a directional pattern in a vertical plane xOz, $F_2(\varphi)$ is a directional pattern in a horizontal plane xOy, θ and φ are angles in a spherical coordinates system, and x, y and z form a rectangular coordinate system (Fig. 4.24). The calculations show, that the expression (4.51) is true only in a small area, limited by a main lobe of a directional pattern.

The proposed method in [73] proceeds by revealing a curve, which is a locus of points with an identical signal magnitude (with an identical

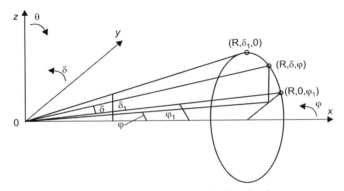

Fig. 4.24. The coordinate system and the directional pattern.

directivity). In this article the angle $\delta = \frac{\pi}{2} - \theta$ is used instead of the angle θ. It follows that the magnitude of the directional pattern is equal to

$$f(\delta, \varphi) = f(\delta_1, 0) = f(0, \varphi_1), \qquad (4.52)$$

where δ_1 and φ_1 are values of coordinates δ and φ at the points of intersection of the mentioned curve with planes xOz ($\varphi = 0$) and xOy ($\delta = 0$) respectively (see Fig. 4.24).

Assume that the direction of the maximum radiation coincides with the x-axis, and the directional pattern is symmetric about the planes xOz and xOy (Fig. 4.25). For example, in-phase array, located in a plane yOz, has such a directional pattern. In this case curves with identical directivity in the first approximation will be have the form of circumferences or ellipses:

$$f(\delta_1, 0) = f_1(\delta_1), f(0, \varphi_1) = f_2(\varphi_1). \qquad (4.53)$$

If for convenience we introduce a new angular coordinate β, measured from axis x (see Fig. 4.25), then, as it is easy to show, the new and the old coordinates will be connected among themselves by the relation:

$$\beta = \cos^{-1}(\cos \delta \cos \varphi). \qquad (4.54)$$

In the case when curves with identical directivity have the form of a circumference, i.e., when the main lobe of the three-dimensional directional pattern has circular symmetry, then

$$\delta_1 = \varphi_1 = \beta = \cos^{-1}(\cos \delta \cos \varphi). \qquad (4.55)$$

More often these curves are ellipses. Let a_1 be the length of its vertical axis, i.e., the length of the segment between the upper and the lower points of the directional pattern, corresponding to the same signal magnitude (to the given magnitude of the directional pattern). In the same way a_2 is the length of its horizontal axis, i.e., the length of segment between the left and right points

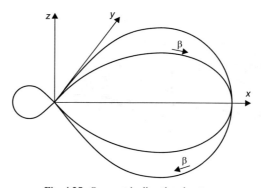

Fig. 4.25. Symmetric directional pattern.

of the directional pattern corresponding to the given magnitude of a signal. Denote the relationship of lengths a_1 and a_2 by $a = a_1/a_2$. Then at $a < 1$

$$\delta_1 = \cos^{-1}(\cos\delta\cos a\varphi), \quad \varphi_1 = \frac{1}{a}\cos^{-1}(\cos\delta\cos a\varphi), \quad (4.56)$$

and at $a > 1$

$$\delta_1 = a\cos^{-1}\left(\cos\frac{\delta}{a}\cos\varphi\right), \quad \varphi_1 = \cos^{-1}\left(\cos\frac{\delta}{a}\cos\varphi\right). \quad (4.57)$$

As is known, maximal directivity of an antenna with the pattern, symmetrical about planes xOz and xOy, is equal to

$$D = \pi / \int_0^\pi \int_0^{\pi/2} f(\delta,\varphi)\cos\delta\, d\delta d\varphi, \quad (4.58)$$

whence

$$D = \pi / \left[\int_0^{\pi/2}\int_0^{\pi/2} f(\delta,\varphi)\cos\delta\, d\delta d\varphi + \int_0^{\pi/2}\int_0^{\pi/2} f(\delta,\psi)\cos\delta\, d\delta d\psi\right]. \quad (4.59)$$

The first summand of the denominator corresponds to a forward half-space, the second summand—to a backward half-space. Here in the second integral the change of variable $\varphi = \pi - \psi$ is performed. At $a < 1$ in accordance with (4.52), (4.53) and (4.56)

$$f(\delta, \varphi) = f_1(\delta_1) = f_1[\cos^{-1}(\cos\delta\cos a\varphi)]. \quad (4.60)$$

At $a > 1$ according to (4.52), (4.53) and (4.57) one can obtain a similar expression. The expression (4.59) in view of (4.60) allows calculating antenna directivity, if its directional patterns are given in main planes.

In Fig. 4.26 the experimental directional pattern of a planar uniform antenna array with in-phase excitation is given at a frequency of 3.4 MHz. As one can see from the figure, the factor a is equal to 1 in intervals 160–180° and 135–145°, to 0.63 in an interval 145–160° and to 0.69 in an interval from 90° to 120° along an azimuthal angle. It means that the main lobe of the three-dimensional pattern has circular symmetry, i.e., the points, located on the main lobe, have a signal of identical magnitude, and the locus of these points is a circumference. Such circular symmetry on some side lobes is absent, and these circumstances should be taken into account for increasing calculation accuracy. In an interval 120–135° the factor a is greater than 1.

Calculation in accordance with the described procedure, gives the magnitude of directivity, equal to 18.6 dB if we take into account intervals with $a = 1$. If we also take into account the interval 120–135° with $a > 1$, we obtain an outcome of 18.2 dB. The measured antenna gain is equal to 18 dB. Thus, one must admit a good conformity of calculated and experimental results.

228   ANTENNAS: Rigorous Methods of Analysis and Synthesis

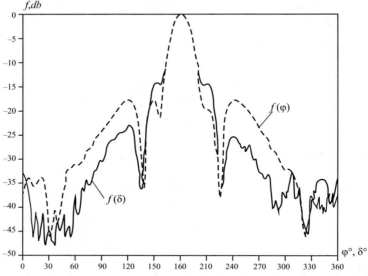

Fig. 4.26. Experimental directional patterns of antenna.

For antennas with one narrow major lobe and small minor lobes the theory ([15]) recommends expression

$$D = 41253/(\theta_1\theta_2), \tag{4.61}$$

where θ_1 and θ_2 are half-power beam widths of the directional pattern (in degrees) in two mutually perpendicular planes. For planar arrays a better approximation is

$$D = 32400/(\theta_1\theta_2). \tag{4.62}$$

The calculation in accordance with these formulas for the directional patterns, presented in Fig. 4.25, gives 20.2 dB and 19.2 dB, i.e., much greater error.

The directivity calculation program is uses two procedures: the procedure of antenna pattern estimation between intermediate angles and the procedure of integrals calculation by summation of numerical magnitudes. These methods can be used too for the solution of other problems, for example, for an estimation of increasing antenna directivity at the expense of decreasing side lobes.

Let is necessary to determine how the directivity will change as a result of diminution of side lobes, be necessary. One must first determine the initial value of the directivity in accordance with (4.60) and next reduce the side lobes to a given level f_0 and calculate the new directivity. For example, one can reduce the side lobes in a vertical plane in the range of angles from φ_{11} to

φ_{12} and in a horizontal plane in the range of angles from δ_{11} to δ_{12}. At $a < 1$ the new directivity is

$$D_1 = \pi \left\{ \int_0^\pi \int_0^{\pi/2} f_1(\delta_1) \cos\delta\, d\delta + \int_{\varphi_{11}}^{\varphi_{12}} \int_{\delta_{11}}^{\delta_{12}} [f_0 - f_1(\delta_1)] \cos\delta\, d\delta d\varphi \right\}^{-1}, \quad (4.63)$$

i.e.,

$$\frac{D}{D_1} = 1 + \left\{ \int_{\varphi_{11}}^{\varphi_{12}} \int_{\delta_{11}}^{\delta_{12}} [f_0 - f_1(\delta_1)] \cos\delta\, d\delta d\varphi \right\} \Big/ \left[\int_0^\pi \int_0^{\pi/2} f_1(\delta_1) \cos\delta\, d\delta \right] =$$
$$1 + D\left[\frac{f_0}{\pi}(\varphi_{12} - \varphi_{11})(\sin\delta_{12} - \delta_{11}) - \frac{1}{\pi} \int_{\varphi_{11}}^{\varphi_{12}} \int_{\delta_{11}}^{\delta_{12}} f_1(\delta_1) \cos\delta\, d\delta d\varphi \right]. \quad (4.64)$$

The program for the calculation of the magnitude of D was performed in MATLAB® and presented in [72].

4.6 Diversity reception

Cellular communications, as well as other forms of communication, suffer from various types of noise and interference. Diversity reception, i.e., mounting two receiving antennas at a finite distance from each other (at a distance of the order of several wavelengths) and the creation of a new signal from two received signals, is an effective measure of the struggle against interferences. As a result, the interference is completely absent or is relaxed substantially. This can be used on a large object, where the interference signal is drastically decreasing. In order to filter the interference, one can use the bridge circuit and the modulator (Fig. 4.27). Receiving antennas A_1 and A_2 are mounted at points 1 and 2 of the bridge.

Let the source of the useful signal be placed at some point P. The signals from this point come to antennas A_1 and A_2 with different phases, caused by the difference in distances. The magnitudes of the received signals with allowance

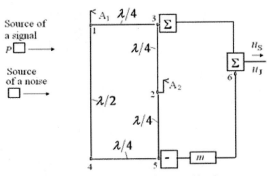

Fig. 4.27. Circuit of interferences filtration.

for changes in their receiving circuits can be written as $u_1 = S_1 \exp(j\varphi_1)$ and $u_2 = S_2 \exp(j\varphi_2)$ respectively. If the distance between the antennas is small, one can assume that the amplitudes of the received signals in the first approximation are the same.

These signals arrive to the point 3 of the bridge circuit in accordance with their phases and sum up in it as:

$$u_1 + u_2 = S \exp(j\varphi_1)[1 + \exp(j\Delta\varphi)], \qquad (4.65)$$

where $\Delta\varphi = \varphi_2 - \varphi_1$. The summary signal comes to an adder, located at the point 6. At point 5 of the bridge circuit the signal comes in anti-phase, as a difference, i.e., one signal is subtracted from the other:

$$u_2 - u_1 = S \exp(j\varphi_1)[\exp(j\Delta\varphi) - 1]. \qquad (4.66)$$

Further the signal difference passes through the modulator M, where it is multiplied by its gain $m = M \exp(j\varphi_m)$, and comes to the adder 6. The useful signal at the output of the adder 6 is

$$u_S = u_1 + u_2 + m(u_2 - u_1) = S \exp(j\varphi_1)[1 - m + (1 + m)\exp(j\Delta\varphi)]. \qquad (4.67)$$

A similar expression can be written for the interference signal. If the distance between the antennas A_1 and A_2 is small, the amplitudes of the interference signal in these antennas are respectively $u_3 = J \exp(j\psi_3)$, $u_4 = J \exp(j\psi_4)$. Interference at the output of the adder is

$$u_J = J \exp(j\psi_3)[1 - m + (1 + m)\exp(j\Delta\psi)], \qquad (4.68)$$

where $\Delta\psi = \psi_4 - \psi_3$. In order for the interference to be equal to zero at the output of the adder, it is necessary that the gain of the modulator be equal to

$$m = \frac{1 + \exp(j\Delta\psi)}{1 - \exp(j\Delta\psi)} = j \coth(\Delta\psi/2). \qquad (4.69)$$

From this expression it follows that the modulator has to change the phase of the transmitted signal. If the amplitudes of interfering signals at the antennas A_1 and A_2 are not the same, i.e., $u_3 = J_3 \exp(j\psi_3)$, $u_4 = J_4 \exp(j\psi_4)$, then assuming that $J_4/J_3 = \exp(\beta)$ we find:

$$u_J = J_3 \exp(j\psi_3)[1 - m + (1 + m)\exp(\beta + j\Delta\psi)],$$

whence

$$m = -\coth\frac{\beta + j\Delta\psi}{2} =$$
$$\frac{1}{\coth^2(\beta/2) + \coth^2(\psi/2)}[\coth(\beta/2)\operatorname{csch}^2(\psi/2) - j\coth(\psi/2)\operatorname{csch}^2(\beta/2)]. \qquad (4.70)$$

In this case, the modulator must change not only the phase but also the amplitude of the signal, passing through.

Accordingly, the useful signal at the output of the structure is equal to

$$u_S = S\exp(j\varphi_1)\left[1 + \coth\frac{\beta + j\Delta\psi}{2} + \left(1 - \coth\frac{\beta + j\Delta\psi}{2}\right)\exp(\alpha + j\Delta\varphi)\right], \quad (4.71)$$

where $\exp(\alpha) = J_2/J_1 = S_2/S_1$. Since $\coth x = [1 + \exp(-2x)]/[1 - \exp(-2x)]$, then

$$u_S = \frac{2S\exp(j\varphi_1)}{1 - \exp[-(\beta + j\Delta\psi)]}\{1 - \exp[\alpha - \beta + j(\Delta\varphi - \Delta\psi)]\}. \quad (4.72)$$

As one can be seen from expression (4.72), unlike the interference, which is equal to zero at the structure output, the useful signal is different from zero, if the value $\exp[\alpha - \beta + j(\Delta\varphi - \Delta\psi)]$ is not equal to 1.

The magnitude $\Delta\varphi$ in the expression (4.72) is the phase difference between useful signals received by different antennas. It is equal to $\Delta\varphi = kd\cos\gamma$, where $k = 2\pi/\lambda$, d is the distance between antennas, γ is the angle between the normal to the segment d and the direction to the useful signal source. Similarly, the magnitude $\Delta\psi$ in this expression is a phase difference $\Delta\psi = kd\cos\delta$ between interfering signals, received by different antennas (δ is the angle between the normal to the segment d and the direction towards the interference source). If the arrival angles of the signal and the interference are close to each other ($\delta \approx \gamma$), then

$$\Delta\varphi - \Delta\psi = kd(\cos\gamma - \cos\delta) \approx kd(\delta - \gamma)\sin\gamma.$$

Obviously, α and β also coincide. In this case,

$$u_S = \frac{2S\exp(j\varphi_1)}{1 - \exp[-(\beta + j\Delta\psi)]}\{1 - \exp[jkd(\delta - \gamma)\sin\gamma]\} = \frac{2jkd(\delta - \gamma)\sin\gamma S\exp(j\varphi_1)}{1 - \exp[-(\beta + j\Delta\psi)]}. \quad (4.73)$$

From (4.73) it follows that the useful signal at the output of the system at similar arrival angles of the signal and the interference is proportional to the difference between the angles of their arrival.

If the signal and the interference come from one and the same azimuth, it is necessary to separate the antennas for height. Otherwise, if the angle in the vertical plane between the directions of arrival of the signal and the noise does not exceed a few degrees with the same polarization, the reception is not possible.

Thus, application of this circuit together with interference attenuation leads to a weakening of the useful signal. Also, the areas of weakened reception may appear. Indeed, suppose that the antennas A_1 and A_2 are the same and have a circular directional pattern in the horizontal plane. The modulator gain in accordance with (4.69) is equal to $m = j\coth(\Delta\psi/2)$. If the phase difference between the interferences, received by the antenna, is equal to $\Delta\psi$, and the phase difference between the useful signals is $\Delta\varphi = \Delta\psi \pm 2n\pi$, where n is a natural number, then in accordance with (4.73) $u_S = 0$. However, if d is not too great, this does not take place. Suppose, for example, that $\Delta\psi = 0.3\pi$, $d \approx 0.35\lambda$, i.e., $kd = 0.7\pi$, $\delta = \cos^{-1} 0.43 = 0.36\pi$, $\gamma = \cos^{-1}(\pm 0.86n\pi + 0.36\pi)$. The absence of real γ values testifies about the absence of these zones.

It is necessary to say a few words about the impact of inaccurate adjustment of the modulator gain. If the gain differs from the optimum value m in magnitude and phase, for example, and is equal to

$$m_1 = m(1 + \varepsilon_1)\exp(j\varepsilon_2), \qquad (4.74)$$

where $\varepsilon_1, \varepsilon_2 \ll 1$, i.e., $m_1 \approx m(1 + \varepsilon_1 + j\varepsilon_2)$, then the interference signal at the output will differ from the zero signal:

$$u_J = -\frac{2J\exp(j\psi_1)}{1-\exp[-(\beta+j\Delta\psi)]}\{1+\exp[-(\beta+j\Delta\psi)]-\exp(\beta+j\Delta\psi)[1+\exp[-\beta-j\Delta\psi)]\}(\varepsilon_1+\varepsilon_2). \qquad (4.75)$$

But, as calculations show, this difference will be significantly lesser than the initial interference in each channel. If, for example, $\varepsilon_1 = 0.01$, $\varepsilon_2 = 1° = 0.017$, then, neglecting the differences of the interferences' amplitudes at two antennas, we obtain

$$u_J = -\frac{2J\exp(j\psi_1)}{1-\exp[-(\beta+j\Delta\psi)]}\sqrt{0.01^2+0.017^2} = -0.0197\frac{2J\exp(j\psi_1)}{1-\exp[-(\beta+j\Delta\psi)]}.$$

Application of diversity reception as an effective measure of a struggle against interferences was known long ago. But its effectiveness essentially enhances if to obtain optimum results we use the method of mathematical programming. Advantages of this approach are visible clearly manifested in the struggle against constant interferences. In this case, the optimization problem is the correct choice of amplitude and phase modulator characteristics, and the variable parameters x which are the modulus and phase of the modulator gain.

4.7 Adaptive array

While receiving a radio signal, possible filtration of interference, i.e., excretion of useful signal and simultaneous suppression of interfering signals, has a great significance. Spatial filtration is one of the most efficient methods of fighting interference. In order to apply it, one must know the direction of the arrival of the useful signal so that the main lobe of the directional pattern could be oriented on the source of the signal, and zeroes of the directional patterns—on sources of interferences and disturbances.

The array of receiving antennas forming a unified system with receiving equipment can act as an efficient spatial filter [74]. Adjustment of such a system is performed with the help of special weighting devices (attenuators), forming the required directional pattern, and adaptive controlling circuit (feedback loop), which uses an iterative procedure to automatically choose optimal parameters of the system and then automatically adapts to changing conditions. For this reason, the described antenna system is called an adaptive one.

An advantage of adaptive processing is the fact that the suppression of interference, as a rule, involves no decrease of the useful signal. The automatic control of parameters is of special significance where constant factors, degrading the antenna performance, often act together with variable factors that exist, for example, aboard ship: running rigging, motion of various steel ropes under the action of wind or pitching and also rolling, turning or tuning of nearby antennas, the weather effects, etc.

An adaptive antenna system operates in the situation, when the spectrum of the useful signal and the direction of its arrival are known, whereas the field structure of the source, which is incorporating noise and interference, and the direction to the source do not know. The system uses an artificially introduced reference signal that is produced in the receiver and has the spectral characteristics and azimuth coinciding with those of the useful signal, known approximately.

The principle of beam forming in the adaptive antenna system with the help of weighting devices is clear from Fig. 4.28.

On multiplying output signals of array elements by weighting coefficients, which are selected to ensure that the main lobe does not undergo changes (i.e., the magnitude of the received useful signal remains the same), and the

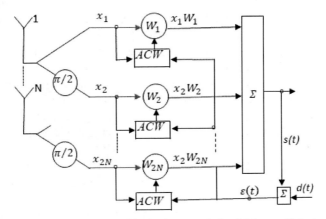

Fig. 4.28. Antennas array with adaptive control of weighting coefficients.

direction of zero reception coincides with the direction of the interference source. A possible variant to implement required weighting coefficient W is using a circuit from two parallel channels at each output element. In one channel the system of amplitude adjustment and phase delay by $\pi/2$ is placed. Such a system is named by a circuit with quadrature channels. Introduction of the phase delay equal to $\pi/2$ is not obligatory, but useful, because it allows obtaining close magnitudes of the weighting coefficients in adjacent channels.

The adaptive control of weighting coefficients (ACW) is performed with the help of a controlling circuit—adaptive processor (AP). It automatically adjusts the weight coefficients by the iterative procedure in accordance with the chosen algorithm. The error signal $\varepsilon(t)$ is used as a controlling signal in adjustment circuits. The error signal is equal to the difference between reference signal $d(t)$ (close to required output signal) and actual signal $s(t)$ at the adder output:

$$\varepsilon(t) = d(t) - s(t), \qquad (4.76)$$

where output signal $s(t)$ is the sum of signals $x_n(t)$ with weighting coefficients W_n:

$$s(t) = \sum_{n=1}^{2N} x_n(t) W_n. \qquad (4.77)$$

Here N is the number of antennas, $2N$ is the number of weighting coefficients. Quantity $x_n(t)$ takes the phase delay into account, and is equal to $\pi/2$ for even values of n.

At the present time three adaptation algorithms are applied: (1) a differential algorithm, or an algorithm of greatest steepness, (2) an algorithm of least mean square of the error, (3) a random search [75]. The first two algorithms are based on the steepest descent method. The adaptation process starts with an installation of several arbitrary coefficients. Then, if the method of the steepest descent is used, the gradient of error function is measured, and the weighting coefficients are set such that a negative gradient is obtained. The procedure is repeated in order to decrease the error and to approach the weighting coefficients to optimal magnitudes. If the differential algorithm is used, the gradient is evaluated directly according to the error function derivatives. In the method of least mean square of the error the error magnitude is squared, and the derivatives of the square are used to calculate the gradient. The random search method includes the measurement of the mean square of the error before and after an arbitrary change of the weighting coefficients and the comparison of the results of these measurements to decide whether to accept the change in the case of error reduction, or to discard it in the opposite case.

The algorithm of least mean square of the error ensures either a most rapid convergence to the same magnitude of the error in all cases or the least error in the same operation time. This algorithm is based on using feedback and adjusting each weighting coefficient according to the law

$$dW_n/dt = \mu \overline{\partial \varepsilon^2(t)/\partial W_n}, \qquad (4.78)$$

where μ is a negative constant governing a convergence rate and a system stability, and the line on top denotes the mathematical expectation. In this case, the number of arithmetical operations is proportional to the number of weighting coefficients, that is, far less than in the direct calculation of the coefficients with the help of a covariance matrix, where the number of operations is in proportion to the third power of the number of weighting coefficients. Therefore, implementation of this algorithm in practice is easier than implementing other algorithms.

Rather than calculating the gradient of the mean square of the error, which requires a great number of statistical samples, it is expedient to use the gradient of a single sample of the squared error (the gradient estimate), i.e., to replace the derivative $\overline{\partial \varepsilon^2(t)/\partial W_n}$ by the derivative $\partial \varepsilon^2(t)/\partial W_n$. Then the law of feedback will take the form,

$$dW_n/dt = -2\mu x_n(t)\varepsilon(t). \qquad (4.79)$$

One can show that expected value of the gradient estimate is equal to the gradient, i.e., the gradient estimation is not displaced.

If the number of iterations increases unlimitedly, the mathematical expectations of the weighting coefficients converge to the Wiener solution, in which gradient $\overline{\nabla \varepsilon^2(t)} = \sum_{n=1}^{2n} \overline{\partial \varepsilon^2(t)/\partial W_n}$ is equal to zero. But this convergence is secured only in the case, when the constant μ lies within certain limits. A practically convenient restriction (although stricter than necessary) is the inequality

$$-1/P < \mu < 0, \qquad (4.80)$$

where P is total power of the input signals.

In accordance with (4.79), input signal $x_n(t)$ and error signal $\varepsilon(t)$, i.e., the difference between output signal $d(t)$ and actual signal $s(t)$ at the adder output, are entered into the processor. If the output signal were equal to the desired signal, then the process of creating the desired signal would be simple and efficient. But there is no such signal in the antenna array. Therefore, an artificially introduced reference signal created in the receiving device is used as the required signal. For this reason, the main lobe of array in the process of adaptation orients in the direction specified by the reference signal. The amplitude characteristic of the antenna system in the band of the reference signal becomes uniform, and the phase characteristic becomes linear.

An introduction of the reference signal can overly distort the useful signal. To resolve this difficulty, two manners of adaptation, single-mode and dual-mode, are developed and used. In the dual-mode adaptation (Fig. 4.29) only one processor is used, i.e., it is more economical. As seen from the figure, the generator of reference signals (RG) creates two signals. One signal goes as the reference signal $d(t)$ into the processing circuit. The second (control) signal imitates the useful signal arrival. It goes through the circuits of delay δ_n to inputs of array channels. Delays δ_n are chosen so that the received input signals are identical to the signals coming from the given direction.

In the first mode, when a switch is set to position I, the control signals are introduced to the inputs of channels of the adaptive processor, and the processor adjusts the weighting coefficients and turns the main lobe of the directional pattern in the given direction. As a result, the output signal does not differ from the reference signal. In the second mode, when the switch is set to position II, the signals from array elements (i.e., from the surrounding space) are introduced to the inputs of adaptive processor channels, and the reference and control signals are removed, lest they distort the external signal. Since there is no reference signal, i.e., $d(t) = 0$, all received signals are suppressed.

Sustained operation in the second mode leads to the self-blocking of the system, when all weighting coefficients tend to zero. But, if the modes rapidly alternate and the weights vary little during operation in each mode, the required direction of the main lobe is retained (at the operation in the first regime), and the power of interference is reduced to minimum (mostly, in the second mode). The useful signal in the second mode (when the switch is set to position II) arrives to the receiver input R.

Digital simulation adaptive processing of signals confirms the procedure convergence and shows that this is an efficient method of the spatial filtration of interference with the useful signal retained. The experimental testing of an

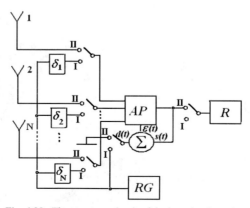

Fig. 4.29. The structure circuit of dual-mode adaptation.

adaptive system confirming its efficiency simultaneously revealed the danger of suppressing the system operation by interferences with frequency close to that of the useful signal. This testing showed the necessity of protection of the adaptive system, for example, by means of modulating the useful and reference signals with help of pseudo noise code [76].

4.8 Struggle with environmental influences

Development of radio engineering and the widespread use of the radio devices in the national economy and in the everyday life of people led to the problem of protecting living organisms and sensitive instruments from strong electromagnetic fields in the near zone of each transmitting antenna. The protection of devices is necessary, since the radiation of nearby devices can disrupt their normal operation, and cause spontaneous switching on and switching off of the device, or a change in operating regimes, among others. The protection of living organisms is required, first, in the vicinity of powerful transmitting centers, where the electrical field strength is great, and, second, near mobile equipment, which is located next to the user and operates in the transmitting mode. A cellular phone, in particular, is such a device.

The undoubted advantage of mobile communication consists in the freedom of its use by everyone, regardless of age, wealth and location. If the radio liberates personal human contacts from fetters of wire systems, the cellular phone allows replacing the radio station, mounted on a truck or on another vehicle, by a small device that may fit in a child palm. As a result, the advent of cellular communication looked like a big bang due to the rapid increase in the number of handsets, the widespread proliferation of phone contacts, and also due to the rapid growth of concern about the potential harm of these devices to human health, in particular, due to anxiety about carcinogenicity.

Together with proliferation of mobile communication systems, there has been an increasing concern about possible hazards for the user's body, especially for the user's head, which is irradiated by an antenna of the handset. During a phone conversation, the personal cellular phone is placed next to the user's head, and its transmitting antenna irradiates sensitive human organs (brain, eyes, etc.). The absorbed energy in the first cell phones was close to half of all radiated energy. In order to minimize any possible health risk, it is necessary to reduce the amount of that energy.

Rumors and the truth about the potential health harm of irradiation require reducing the Specific Absorption Rate (SAR) in the user's head. This is an extremely complex problem, since one must reduce the near field of the transmitting antenna without decreasing the field in the far zone and at that do not spoil the directional pattern of the antenna. In addition, the problem is not limited by the cellular phones, since a man often works with a low-power transmitter or uses it. This transmitter can be located nearby in a production area or in a vehicle.

The rather obvious idea of head protection by means of screening, i.e., by shading effect, is unrealizable. The near field has no ray structure, and hence the shadow behind a metallic screen can cover only an area approximately equal to the screen size. In order to protect the head of the cellular phone user, the screen must be much larger than the cross-section of the handset housing. For similar reasons, one should discard the idea of using an absorber, i.e., a dielectric shield that absorbs some part of energy. The distortion of the antenna directional pattern is still another obvious disadvantage of using screens and absorbers.

The protective action of the compensation method is based on a different principle—on the mutual suppression of fields, created by various radiating elements in a certain area. The diverse variants of using this principle posseses an opportunity to reduce irradiation of the user organism, especially his head, without distorting the antenna pattern in the horizontal plane. In accordance with the key variant of the compensation method, as shown in Fig. 4.30, main radiator 1 is supplemented with second (auxiliary) radiator 2, situated in the plane, passing through the center of a human head and the excitation point of the main radiator. The second radiator is placed between the head and the main radiator and is excited approximately in anti-phase with it (not exactly in anti-phase, because the phase of the field is changed along the interval between the radiators). So, the fields of two radiators will compensate each other at a certain point inside the head, and the point will be surrounded by a dark spot, i.e., by a zone of a weak field.

The dipole moment of the auxiliary radiator must be smaller than that of the main radiator, since the field of an electrical radiator, decreases quickly. In order to get the same magnitude of field at the compensation point, if the currents of radiators differ substantially, it is enough to place the radiators at a distance of a few centimeters from each other (1–2 cm at frequency 1 GHz). Therefore, the common far field is little different from the main

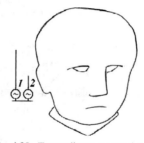

Fig. 4.30. Two radiators next to head.

radiator field, i.e., the directional pattern stays close to circular. Since the field of the auxiliary radiator is relatively small, the common field in the far region does not almost change, and the directional pattern remains close to the directional pattern in the absence of compensation.

Design options for personal cell phones that use the compensation principle were described in [10]. The obtained results confirmed a significant decrease in the power absorbed by the user's head as a result of its irradiation.

Application of the compensation method under realistic conditions often meets with difficulties because of a changing surroundings, since the operation of a complicated radiating structure often is disrupted by various external actions. The disturbances are to be counteracted. In particular, metallic objects approaching the antenna or relative displacement of the antenna elements as a consequence of user's movement can result in a tuning disturbance. In such cases, the compensation point may be displaced and the field inside the dark spot can grow significantly. Consider the impact of metallic objects approaching an antenna as an example of the external action.

In order to eliminate the consequences of these influences, one can try to retain the amplitudes and phases of the driving voltages and currents (the first method) or prevent the change of the radiator fields (the second method). But it should be noted, first, that the appearance of a metal body near a radiator changes the current at all points along the wire of the radiator, whereas the feedback circuits are capable of adjusting the current only at a single point in each radiator, e.g., at its input. And, second, one must observe that a metal body causes different changes of the radiator fields in the entire space, whereas the feedback circuits are capable of adjusting the field only in one or in two points. Therefore, the efficiency of both methods, especially the first one, is inherently limited. This Section is devoted to the comparative analysis of efficiency of both methods. It should be emphasized that the analysis considers cases of severe distortion in antenna systems.

The circuit of the antenna system is given in Fig. 4.31. The system consists of two monopoles (A is the main radiator; B is the auxiliary radiator) mounted on a metal plate near the model of a human head. The compensation point is located inside the head, near to its front boundary. Dimensions in the figure are indicated in centimeters. Figure 4.32 shows the same circuit for the case, when a vertical metal sheet in the shape of a square is placed not far from the antenna system (at a distance of 0.5 m from the radiators). Presence of a metal sheet in the proximity of a cellular phone may occur, e.g., when a phone's user enters an elevator or a car. The metal sheet affects the antenna system and causes the growth of fields in the zone of a weak field.

Let us start with a relatively simple system with a single radiator. The input current of a single radiator in the absence of any metal sheet is $J_{A1} = e_A/Z_A$, where e_A is an electromotive force, and Z_A is the input impedance

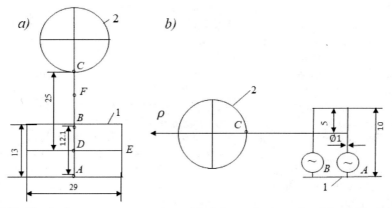

Fig. 4.31. Circuit of antenna system near the model of a human head: top view (*a*) and side view (*b*).

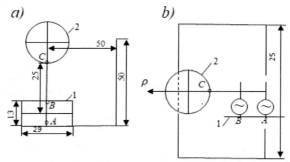

Fig. 4.32. A metal sheet is placed not far from the antenna: top view (*a*) and side view (*b*).

of the radiator. The metal sheet changes the input impedance of the radiator and makes it equal to Z'_A. To avoid the current change in the radiator base, one must change the emf at its input to the to the magnitude,

$$e'_A = J_{A1} Z'_A = \frac{e_A Z'_A}{Z_A}. \qquad (4.81)$$

For a single radiator of height 10 cm and diameter 1 cm we have: $e_A = 1$, $Z_A = 38.6 + j12.8$, $Z'_A = 41.7 + j14.3$, $e'_A = 1.09 \exp(j0.0145)$. Figure 4.33 shows the field of single radiator A along the horizontal line, passing through the radiator and the center of the human head in the absence (1) and presence (2) of a metal sheet and after adjustment of the emf (3). Figure 4.33 and several others figures were divided into two parts in order to use the different scales.

In the case of two radiators one must first calculate the amplitude and the phase of the emf of the second radiator, which ensures the field compensation at a given point. For this purpose one must take turns excite both radiators by

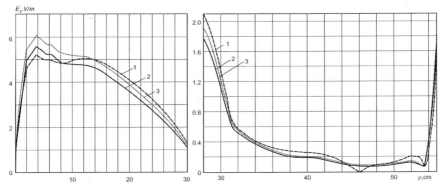

Fig. 4.33. The field of the single radiator A in the absence of (1) and the presence of (2) the metal sheet and after adjustment of the emf (3).

emf e_1, calculate the fields E_1 and E_2 of each radiator at the compensation point and use emf $e_2 = -e_1 E_1/E_2$ as the emf of the second radiator. It should be noted that the field and the input impedance of each radiator are calculated in the presence of another radiator, whose input is grounded.

Calculations were performed with the CST program, which permits to simulate of the total circuit of both generators and to find the self-impedances Z_{11} and Z_{22} of each radiator and their mutual impedance Z_{12}. The solution of the set of two equations $e_1 = J_1 Z_{11} + J_2 Z_{12}$, $e_2 = J_2 Z_{22} + J_1 Z_{12}$ allows us to find currents J_1 and J_2. The metal sheet changes the self- and mutual impedances of in radiators bases. To avoid the change of the currents at radiators bases, the emfs must be changed to

$$e'_1 = J_1 Z'_{11} + J_2 Z'_{12}, \; e'_2 = J_2 Z'_{22} + J_1 Z'_{12}, \qquad (4.82)$$

where the new impedances are marked by sign prime. Calculating the new emf's, we may find the new fields and ascertain the adjustment results.

For two radiators of the same dimensions we initially obtain: $e_1 = 1$, $E_1 = 0.48 \exp(j1.11)$, $E_2 = 1.09 \exp(j2.54)$, i.e., $e_2 = 0.44 \exp(j1.71)$. Accordingly in the compensation mode we obtain for impedances: $Z_{11} = 37.3 + j16.2$, $Z_{22} = 39.5 + j16.8$, $Z_{12} = 0.67 - j28.9$, and for currents: $J_1 = 0.013 - j0.0008$, $J_2 = 0.0064 + j0.018$. The adjustment results for impedances are follow: $Z'_{11} = 40.6 + j17.1$, $Z'_{22} = 42.0 + j18.9$, $Z'_{12} = 2.04 - j28.02$, and for emfs: $e'_1 = 1.04\exp(j0.04)$, $e'_2 = 0.51\exp(j1.70)$. The calculated curves for the fields of the two-radiator structure are presented in Fig. 4.34.

The verification of adjustment results involves computation of fields along the horizontal line, passing through the radiators and the center of the head, as well as calculation of the total SAR and the local SAR (in 1 g). The results, obtained for the single radiator and for the structure from two radiators in the absence of and in the presence of a metal sheet and after adjustment of

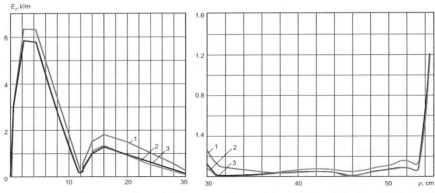

Fig. 4.34. The field of two radiators in the absence of (1) and the presence of (2) the metal sheet and after adjustment of the emf (3).

the emfs in accordance with expressions (4.81) and (4.82), are compared with each other.

The total SAR and the maximal local SAR for the corresponding cases are given in Table 4.2. The calculated amplitudes of field at the compensation point are also given in Table 4.2. The maximal level of SAR and the fields are presented in W/kg and in V/m, respectively. As one can see from the table and figures, the results indicate the rather low efficiency of the correction method, based on retaining the driving currents of radiators.

The implementation of the second method calls for using one radiator as a measuring antenna and the second radiator as a local transmitting antenna (i.e., as a field source), or using both radiators in turn as a measuring and a transmitting antenna, or using the third radiator as a measuring antenna. The third radiator may be mounted at any suitable and convenient place. The adjustment is performed in the following way. The field at the receiving point is measured in the presence of a metal sheet and is compared with its magnitude in the absence of the metal sheet (i.e., in the compensation mode), and afterwards the emf of transmitting antenna is varied until the initial value of the field.

The results of such adjustments are given in Fig. 4.35 and Table 4.3 for the following variants: (1) when the main antenna is used as the transmitting and the auxiliary antenna is used as the receiving, (2) the opposite case: the main antenna is used as the receiving, whereas the auxiliary antenna is used as the transmitting, (3) both emfs are replaced. Figure 4.35 and Table 4.3 present the results of field adjustment, using the third antenna located in the center of a metal plate, i.e., at point D (see Fig. 4.31) for the following variants: (4) when the emf and the field of the main antenna is changed so that its field at point D becomes equal to its original field (before distortion), (5) when the

Table 4.2. SAR and field at adjustment of emf in accordance with currents.

Characteristic	Total SAR	Local SAR	Field	Total SAR	Local SAR	Field
	one radiator			two radiators		
Sheet is absent	$4.5 \cdot 10^{-5}$	$1.54 \cdot 10^{-3}$	0.79	$1.4 \cdot 10^{-5}$	$0.3 \cdot 10^{-3}$	0.0029
Sheet is present	$2.8 \cdot 10^{-5}$	$0.29 \cdot 10^{-3}$	0.68	$1.7 \cdot 10^{-5}$	$0.11 \cdot 10^{-3}$	0.0044
With adjustment	$3.4 \cdot 10^{-5}$	$0.34 \cdot 10^{-3}$	0.74	$2.1 \cdot 10^{-5}$	$0.18 \cdot 10^{-3}$	0.1214

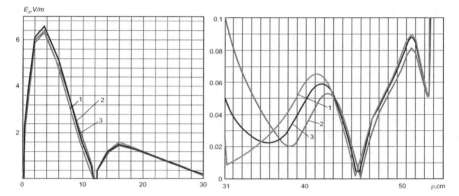

Fig. 4.35. The fields of the structure from two radiators after emf adjustment, based on the fields, received from by these radiators.

Table 4.3. *SAR* and field at adjustment of emf in accordance with fields.

Variant	Total *SAR*	Max. Local *SAR*	Field
1	$1.99 \cdot 10^{-5}$	$0.137 \cdot 10^{-3}$	0.035
2	$1.83 \cdot 10^{-5}$	$0.151 \cdot 10^{-3}$	0.070
3	$2.02 \cdot 10^{-5}$	$0.152 \cdot 10^{-3}$	0.052
4	$1.57 \cdot 10^{-5}$	$0.102 \cdot 10^{-3}$	0.070
5	$1.45 \cdot 10^{-5}$	$0.125 \cdot 10^{-3}$	0.130
6	$1.18 \cdot 10^{-5}$	$0.090 \cdot 10^{-3}$	0.047

field of the auxiliary antenna is changed for this purpose, (6) when the fields of both antennas are changed.

As one can see from Table 4.3 and Figs. 4.35 and 4.36, the method, based on the measurement of the fields, demonstrates a higher efficiency. But acceptable results are obtained only on application of variant 6. Other variants, including placement of the third antenna at points E and F (see Fig. 4.31), do not give satisfactory results. And it should be pointed out that the monitoring signal is a weak signal that serves as the signal of feedback

244   ANTENNAS: Rigorous Methods of Analysis and Synthesis

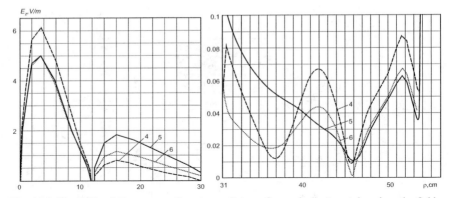

Fig. 4.36. The fields of the structure from two radiators after emf adjustment, based on the fields, received by a third radiator.

for both the main and auxiliary radiators. Therefore, proper adjustment by both methods requires that no signal from the transmitter impinges on the measuring antenna during the measurements of field.

In order to prevent the changes of the field under external actions (e.g., at the approach of a metallic object), one may use a manual or an automatic adjustment. Figure 4.37 gives the block diagram of an automatic adjustment. It contains transmitter 1, main radiator A_1 and auxiliary radiator A_2, connected to the transmitter through the power divider 2, the amplitude controller 3 and the phase shifter 4. The amplitude controller and the phase shifter provide the initial tuning and the field compensation at a given point. The amplitude controller is usually implemented by means of the potentiometer, and the phase shifter is implemented by means of the delay line, the low-pass filter or the high-pass filter. Two circuits, consisting of the amplitude controllers 13 and 15 and the phase shifters 14 and 16, provide two reference signals for the radiators A_2 and A_1, respectively.

The external action (e.g., an approach of a metal body) changes the phases of the signals of both the radiators, received by antenna. The phase detector 7 in turn compares these phases with phases of the reference signals and produces an error signal, proportional to the difference of the said phases. Low-pass filter 8 removes short-term fluctuations of the error signal. The error signal passes through amplifier 9, controls the phase shifters 4 and 6 and brings up the optimal phase differences. As it can be seen from Fig. 4.37, a feedback circuit is constructed, and it provides a phase self-tuning action, similar to action of a phase locked loop (PLL).

The second feedback circuit is used for predicting the optimal signal amplitudes. It is similar to an automatic gain control (AGC) circuit. The input signal of radiator A_2 is compared in amplitude by comparator 10 (operational

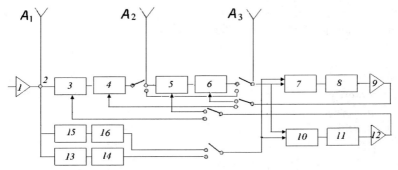

Fig. 4.37. Block diagram of the proposed automatic adjustment circuit, based on the constancy of fields.

amplifier) with the reference signal. Low-pass filter 11 removes short-term fluctuations of the signal at the comparator output. The error signal passes through the amplifier 12, controls the amplitude controller 3 and brings up the amplitudes' relationship to the optimal ratio.

As a result, two feedback circuits allow optimizing the amplitude relationship and the phases of the emf, feeding main radiator A_1 and auxiliary radiator A_2. In contrast to the conventional automatic gain control circuits, the amplitude difference and the phase difference of the two different radiator signals are not zero in this case.

The obtained results show that it is possible to automatically adjust a complicated antenna system under conditions of intensive disturbances. The proposed method, based on measurement of the fields, demonstrates a higher efficiency. It allows obtaining acceptable practical results under severe disturbance of the antenna system operation even in cases, when (as in our example) the zero-field point is not achievable.

4.9 Turn of the directional pattern of a cellular base station as concrete task

One of the ways to improve the quality of service for cellular communication users, as well as its profitability, is to create base station antennas with a guided radiation pattern [77]. In this case, the simplest and most essential requirement to the radiation pattern of its antenna is the ability to turn its main lobe into the horizontal and vertical planes. That permits to providing high-quality communications to a maximum number of users depending on the time of day, and on the occasion of organized events, etc. Important conditions when fulfilling the stated task are, first, the electrical and not mechanical adjustment of the beam and, second, the preservation, if possible, of the available equipment of base stations.

The antenna of the base station is as a rule a uniform linear phased array located along the vertical axis. The angular width of the directional pattern in the vertical plane is from 5° to 15°, and in the horizontal plane it is equal to 60° or more. The direction of the maximum radiation coincides with the perpendicular to the array axis (at zero phase shift between the elements) or differs somewhat from it in accordance with the required angle of inclination.

In most cases turning the directional pattern in a vertical plane is easier than turning it in a horizontal plane. For this, a phase shift between the array elements should be provided by introducing phase-shifting circuits into the distribution scheme. Changing the phase shift during operation (for example, using the switches), one can change the angle of inclination of the main lobe of the directional pattern. However, this requires a partial replacement of the base station equipment, primarily antennas.

To avoid this and not to make changes to the existing antenna array, one can install an additional radiator along its vertical axis, exciting it with a given phase shift. In fact, this is an increase in the length of the antenna array by one element. Depending on the magnitude of the phase shift, one or another inclination angle of the general directional pattern can be obtained. The power supplied to this element may be close to the excitation power of each element of the existing antenna. In order to change the angle of inclination of the main lobe of the general directional pattern during operation, one can either switch on or switch off the additional radiator or switch phase-shifting circuits in it.

The possibility of implementing such an antenna system is confirmed by the following example. Let the existing antenna be a vertical array of N identical in-phase radiators located at a distance b from each other. The array multiplier is

$$f_1(\theta) = \sin\frac{Nkb\cos\theta}{2} \bigg/ \left(\sin\frac{kb\cos\theta}{2}\right), \qquad (4.83)$$

where θ is the angle measured from the array axis (Fig. 4.38), k is the wave propagation constant in air.

If the directional pattern of the additional radiator coincides with the directional pattern of the array element, then, considering the new antenna system as a combination of two radiators: the array of N elements with an input current I_1 and an additional radiator $A_{(2)}$ with the input current I_2—we obtain the overall multiplier of the system:

$$f(\theta) = f_1(\theta) + a\exp[j(kd\cos\theta + \psi)], \qquad (4.84)$$

where $a = I_2/I_1$, d is the distance between the centers of two radiators, ψ is the phase shift in the second radiator. From here

$$f(\theta) = \sqrt{[f_1(\theta) + a\cos(kd\cos\theta + \psi)]^2 + a^2\sin^2(kd\cos\theta + \psi)}.$$

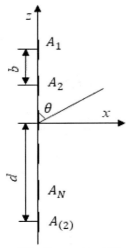

Fig. 4.38. Array of N radiators with additional antenna $A_{(2)}$.

At a point where the directional pattern of the antenna array has a maximum

$$\frac{d|f(\theta)|}{d\theta} = \frac{1}{|f(\theta)|}\{[f_1(\theta) + a\cos(kd\cos\theta + \psi)]\left[\frac{df_1(\theta)}{d\theta} + akd\sin(kd\cos\theta + \psi)\sin\theta\right] - \frac{a^2 kd}{2}\sin2(kd\cos\theta + \psi)\} = 0.$$

(4.85)

Let $\theta = \frac{\pi}{2} - \delta$. We will assume that the angle of deviation of the maximum radiation from the horizontal is small, i.e., $\delta \ll 1$. Then, in a first approximation $\sin\theta = 1$, $\cos\theta = \delta$, $f_1(\theta) = 1$, i.e.,

$$[1 + a\cos(kd\delta + \psi)]\left[\frac{df_1}{d\theta} + akd\sin(kd\delta + \psi)\right] = \frac{a^2 kd}{2}\sin2(kd\delta + \psi).$$

We introduce the notation $\alpha = \frac{kb\cos\theta}{2}$, i.e., $f_1(\theta) = \frac{\sin N\alpha}{N\sin\alpha}$. It follows that,

$$\frac{df_1}{d\theta} = \frac{df_1}{d\alpha}\frac{d\alpha}{d\theta} = -\frac{kb\sin\theta}{2N^2\sin^2\alpha}[N^2\sin\alpha\cos N\alpha - N\sin N\alpha\cos\alpha],$$

and $\alpha = \frac{kb\delta}{2} \ll 1$. Replacing the trigonometric functions of a small argument by the initial members of the series, we find:

$$\sin\alpha = \alpha\left(1 - \frac{\alpha^2}{6}\right), \quad \sin N\alpha = N\alpha\left[1 - \frac{(N\alpha)^2}{6}\right], \quad \cos\alpha = 1 - \frac{\alpha^2}{2},$$

$\cos N\alpha = 1 - \dfrac{(N\alpha)^2}{2}$, i.e.

$$\frac{df_1}{d\theta} = \frac{k^2 b^2 (N^2 - 1)\delta}{12}.$$

This magnitude at the maximum point of the directional pattern of the antenna system is equal to,

$$\frac{df_1}{d\theta} = [a^2 k d \sin 2(kd\delta + \psi)/2]/[1 + a\cos(kd\delta + \psi)] - akd\sin(kd\delta + \psi),$$

and at a small current I_2 of the additional radiator, assuming that a is small, and $\dfrac{df_1}{d\theta} = -akd\sin(kd\delta + \psi)$, we obtain

$$I_2 \sin(kd\delta + \psi) = -(N^2 - 1)kb^2\delta/12ad. \qquad (4.86)$$

Let, for example, $N = 6$, $d = 4b$, $a = 0{,}2$, $b = 0{,}8\lambda$. Then $\sin(20\delta + \psi) = -17{,}5\delta$, i.e., at $\delta = 0{,}05$ ($\sin^{-1}\delta = 3°$) $\psi = -118°$. The above example shows that the phase shift between the currents in the secondary radiator and the main array should be sufficiently large. Thus more is the shift between currents, the more is the difference between the powers in these two parts of the total array. However, vertical turn of the radiation pattern is possible.

In contrast to the turning in the vertical plane, the ability to turn the directional pattern in the horizontal plane is problematic. Installing a single additional radiator in the plane of the array along the horizontal axis does not solve the problem, not only because the additional antenna does not have a narrow directional pattern in the vertical plane, and from that the gain of the system drops sharply. A significant drawback of such a device is the fact that to turn the general directional pattern, the power in the additional radiator must be G times greater (G is gain of the array) than the power of existing antenna so that the signals of both radiators at the receiving point are equal. A competent technical solution is the installation of a second antenna array parallel to the first, with a given phase shift between the currents of both arrays. However, the implementation of this solution significantly increases the complexity and cost of the equipment.

As is known, an additional radiator with characteristics similar to the characteristics of the main radiator can be created by mirroring the main radiator in a perfectly conducting metal plane. In this case, the signal of the additional radiator differs from the signal of the main radiator by a fixed phase. The total directional pattern of a system of two such radiators is

$$F(\theta, \varphi) = f_1(\theta, \varphi) f(\varphi),$$

where $f_1(\theta, \varphi)$ is the directional pattern of a single array, $f(\varphi) = \sin\dfrac{kd\cos\varphi}{2}$ is the array multiplier of two radiators, $d = 2d_1$ is the distance between the axes of the radiators. It is equal to twice the distance between the main radiator and the metal plane. As can be seen from this expression, even the turn of the metal plane does not lead to the turn of the radiation pattern, if the direction of the maximum radiation of the main antenna does not change (the signal magnitude simply decreases in the direction of the minimum of the array multiplier). Mechanical (or electromechanical) turn of the main antenna around a vertical axis allows to turn the radiation pattern around this axis and makes the metal plane unnecessary. But such an adjustment of the beam by means of a mechanical turn is expensive and impractical.

The analysis shows that the turn of the radiation pattern in the vertical plane is real and may be accomplished by fairly simple means. The turn in the horizontal plane by the simple means is impossible.

Chapter 5
Problems of Design and Placement of Antennas

5.1 Ship antennas of medium-frequency waves

Antennas are widely used for a variety of purposes in diverse areas of our lives. These purposes and areas of use necessitate the development of products that correspond to an aim and to a placement of antennas. One of the most important applications of antennas is a radio communication and broadcasting. A peculiar installation place for antennas are ships and vessels for various purposes. This chapter is devoted to the analysis of the features of antennas used in ships for radio communication and broadcasting.

The main ship's antenna is an antenna operating in the range of medium frequencies (of medium-frequency waves). With allowance for peculiarity of their propagation, these electromagnetic waves should be directed along the earth's surface and have a vertical polarization. In the horizontal plane, the directional pattern of the antenna should be close to circular. The efficiency of the main antenna should be sufficient to create an electric field with a voltage of 50 µV/m at a distance of 150 miles from the ship when the main radio transmitter of the ship is connected to the antenna.

As a rule, the height of the ship's antenna is small in comparison to the wavelength, i.e., the resistance of radiation is small, and that leads to low efficiency. Accordingly, the main task of developing new medium-frequency antenna is the increase of its effective height. Therefore, the antennas with capacitive loads at the upper end are widespread. Their load is fabricated in the form of a horizontal wire structure and permits to improve (to make more uniform) a current distribution along the vertical wire of the antenna in order to increase its effective height and resistance of radiation.

Such an antenna is excited at the base as a rule. It is named inverted-L antenna. Its circuit and a distribution of a current amplitude along wires are given in Fig. 5.1a. The vertical segment may be single-wire, multi-wire or

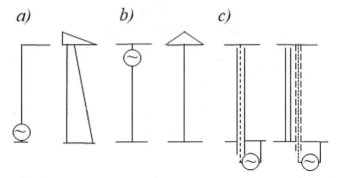

Fig. 5.1. Ship's wire antennas: a – antenna with excitation at a bottom point, and the current distribution along it, b – antenna with an upper excitation, and the current distribution along it, c – placement of the feeder inside and outside the mast.

fan-shaped, i.e., consist of several wires placed in one vertical plane, located in parallel or convergent to an excitation point. This segment is connected to an end or to the middle of a horizontal load. Accordingly, the antenna has inverted-L or T-shape.

Along with the antenna excitation at a bottom point (in the base), one can use the circuit of the upper excitation, when a radio transmitter is connected to the bottom end of a wire located inside the mast or inside a special feeder. The upper end of the wire is connected to a horizontal load (Fig. 5.1b, c). In this case, the outer surface of the mast acts as a radiator, and the internal surface of the mast and a central wire form the feeder. Such a circuit of antenna excitation does not give particular advantages, although the effective height of the antenna in this case significantly increases and is close to the height of the mast. But a long feeder reduces the load resistance of a transmitter, i.e., decreases the positive effect created by the upper excitation. To weaken the feeder influence one must increase the mast diameter.

The effective antenna height increases with increasing load length and width. It is possible to increase the load efficiency without changing its length by replacing the usual structure of the parallel wires with the meander load described in Section 1.5. A calculating method for electrical characteristics of an antenna with the meander load and results of applying such a load are given in [27].

Disadvantages of wire antennas are obvious: they require a second mast, which hinders cargo operations, and they are torn in stormy conditions. In addition, in a circuit with excitation at the bottom point, the mast, near which the vertical antenna wire is suspended, not only plays the role of a support but also creates an additional parasitic capacitance between the antenna and the ground, reducing the resistance of a radiation.

The calculating procedure for the resistance of a radiation for an antenna with a bottom excitation, the wire of which is parallel to the mast, is described in [35]. It is based on the theory of a folded radiator. Figure 5.2 shows the results of calculating the radiation resistance for antennas 6.5 and 13 m long at a frequency 0.46 MHz depending on the distance b between the antenna and the mast for different radii a_2 of the mast (all dimensions are indicated in meters). The radius of the antenna wire is equal to $a_1 = 3.7 \cdot 10^{-3}$ m. The points in the figure give the results of an experimental check, when the mast radius is equal to $a_2 = 0.4$ m. For comparison, the graphs show the radiation resistance R_{11} of an isolated antenna. It can be seen from the figures that when the distance between the antenna and the mast decreases, the resistance of radiation drops sharply.

It is seen from the figure that active component R_A of the antenna input impedance drops sharply as the distance between the antenna and the mast is small. Horizontal wire loads weaken the influence of the mast. But in this case also it is necessary to move the antenna away from the mast as far as possible (from 4 to 8 m, depending on the mast height).

Low radiation resistance and lower efficiency are not sole drawbacks of wires antennas. Another drawback is a wide variation range of the input impedance, which hampers standardization of antennas types and complicates the onboard equipment. Besides, an antenna often hinders cargo handling. A storm or ice formation can tear the vertical segment or the horizontal load (curtain) of the antenna. The antenna may require mounting a second mast, which is not necessary for a contemporary ship.

For this reason, antenna-masts have found their use as the main ships' antennas. At first, three variants of such antennas appeared: (1) with guy ropes, (2) free-standing and (3) placed on the mast. The first variant was dropped

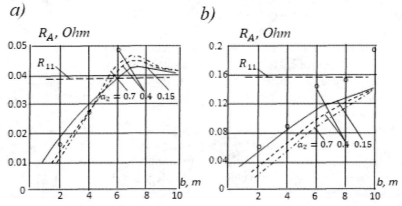

Fig. 5.2. Dependence of the radiation resistance of the wire antenna 6.5 m high (*a*) and 13 m high (*b*) on the distance to the mast.

soon because of the large area occupied by the antenna. Therefore, in the first stage free-standing (self-supporting) antenna-masts were manufactured, but they were used only on big ships. Further, in order to reduce the cost of the antenna and to use it on the ships of small and medium tonnage, the antenna-mast with an inductive-capacitive load was designed. It was mounted on the existing ship mast.

The circuit of this antenna is presented in Fig. 5.3a. The circuit corresponds to an option with bottom excitation and an open vertical wire. This means that this antenna differs from an inverted-L antenna only by the type of load. An antenna option with an upper excitation is also possible. The load is created in the form of a vertical structure, which is the extension of the mast. The mast acts as a support. As a result, the geometric height of the antenna is significantly greater than the mast height. Increasing the effective height of the antenna caused by increasing its geometric height improves all its electrical characteristics. Besides, this antenna can be installed on board of an exploited ship.

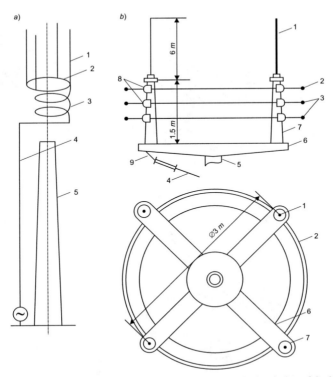

Fig. 5.3. Antenna with inductive-capacitive load: a – circuit, b – design of the load.
1 – whip antenna, 2 – conducting ring, 3 – two-turn volumetric spiral, 4 – open vertical wire, 5 – mast, 6 – work platform, 7 – dielectric column, 8 – base insulator, 9 – rod insulator.

As shown in Fig. 5.3b, the antenna load consists of four whip antennas connected on the base by a conductive ring, and a two-turn volumetric spiral connected in series with the system of whip antennas. The system of four whip antennas is equivalent to a thick metal radiator with a great capacitance and a low wave impedance. The spiral increases the electrical length of the antenna. The system of whip antennas creates the capacitive component of the load. The two-turn volumetric spiral creates the inductive component of it. Both elements decrease the input reactance of the antenna and increase its effective height and the radiation resistance. Use of whip antennas allows, if necessary, to incline them and decrease the total height of the structure. A necessity for manual access to elements of the antenna load is foreseen at the time of parking the ship in a port and in a calm weather. The lightning-conductor (spark gap) is installed at the antenna wire upstream of the transmitter room.

The antenna-mast with inductive-capacitive load was proposed in 1967 [78] and was improved in 1970 [79]. The antenna prototype was mounted on the board of cargo ship "*Konstantin Shestakov*" with a displacement of 3500 ton (Fig. 5.4). The antenna was placed on the ship's upper bridge, on the mast of height 9.5 m and diameter 0.3 m. The total antenna height was about 16 m. The static capacitance of an antenna is equal to 442 pF, the self wave length is 240 m, and the resistance at frequency 400 kHz is 4.3 Ohm. In 1968–1969, the antenna passed comprehensive operational tests. They showed that the antenna fully meets the Maritime Register requirements, including a level of intensity of the created field.

In subsequent years, antennas of a similar type were built in other countries (Fig. 5.5, dimensions are given in meters). They include Norwegian antenna

Fig. 5.4. Antenna-mast with inductive-capacitive load on the upper bridge of cargo ship "*Konstantin Shestakov*".

Problems of Design and Placement of Antennas   255

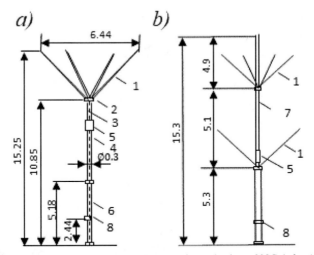

Fig. 5.5. Antenna-masts with inductive-capacitance load: *a* – 938G-1, *b* – AS9.
1 – whip antenna of fiberglass with wicker copper mesh, 2 – bronze cap, 3 – central copper wire, 4 – fiberglass mast, 5 – inductance, 6 – aluminum tube, 7 – fiberglass tube, 8 – antenna lead-in.

AS9 (the antenna option with possible inclination and is named AS9ST) and antenna 938G-1 of the firm *Collins*, USA. The capacitive loads in them, as in the described antenna, is made in the form of whip antennas installed on the mast top. The inductive load is made by means of a coil or spiral, connected in series with a system of whip antennas and a vertical wire.

They are built as free-standing structures of fiberglass and have a total height of about 15.3 m. Combining these antennas with a conventional ship mast is not provided.

The calculating method of the electrical characteristics of the antenna-mast (using the example of an antenna-mast with inductive-capacitance load) is described in [35]. Also, the antenna test results are given.

In principle a different version of a main antenna was developed for ships of another type, where the masts are almost always absent and replaced by tower-type superstructures. The antenna general view is shown in Fig. 5.6. This is the antenna with an upper excitation. The antenna consists of an excited metal structure (tower) and two wire cylinders 13–15 m long and about 0.5 m in diameter. The cylinders play the role of an antenna load and are made of six parallel wires located along the cylinder generatrixes and converging at the ends. Cylinders can be located at the same height at an angle of 0° to 20° to the horizontal and 20° to 50° to each other. If there is no space for such placement, then the cylinders can be placed one below the other with an angle from 10° to 30° between them (see Fig. 5.6).

256 ANTENNAS: Rigorous Methods of Analysis and Synthesis

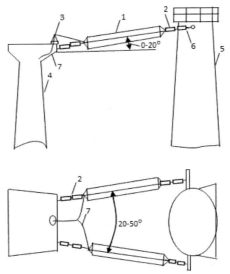

Fig. 5.6. Antenna with load in the form of two wire cylinders.
1 – wire cylinder, 2 – insulator circuit, 3 – coupling, 4 – excited structure, 5 – support tower, 6 – connector, 7 – cable.

The upper ends of the cylinders are fixed to a support cross-beam or bracket through insulators. The lower ends are fixed to the excited structure (also through insulators). Using flexible jumpers, the ends of the cylinders are connected to each other and to a cable. To reduce losses in the feeder, an end cascade or remote output resonant circuit of the radio transmitter should be located near the excitation point of the antenna. The distance between the side surface of the cylinder and the metal superstructure must not be less than 3 m.

The described antenna was installed on one of the navy ships. The tests carried out confirmed its prospects and showed that the antenna has a fairly high level of matching and an acceptable directional pattern, which were realized when the above restrictions are met during installation of the antenna on the ship. The advantage of the antenna is the absence of resistive elements in its design, and resulting in an antenna efficiency close to 1.

To ensure the safety of navigation, ships in addition to stationary antennas must be equipped with antennas that can be deployed and used in emergency situations. Such an emergency antenna should be kept fully prepared for immediate installation. For this purpose, pneumatic and ballistic antennas can be used. In both cases, the antenna is a vertical radiator excited at the base.

In a pneumatic antenna [80] the radiator is a flexible wire that is supported in a vertical position using one or more rubber containers (Fig. 5.7a). Containers are made of a flexible, durable and airtight material. When the compressor delivers compressed air to the container, then, thanks to the guiding metal

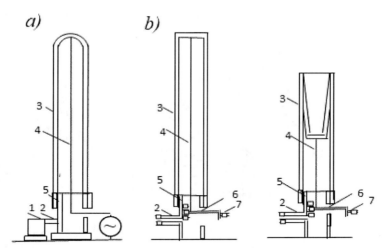

Fig. 5.7. Pneumatic antenna (*a*), its deployment mechanism (*b*), and a height adjustment (*c*). 1 – compressor, 2 – tube, 3 – container, 4 – flexible wire, 5 – base, 6 – drum, 7 – handle of winch.

pipe, the antenna is promoted and assumes a vertical position. After this, it is only necessary to maintain the desired pressure in the container according to the signal of a special sensor (Fig. 5.7*b*). The air pressure must be sufficient to maintain the antenna upright. In the folded position, the shell of container is wound on a drum of a winch. Upon transition to the operating position, the compressor increases the air pressure inside the container, forcing the shell to rise upward and gradually increase the antenna height. A height adjustment is made by turning the drum (Fig. 5.7*c*).

To operate in the medium wave range, the pneumatic antenna should have a height of about 20 m. In order to improve the electrical characteristics and increase the stability of the shell, one can overlay a metal mesh on the shell or replace the vertical wire with several wires.

Another variant of the deployable emergency antenna is a ballistic antenna [81]. An antenna circuit is shown in Fig. 5.8. A flexible filament closed in a ring is throwing away with the help of a catapult (a rotating pulley mounted on the motor axis), forming an ascending branch of the antenna.

Another pulley accepts the filament arriving in a descending branch. The catapult and pickup pulley are isolated from the ground. A high-frequency energy is transmitted to the antenna by means of an inductive coupling.

One can use a metal hawser with washers or discs mounted on it as a filament. Such a filament has an increased aerodynamic drag, which allows the reduction of the speed of the filament, without threatening the stability of the antenna. In another embodiment, the radiator is made of a high strength elastomer with a complex cross-sectional profile. It is twisted along the

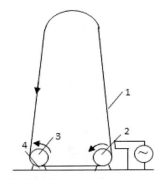

Fig. 5.8. Circuit of ballistic antenna.
1 – filament, 2 – motor with a pulley (catapult), 3 – pickup pulley, 4 – insulator.

longitudinal axis by an angle $n\pi$ (of the Möbius loop type) and is covered with a metal layer with a thickness of 0.01–0.1 mm ($10^{-5} - 10^{-4}$M). This design permits the implementation of an antenna with a height of several tens of meters or more. The design provides fast ascent and descent. Changing the height of the antenna is done by changing the relative position of the additional rollers along which the filament is moved.

Concluding the section devoted to medium wave antennas for the ships, it is necessary to summarize what has been said. At first glance, this Section is devoted to the relatively narrow field of antennas application, and these antennas are used to solve a simple problem. But, as follows from the foregoing, this task entails a variety of side effects associated with the placement of antennas on the object and the influence of operating conditions. To solve problems due to these side effects, sometimes a strict approach is required at the highest scientific level. The issue of the interaction of antennas of different types and ranges placed together on objects is one of the most serious causes of problems caused by side effects and will be considered further.

5.2 Ship antenna of high-frequency waves

In ship radio communications, a range of short (decameter) waves is widely used. Its peculiarity is the operational change of working frequencies. The need for such a change is caused by daily and seasonal changes of the state of the ionosphere. In addition, the choice of the operating frequency depends on the ship's location, the length of the radio communication, the level of the sun's activity, etc. Therefore, it is necessary to operate practically in almost the entire high-frequency range.

The use on ships in the range of decameter waves of wide-range radio transmitters based on power amplifiers with distributed amplification simplifies the operation of radio transmitters and the automation of their control. But at the same time, for them it is necessary that the load impedance of the transmitter lies within the specified limits. These limits are much narrower than the limits necessary for the operation of transmitters with resonant amplification.

Therefore, the use of wide-range transmitters is impossible without wide-range antennas. Also, an automation of radio communication is not realized without wide-range antennas. Thus, the first requirement of a ships' shortwave antennas is a wide operating range.

Simultaneously with operation in a wide frequency range, it is necessary to ensure high electrical characteristics at each operating frequency. This requirement is associated not only with the use of wide-range radio transmitters, but also with the need to increase the power transmitted by the cable and to reduce cable losses.

The directional pattern of antenna in the horizontal plane should be close to circular. As for the vertical plane, the greatest range of radio communication during signal propagation by a surface (terrestrial) and spatial beam is obtained in the case of a maximal radiation in the horizontal direction or at small angles to the horizon, i.e., for a pattern factor, close to 1. In this direction, maximal radiation should be ensured. In [82], the results of statistical processing of various communication lines are presented. On their basis the optimal relationships between the signal frequency, the angle of maximum radiation, and the width of the directional pattern in the vertical plane are established (Fig. 5.9). Each curve in the figure shows the probability p of the event when the elevation angle θ_1 hit into the interval $\theta \leq \theta_1 \leq 90°$. As can be seen from the figure, the hatched region covers 80% of the angles θ_1. Therefore, the width of the main lobe of the directional pattern should correspond to the width of the shaded area, and the direction of maximal radiation should be defined by a curve passing through the middle of this region ($p = 50\%$). With increasing frequency, the angle of rise of the main lobe above the earth's surface should decrease. Of course, while using a real antenna it is difficult to obtain such a

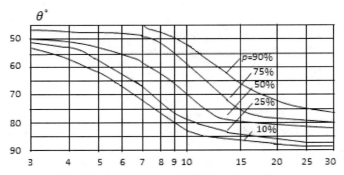

Fig. 5.9. Optimal angles of radiation and reception.

dependence of the radiation pattern on frequency. However, the general trend is clear: the main radiation in the decameter range should be directed under angles from 0° to 40° to the horizon.

The choice of polarization of the radiated signal is associated with the shape of directional pattern. Horizontal antennas create weak radiation in the horizontal direction. Mirroring in the ground multiplies the signal of a horizontal antenna located at a height h above the ground by the value $\sin(kh \cos \theta)$, i.e., sharply attenuates the signal. Horizontally polarized waves during propagation along the sea surface decay faster than vertically polarized ones. Therefore, the ships primarily use vertical polarized antennas. An exception is log-periodic antennas, which are installed at a great height above the deck, that weakens the influence of the earth.

The low efficiency of the antenna leads not only to useless energy losses, but also limits the allowable power. As a result, simultaneous operation of several transmitters on one antenna becomes impossible, and that does not allow us to decrease the number of antennas. A wide-range antenna with high characteristics at all frequencies solves this problem.

Thus, the second requirement of a ship's shortwave antenna is the presence of high electrical characteristics at each operating frequency.

The third requirement to a ship's shortwave antenna is necessity to have a free-standing antenna design. As a rule, only a free-standing antenna design can provide parameters independent of the type and size of the ship. Therefore, a wide-range shortwave antenna should have such a design. Based on the real possibilities of placement, the maximum height of the structure should not exceed 10–12 m.

Consider the specific options for the antennas used. Although whip antennas cannot be called wide-range, they are widely used for short-wave radio communication. The antennas used in the US Navy are described in detail in [83]. Aluminum and fiberglass rod antennas are widely used on ships. The aluminum rods *NT-66046* and *NT-66047* consisting of four or five elements, are mounted on feedthrough insulators and have a total height of 8.5 and 10.6 m (Fig. 5.10*a*). Plastic glass antennas *AS-2537/SR* and *AS-2537A/SR* consist of two elements inserted or screwed into each other (Fig. 5.10*b*). The total antenna height is 10.5 or 10.6 m. The radiator is made in the form of a wire or a metal strip of beryllium bronze, reinforced with fiberglass. The lower element is mounted on an insulator inserted in a metal base. Power is supplied to the contact located above the insulator.

Antennas have a small diameter and a high impedance. For example, the diameter of the *NT-66047* antenna is 7.6 cm at the base. The transverse dimension of the *AS-2807/SRC* antenna is increased. The antenna is made in the form of three hollow aluminum boxes of square cross-section, the lower one of which is mounted on a base insulator inserted into a metal base

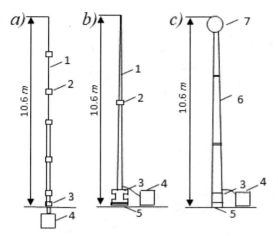

Fig. 5.10. Ship antennas of HF range *NT-66047* (*a*), *AS-2537A/SR* (*b*) and *AS-2807/SRC* (*c*). 1 – element, 2 – screw tie, 3 – base insulator, 4 – matching device, 5 – base, 6 – aluminum box, 7 – damper, 8 – feedthrough insulator.

(Fig. 5.10*c*). The antenna height is 10.6 m, the width at the base is 0.3 m, and at the top it is 0.025 m. The operating range in the transmission mode is from 2 to 32 MHz. At the base of each antenna, a matching device is installed that provides *TWR* in the cable of at least 0.3.

Whip antennas of the indicated length do not have the desired directional pattern in the upper part of high-frequency range, in the vertical plane. A tunable matching device allows us to obtain the required *TWR* in the cable, but this device takes more time for tuning and does not allow a quick change of frequency of the radio transmitter. As already mentioned, whip antennas with such properties cannot be considered wide-range ones.

The task of creating wide-range antennas with a high level of matching requires a reduction of the antenna wave impedance. As follows from (3.49), *TWR* is maximal, if the reactive component of an input impedance is small and an active component is close to a wave impedance of a cable. It is possible to reduce the antenna wave impedance by increasing its transverse dimensions. To the mass and windage of the resulting design were not excessive, one may replace the cylindrical antenna by the antenna of several wires. A wire system can be installed around the rod antenna. An example of such a volume radiator is the *AS-2805/SRC* antenna, described in the book [83]. It is built on the basis of the *AS-2807/SRC* rod radiator, around which a four-wire construction is installed (Fig. 5.11*a*). The wires are stretched between the top (upper boundary) of the second aluminum element and the spacers. The antenna height is 10.6 m, and the width at the base is 2.45 m. The operating range is

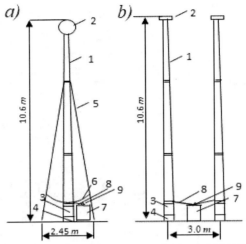

Fig. 5.11. Volumetric radiators *AS-2805/SRC*, based on one rod (*a*), and *OE-214/U*, based on two rods (*b*). 1 – aluminum element (box), 2 – damper, 3 and 9 – insulators, 4 – base, 5 – wire, 6 – connecting wire, 7 – matching device, 8 – jumper, 10 – spacer.

from 4 to 12 MHz. In the housing with a side 0.3 m long, near the antenna, a matching device is installed, connected to it with a flexible jumper.

To increase the matching level, one can use two rods. For example, the *OE-214/U* antenna consists of two *AS-2807* rod antennas (Fig. 5.11*b*) installed at a common site at a distance of 3 m from each other and connected by flexible jumpers to a common matching device.

Bulkier versions of volumetric antennas with different heights are created on the basis of a metal mast with a disk, on which a cylindrical or conical multi-wire radiator is suspended. The radiator can be installed around the ship's mast. It is fabricated of insulated mast wires, joined with each other, and connected to the central conductor of a cable (Fig. 5.12*a*). If necessary, such an antenna can be made separately from the mast and adjusted in height [84]. For this, wires of variable length are stretched along the generatrixes of a cylinder or cone between the disk mounted on the top of the central rod and the drum on the winch, to which an internal cable conductor is connected using a spring contact. The central rod itself can also be made expanding, for example, telescopic (Fig. 5.12*b*).

In the idle position, the telescopic elements are inserted into each other, and the wires are wound on the drum. When lifting the antenna, the rod is extended, and the wires are wound from the drum. When lowering the antenna, the wires are wound on the drum and wherein pull the upper element of the telescopic structure. The tracking drive allows to extend the antenna to the desired height in accordance with the operating frequency.

Problems of Design and Placement of Antennas 263

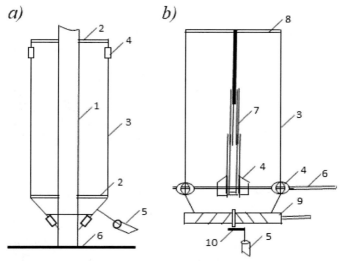

Fig. 5.12. Stationary (*a*) and expanding (*b*) volumetric radiators.
1 – mast, 2 – spacer, 3 – wire, 4 –insulator, 5 – cable, 6 – deck, 7 – telescopic element, 8 – disk, 9 – drum, 10 – spring contact.

This antenna is an example of a tunable radiator, whose height changes with a change of the operating frequency that allows optimal electrical characteristics at each frequency. In the idle position, the antenna dimensions are small, simplifying the task of storage and transportation. An expanding antenna can be also built on the basis of thin-walled elastic profiles or telescopic structures. Elastic profiles are able to forcefully take the form of a flat tape and in this form are twisting into a roll. Therefore, the antennas, in which elastic profiles are used are named tape antennas. If necessary, the profiles are deployed due to the energy stored during twisting and acquire a given shape. Therefore, a thin-walled profile can be used as an expanding transformable elastic element. An expanding antenna created on the basis of such an element is described in [85] and is presented in Fig. 5.13*a*. One end of the elastic element is rigidly fixed to the insulator, and the other is twisted on a drum equipped with a locking device (pin). In pulling out the pin, the antenna unfolds and assumes the operating position.

As shown in Fig. 5.13*b*, the elastic element may have a spiral profile. The drum, on which the elastic element is twisted can be connected through the gearbox with the drive and, with the help of the drive, the height of the antenna can be adjusted. To increase the rigidity of the structure, the antenna can be fabricated from two or more elastic elements, twisted on different drums (Fig. 5.13*c*). The diameter of the profile of the internal tape must be greater than the diameter of the profile of the outer tape [86].

264   ANTENNAS: Rigorous Methods of Analysis and Synthesis

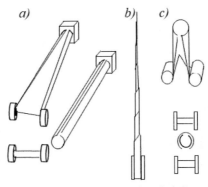

Fig. 5.13. Self-expanding antennas with a grooved (*a*) and a spiral (*b*) profile, from two elements (*c*).

Another variant of the extendable antenna is a telescopic design. This is a mounting on an insulator system of hollow telescopic elements inserted one into one another. The central conductor of the cable is connected to the lower (fixed) element of the antenna, and the outer shell is connected to the body of the object. Locking mechanisms and contact brushes are installed on the upper ends of the elements. To raise and to lower the telescopic antenna one can use various types of drives: cable, rack, pneumatic and other types. The circuit of a pneumatic drive is shown in Fig. 5.14. This drive uses compressed air for lifting the antenna. To lower the antenna, one can use a hawser, one end of which is connected to the upper element of the antenna, and the other is wound on a drum [87].

A smooth change of height when changing the operating frequency complicates the design of the antenna and its control system, and increases the tuning time and reduces the operational reliability. Therefore, in practice, a stepped extension system is often used with several fixed heights selected so as to reduce dips of the *TWR* curve in the cable. Unfortunately, this approach helps poorly. A significant improvement of electrical characteristics can be achieved by creating an expanding impedance radiator on the base of a telescopic design. For this, a metallic disk is mounted on the top of each telescopic element. Figure 5.15 demonstrates the positive effect of such a replacement, comparing the level of matching with the cable of three antennas: an impedance antenna 3.5 m high with a central rod of a diameter 0.02 m and a disk of a radius 0.3 m (curve 1); a rod antenna 6 m high and a diameter 0.02 m (curve 2) and a rod antenna 3,5 m high and a diameter 0.02 m (curve 3). The circles in the figure show the results of an experimental check of the first antenna. From the figure it follows that the *TWR* of the antenna with the disks is significantly higher than the *TWR* of the rod with the same height and approaches the *TWR* of the rod of double the height.

Problems of Design and Placement of Antennas 265

Fig. 5.14. Pneumatic mechanism for lifting telescopic antenna.
1 – element, 2 – insulator, 3 – cable, 4 – tube for air supply, 5 – hawser, 6 – drum.

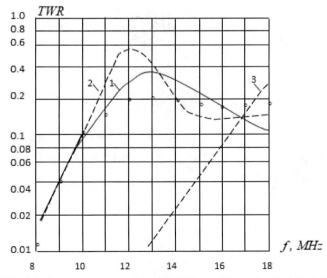

Fig. 5.15. *TWR* of the impedance antenna 3.5 m high (1) and of the rods with the height 6 m (2) and 3.5 m (3).

Antennas with discs can be used on small ships, including river passenger ones. They are made in the form of a lightweight, durable and, if necessary, folding design.

Volumetric and expanding constructions permit improve the electrical characteristics of antennas, in particular ship ones, but they do not solve the problem of creating non-tunable wide-range antennas. In the course of solving this problem, a method based on the use of concentrated loads was considered. First, resistive loads were examined [88], followed by complex impedances in the form of a resistor and an inductance coil, connected in parallel. The reason for the transition to a complex load was the hope of reducing losses and increasing efficiency. One of the design options for an antenna with such a load is a broadband antenna [89], whose general view is shown in Fig. 5.16.

The antenna load is made in the form of a broadband absorber with a maximum dissipation power of 1.5 kW. It created in the form of a coaxial line with spiral wires of resistive materials. Its structure permits a uniform power dissipation. The absorber input is located at a height of $\frac{2}{3}L$ from the ground, where L is the height of the antenna. The purpose of its inclusion is to provide a traveling wave mode on a lower segment of the antenna. A spiral-exponential transformer is installed at the base of the antenna to convert the active component of its input impedance to cable wave impedance.

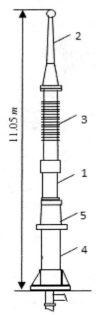

Fig. 5.16. Antenna with a complex load.
1 – trunk, 2 – rod, 3 – absorber, 4 – transformer, 5 – surface insulator.

Problems of Design and Placement of Antennas 267

The operating frequency range of the described antenna is from 4 to 25.6 MHz. A similar antenna circuit designed to operate at frequencies from 2 to 12 MHz was proposed in [90]. Low efficiency is a very serious disadvantage of antennas with resistive load. Therefore, they became quickly outdated in antenna technology. In order to increase their efficiency, a multi-radiator antenna was developed, in which around a central rod with a broadband absorber a structure of several radiators of different lengths connected to this rod at the base was installed. The circuit of such an antenna is given in Chapter 1 (see Fig. 1.12). Unfortunately, this did not significantly improve its electrical characteristics. Attempts to create a wide-range of antenna from several radiators in the absence of resistive load were also futile. The point is not that the diameter of such an antenna was unacceptable. The characteristics of the antenna with such a diameter remained unsatisfactory.

The way out was found in Hallen's proposal to use purely reactive capacitive loads included along the entire length of the linear antenna, and to create an in-phase current distribution with help of these loads. The results obtained during the analysis and development of this proposal are described in detail in the Chapter 3.

A general view of the antenna K674-3, based on a circuit of a linear radiator with capacitive loads, is shown in Fig. 5.17. The antenna is made in

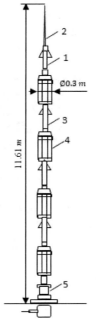

Fig. 5.17. Antenna *K674-3* with capacitive loads.
1 – tube, 2 – rod, 3 – dielectric insert, 4 – system of wires, 5 – insulator.

two versions: tunable and non-tunable. The first option provides a higher level of matching and an increased pattern factor. In this option switches used for changing of capacitive loads without changing the dimensions of the antenna and its components, are used. Loads are performed in the form of lumped elements (capacitors). The antenna height is close to 12 m. Its trunk consists of steel tubes. To increase the antenna diameter, around each steel tube the wires systems located along the generatrixes of the cylinders are installed. They are fabricated from stranded wires and connected to the ends of the tubes. As a result, the antenna diameter increases to 0.3 m, that makes it possible to increase the *TWR* to 0.4 at frequencies from 6 to 30 MHz and to 0.45 at frequencies from 30 to 60 MHz.

Loads and switches are placed inside the dielectric inserts connecting the metal tubes of the trunk. Switching is done remotely. In the base of the antenna and inside the inserts connectors of control circuits are installed. They are open during operation of the antenna with the connected radio transmitter, and when switching capacitors, are closed sequentially starting from the base of the antenna. That prevents shunting of loads and the insulator by control wires.

By reducing the requirements for electrical characteristics, the number of sub ranges can be reduced or tuning cancelled altogether. A non-tunable antenna with capacitive loads provides *TWR* of at least 0.2 at frequencies from 8 to 20 MHz and at least 0.3 at frequencies from 20 to 60 MHz. In the middle and upper part of the range (from 24 to 30 and from 45 to 60 MHz at small angles to the horizon, the signal decreases significantly.

At the end of the section, one must say a few words about the great folded antennas, which are sometimes called loop antennas. They are used on longer waves. The principle of their action is based on the excitation of individual elements of ship structures: masts, towers, various columns, among others. Their application facilitates the solution of the problem of placement of radiators on ships, whose relevance is caused by free territory reduction for the placement of antennas.

A great folded antenna consists of a grounded design (for example, a superstructure) and a system of thin conductors located in parallel to this design or at a slight angle to it. In fact, this is an asymmetric folded radiator of conductors with different diameters. The exciting conductor is made from wires, and the second wire (closed to the ground) has the shape of a metal superstructure. The study of electromagnetic waves in systems consisting of parallel thin conductors is the subject of article [28]. In [28] it is shown that the current in each conductor of such a structure can be divided into in-phase and anti-phase, and the entire structure can be considered as a combination of a linear radiator and non-radiating long lines. An example of a more complex structure of this type is an asymmetric multi-folded radiator.

A general view of one variant of such an antenna with bottom excitation is shown in Fig. 5.18. The conductors are made of metal tubes or wires and are placed on insulators along the metal superstructure at a distance of 0.4–1.0 m from it. Their upper ends are connected to this structure. The bottom ends are combined and with help of flexible jumper connected to the cable through a coupling. As is seen from figure, two antennas (A1 and A2) are installed on one superstructure.

Fan-shaped antennas built according to the circuit with upper excitation are designed for operating with wide-range radio transmitters. Using a similar scheme, dual fan-shaped antennas, built in the USA and described in [83], are manufactured. One of them is antenna *AS-2803/SRC*, made of two fans. Each fan consists of three wires located in one plane. A general view of antennas of this type is shown in Fig. 5.19.

The previous two sections of Chapter 5 are devoted to ship antennas of medium-frequency and high-frequency waves and also to questions about development of these antennas', taking conditions of their use and the present level of their characteristics, into account. From the point of view of a logical sequence of exposition, we should go further to directional antennas of high-frequency waves, taking into account the diversity of these antennas and the importance of their use. The use of directional antennas on ships is an effective method of increasing the range and reliability of short-wave radio communications. These antennas create predominant signal radiations in a given direction.

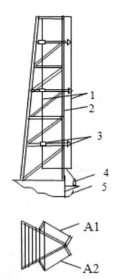

Fig. 5.18. Great folded antenna.
1 – exciting conductor, 2 – mast, 3 – insulator, 4- coupling, 5 – cable.

270 ANTENNAS: Rigorous Methods of Analysis and Synthesis

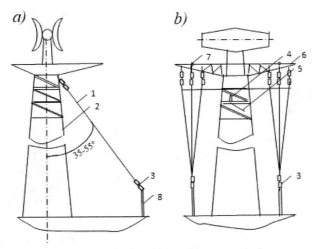

Fig. 5.19. Great folded antenna with upper excitation.
1 – exciting conductor, 2 – tower, 3 – insulator, 4 – coupling, 5 – cable, 6 – ray, 7 – connector, 8 – rod.

To provide radio communications at different times and at any latitude, the directional antennas must operate in a wide frequency range and have high electrical characteristics: an appropriate level of matching and acceptable efficiency. In the vertical plane, the main radiation should be directed at angles from 0° to 40° to the horizon. In the horizontal plane, the radiation pattern should not be too narrow so as not to lose communication due to the pitching of the ship in stormy weather or because of the deviation of the beam from an arc of large radius connecting the position points of the correspondents. The level of the side lobes determines the amount of interference to closely located receiving antennas, as well as the reverse echo, which troubles communication in the shortwave range.

To provide the ability to change the direction of the main radiation in accordance with the direction to the correspondent and the course of the ship, two options for changing this direction are used. The first of them is the mechanical rotation of a directional antenna (for example, log-periodic one) around a vertical axis. The second option is the electrical control of the directional pattern of a stationary antenna by changing the amplitudes and phases of the currents in its elements (phased array antenna).

Many pages already were devoted to directional antennas, due to their importance. In particular, methods of calculating directional characteristics and methods of increasing directivity are discussed in Chapters 3 and 4 of this book. The most effective and at the same time the most complex directional antenna of the decameter range is the log-periodic antenna. The book [10] details the new and almost unknown results obtained by A.F. Yakovlev on the way to improve the performance and reduce the dimensions of this antenna. His main achievement is the development of an asymmetric coaxial log-periodic antenna, which the author named the log-periodic monopole antenna (*LPMA*) by analogy with the *LPDA* (log-periodic dipole antenna). Chapter 3 of this book proposes a method for synthesizing the optimal characteristics of directional antennas, mentioning the electrostatic analogy method, and discusses the results of its application to log-periodic antennas.

The antenna array has got less attention. This array allows on the one hand the increase of the electromagnetic field strength in the far zone without increasing the power of a radio transmitter, and to increase the total power in the antenna array without exceeding a maximal allowable power in a separate radiator and a feeder, on the other hand. In ship conditions, it is difficult to place an antenna system that occupies a significant area. Therefore, large flat antenna arrays are almost never found on ships. But the linear array of radiators can be placed along the ship board. Application of such an array is very useful.

In principle, the array may consist of active and passive radiators. But the use of passive radiators requires the development of switched reactances, which must be designed for high power and also tune in the frequency range. At the same time, phasing of active radiators can be performed in pre-amplification stages, i.e., at a low power level. Therefore, preference should be given to the array from active radiators. It is also advisable to use uniform arrays, in which the amplitudes of the currents in the radiators are the same, and the phases vary linearly. It is known that one can optimize the radiation pattern of the antenna array, i.e., to select the amplitudes of currents in separate radiators to reduce the width of the main lobe of the array and to lower the level of the side lobes. However, the implementation of arrays with optimized directional patterns in a ship is difficult, since the influence of the surrounding metal bodies significantly distorts their shape.

As radiators in the phased array the wide-range antennas should be used.

In Fig. 5.20 the recommended layout of the ship's antenna array from eight radiators is shown. Along with the linear array an array of randomly located radiators, created on the basis of the existing fleet of antennas by converting

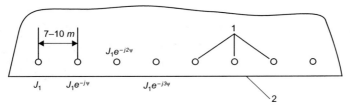

Fig. 5.20. Ship's antenna array.
1 – radiator, 2 – board of ship.

them into a single antenna system, can be implemented on the ship. Under the conditions of the ship the calculated phase shifts should be adjusted, taking into account the actual placement of the antennas in accordance with the results of measurements on the model or in natural conditions.

5.3 Antennas of meter and decimeter waves

At high frequencies, radio communication is provided mainly in conditions of direct visibility (radio waves decay rapidly beyond the horizon line). Such radio communication is of a high quality and reliability. Compared to lower frequencies, communication in the meter and decimeter ranges has several advantages. This results in an increase of the number of channels, a decrease of the power and dimensions of radio equipment, as well as a simplification of its design.

Currently, two options of radio communication are used in the considered ranges. The first is a direct communication between two objects, the second is a communication using a beam, reflected from a highly flying object. The requirements of the antennas used in both communication options have their own characteristics. If radio communication is carried out in conditions of direct visibility, then to increase the distance the antennas must be raised above the ground. As a rule, they are placed on superstructures or masts and are connected to radio stations by long cables. Based on the conditions of placement, the antenna design should be as simple as possible, lightweight and durable, i.e., insensitive to mechanical loads—pitching, vibration, shock, etc. In this case, antennas with vertical polarization are mainly used.

Radio communications through an artificial satellite primarily requires the antennas to provide a comprehensive azimuth and elevation overview. The second condition is to ensure a high gain. Both of these requirements must be met through the use of rotary antennas with a large directivity. Antenna for satellite communications is installing on the deck or on a special base, providing gyro stabilization. It must radiate and receive electromagnetic waves with a circular polarization.

The following typical designs of ship antennas for various purposes, operating in the meter and decimeter ranges are given further. In Fig. 5.21 the variants of symmetric radiators (dipoles) are shown. In Fig. 5.22 variants of asymmetric radiators (monopoles) are presented. Discone antennas are presented in Fig. 5.23, coaxial radiators are shown in Fig. 5.24. Figure 5.25 demonstrates linear arrays with vertical polarization, Fig. 5.26 demonstrates antennas with elliptical polarization. In Figs. 5.27 and 5.28, a general view of two variants of radio buoy is given—with a whip antenna and with a multi-wire radiator in an inflatable shell.

Problems of Design and Placement of Antennas 273

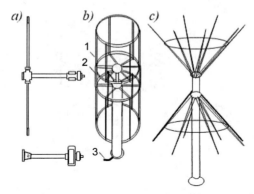

Fig. 5.21. Symmetric radiators of the meter range: a – NT-66095, b – AS-2231, c – AS-2811.
1 – antenna wire, 2 – dielectric insert, 3 – cable.

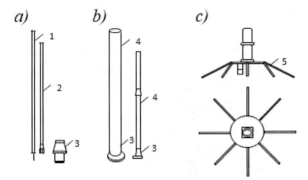

Fig. 5.22. Thin asymmetric radiators (monopoles): a – AS-1729, b – AS-3226, c – AS-390.
1 – upper rod, 2 – lower rod, 3 – insulator, 4 – fiberglass body, 5 – counterpoise.

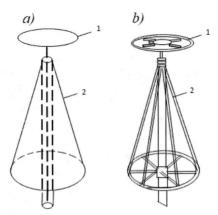

Fig. 5.23. Discone antennas: a – circuit, b – multi-wire construction.
1 – disc (counterpoise), 2 – cone.

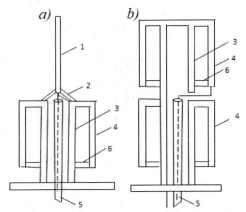

Fig. 5.24. Coaxial radiators with one (*a*) and two (*b*) quarter-wave arm.
1 – a rod, 2 – an insulator, 3 – rack, 4 – arm, 5 – cable, 6 – dielectric insert.

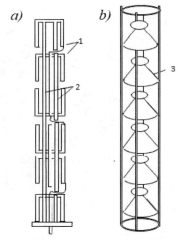

Fig. 5.25. Vertical linear arrays of coaxial (*a*) and discon (*b*) radiators.
1 – coaxial antenna, 2 – rack, 3 – housing of discon antenna.

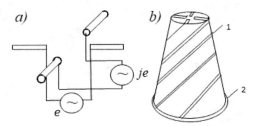

Fig. 5.26. Antennas with elliptical polarization: *a* – turnstile, *b* – four-way conical spiral.
1 – spiral, 2 – counterpoise.

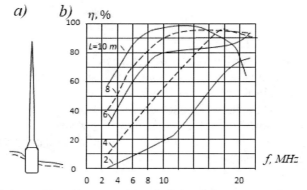

Fig. 5.27. Radio buoy with a whip antenna (*a*), and efficiency of an antenna and antenna tuner (*b*) as function of the whip height *L*.

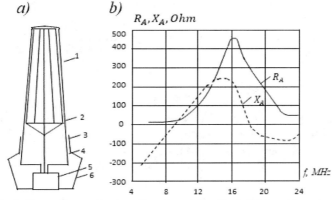

Fig. 5.28. Radio buoy with a multi-wire radiator in an inflatable shell (*a*), and input impedance of a multi-wire radiator (*b*).
1 – radiator, 2 – shell, 3 – screen, 4 – cup, 5 – radio transmitter, 6 – housing.

5.4 Influence of ship designs on antenna characteristics

As noted at the end of Section 5.2, the description of new results in the development of log-periodic antennas has deprived this topic of particular relevance.

Therefore, for the next Section the topic of the influence of metal structures on antennas characteristics, in particular on directivity characteristics, was selected. This topic is no less important for ship conditions. The dependence of the antennas characteristics on nearby metal bodies takes place for any antenna located on any object. But, as a rule, this influence is especially great and the task to reduce this effect is especially difficult when placing antennas on moving objects, in particular on ships. One of the most important and most difficult tasks of analyzing the antennas' characteristics is to take into account the influence of three-dimensional metal structures. A feature of the ship's antenna field is, as already mentioned, the limited space for placing antennas, in connection with which antennas are installed near various metal structures— masts, superstructures, chimney, rooms among others. These structures and a ship hull participate in the formation of electromagnetic radiation, and their influence on the characteristics of the antenna is often decisive. Such designs surround any antenna and are included in the structure of the antennas themselves. Knowledge of the laws of this influence is extremely important for predicting the actual characteristics of ship's antennas.

The analysis of the influence of these objects is complicated even in the simplest cases when, for example, only one superstructure is considered— in the form of a circular cylinder of infinite length or an elliptical cylinder. One can take into account the influence of the superstructure, considering it an additional passive antenna, similar to a thin linear radiator. But this approximation is too crude.

Over time, the influence of the hull and superstructures on the characteristics of antennas has been studied experimentally. But the practical possibilities of a comprehensive analysis with the subsequent selection of optimal solutions in this case are small. An analysis method based on replacing a metal body with a system of intersecting thin wires is more effective [25, 47]. In such a formulation, the problem is reduced to calculating the current distribution in a wire structure consisting of randomly oriented wire segments. The radii of the wires are considered small in comparison with the lengths of the segments and the wavelength. Knowing the currents along the wires, we can calculate all the electrical characteristics of the radiator.

A widespread use of a calculation program based on the above replacement and the Moments method substantially changed the situation. As a rule, the used algorithm includes one of the integral equations, the Galerkin method and a system of piecewise sinusoidal basis functions. In general, the new method of analysis has become part of the generalized method of induced emf.

One of the first problems considered using the new method was the problem of the influence of a cylindrical reradiator on the parameters of the antenna and antenna array [91]. The content of the problem is clear from Fig. 5.29. This figure shows a wire structure equivalent to a thick superstructure in the shape of a circular cylinder of finite length, near which a whip antenna is placed. As it follows from the figure, the metal cylinder is replaced by a wire structure with equally spaced wires located along its generatrixes and radii of a butt-end. The origin for the convenience of calculations is combined with the center of the cylinder base. The wire diameter for simplicity is taken equal to the diameter of the whip antenna. The surface of the earth is considered to

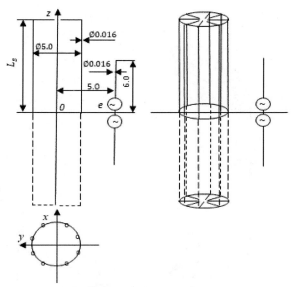

Fig. 5.29. Whip antenna near a metal superstructure.

be perfectly conductive, and the structure under consideration is symmetrical with respect to this surface. The mirror image is given in the figure by dotted lines.

The figures show the directional patterns of a whip antenna 6 m high and 0.016 m in diameter near superstructures of different heights L_s at three frequencies of the HF range in the horizontal (Fig. 5.30) and vertical (Fig. 5.31) plane. The calculated values are represented by curves, and the experimental values are represented by points of various shapes.

The calculation results show that the presence of the superstructure changes all the characteristics of the antenna. A good agreement between the calculation and the experiment confirms the correct choice of the wire structure. Calculations showed that the lengths of the segments into which the conductors are divided should not exceed 2λ, and the number of wires of the selected structure in such problem should be chosen based on the condition that the distance between the parallel wires is of the order of $(0.6–0.8)\lambda$. When these conditions are met, the shape of the radiation pattern becomes stable.

Calculations allow us to find the minimum height of the cylindrical superstructure, at which the characteristics of the antenna located close to it coincide with the characteristics of the same antenna located next to the infinitely long reradiator. From the calculations it follows that for this the height of the reradiator should be greater than the height of the antenna by about a quarter of a wavelength.

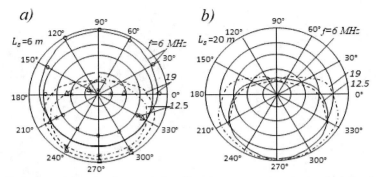

Fig. 5.30. Horizontal directional patterns of a whip antenna near a superstructure high 6 m (*a*) and a superstructure high 20 m (*b*).

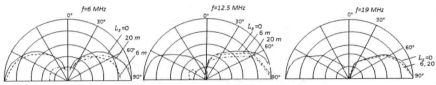

Fig. 5.31. Vertical directional patterns of a whip antenna near a superstructure.

At the next stage of work, similar results were obtained for antennas located near superstructures of complex shape. Comparing the influence of similarly sized superstructures on the properties of antennas equally distant from the axes of differently shaped superstructures showed that the results of this influence are basically the same.

Different authors mentioned different magnitudes for the minimal number of conductors, which ensure the equivalence of electrodynamic properties of a metal body and its model. These differences are caused by such circumstances as solving different problems and using different basis functions. It is advisable to arrange the conductors of the structure along the most probable directions along which current flows. In this case, the number of conductors and the amount of computation will be minimal.

The discussion results of this issue were later summarized in [47]. In this paper, three different types of problems are considered: calculation of directional patterns, mutual coupling, and input impedances. The directional pattern is not affected by minor changes in the distribution of a current, and accurate results are obtained with a small number of basis functions. Calculation of the interconnection and the input impedances requires an accurate representation of the current, and more basis functions are required to obtain accurate results. The presented table describes the various options of structures. For a single wire, when calculating the directional pattern, it is required to use from 6 to 10 basis functions per wavelength. Thick wires require more features than thin wires—up to 15 per wavelength. For wire meshes and surfaces the number of basis functions is given per square with a side equal to the wavelength and depends on the surface shape: for a flat one—about 40, for a three-dimensional one—more. For bodies of revolution, the number of functions per wavelength along the generatrix should be of the order of 20 and depends on the maximum circumference length. Active elements require more functions than passive ones.

An excessive increase in the number of functions can worsen the result, since rounding off the calculated magnitudes will lead to a significant error.

The goal of another task was to analyze the characteristics of the antenna depending on the construction, on which it is installed. In real conditions, ship's short-wave antennas are installed on various objects and elevations—racks, masts, crossbars, brackets, etc. The characteristics of such antennas differ significantly from the characteristics of antennas mounted on an infinite ideally conducting plane. When placing antennas on an object, much attention needs to be paid to the shape of their directional patterns in the vertical plane. The input impedance, and consequently, the operation mode of the cable and the load impedance of the radio transmitter, also depend on dimensions of a rack or mast, on which the antenna is placed.

The antenna mounted on a rack can be considered as a radiator with an excitation point shifted from the base. For generality, we assume that the propagation constants of the wave along the antenna k_1 and the rack k_2 have different magnitudes and differ from the propagation constant of the wave k in air. In this case, the result can be used both for a metal antenna and for an antenna with impedance boundary conditions. In addition, different values of the propagation constants make it possible to take into account the decrease in wave velocity along a thick structure.

We define the electric field in the far zone, considering a radiator of finite length as the sum of elementary electric radiators. In accordance with (1.55) we write:

$$E_\theta = j\frac{30k\exp(-jkR)\sin\theta}{\varepsilon_r R}\int_{-L}^{L} J(z)\exp(jkz\cos\theta)dz = j30kJ(h)\frac{\exp(-jkR)}{\varepsilon_r R}H(\theta),$$

(5.1)

where,

$$H(\theta) = \frac{\sin\theta}{J(h)}\int_{-L}^{L} J(z)\exp(jkz\cos\theta)dz. \quad (5.2)$$

Here $J(z)$ is the generator' current, R is the distance from the origin (from the rack base) to the observation point, $J(z)$ is the current along the radiator. The integral is calculated both along the radiator (along the antenna and rack), and along its mirror image. The current distribution along the antenna is similar to the current distribution along the long line open at the end:

$$J(z) = J(h)\frac{\sin k_1(L - |z|)}{\sin k_1(L - |h|)}, h \le |z| \le L. \quad (5.3)$$

The distribution of current along the rack is similar to the distribution of current in a short circuit (to a ground):

$$J(z) = J(h)\frac{\cos k_2 z}{\cos k_2 h}, 0 \le |z| \le h. \quad (5.4)$$

Then,

$$H(\theta) = H_1(\theta) + H_2(\theta), \quad (5.5)$$

where,

$$H_1(\theta) = \frac{2k_1 \sin \theta}{\sin k_1 L_1 (k_1^2 - k^2 \cos^2 \theta)}.$$

$$[\cos(kL \cos \theta) - \cos k_1 L_1 \cos(kh \cos \theta) + k \cos \theta \sin k_1 L_1 \sin(kh \cos \theta)],$$

$$H_2(\theta) = \frac{2 \sin \theta}{\cos k_2 h (k_2^2 - k^2 \cos^2 \theta)} [k_2 \sin k_2 h \cos(kh \cos \theta) - k \cos \theta \cos k_2 h \sin(kh \cos \theta)].$$

The quantity $H(\theta)$ is sometimes called the generalized effective height of the antenna. It connects the electric field in an arbitrary direction with the generator' current:

$$H(\theta) = h_e F(\theta, \varphi), \quad (5.6)$$

where h_e is the effective height of the symmetric radiator, and $F(\theta, \varphi)$ is the directional pattern. Equalities (5.1), (5.2) and (5.5) allow us to determine the field in the far zone and the directional pattern of the antenna mounted on the rack. For $k_1 = k_2 = k$ the formulas are significantly simplified:

$$E_\theta = j60J(h)\frac{\cos kh \cos(kL \cos \theta) - \cos kL \cos(kh \cos \theta)}{\sin k_1 L \cos kh \sin \theta} \frac{\exp(-jkR)}{R}. \quad (5.7)$$

Putting $h = 0$, we obtain an expression for the directional pattern of the impedance radiator:

$$E_\theta = j60J(h)\frac{k_1 k \sin \theta [\cos(kL \cos \theta) - \cos k_1 L]}{(k_1^2 - k^2 \cos^2 \theta) \sin k_1 L} \frac{\exp(-jkR)}{R}.$$

The effective height h_e of the antenna on the rack is determined from expression (5.5) by substitution $\theta = \pi/2$:

$$h_e = H(\pi/2) = h_{e1} + h_{e2}, \quad (5.8)$$

where $h_{e1} = \dfrac{1}{k_1}\tan\dfrac{k_1 L_1}{2}$, $h_{e2} = \dfrac{1}{k_2}\tan k_2 h$. The radiation resistance of an asymmetric radiator with a shifted excitation point is,

$$R_\Sigma = 40 k^2 h_e^2 = 40\left(\dfrac{1}{k_1}\tan\dfrac{k_1 L_1}{2} + \dfrac{1}{k_2}\tan k_2 h\right)^2. \tag{5.9}$$

The reactive component of the input impedance is,

$$X_A = X_{A1} + X_{A2}, \tag{5.10}$$

where $X_{A1} = -60\ln\dfrac{2L_1}{a_1}\cot k_1 L_1$ is the reactive component of the antenna input impedance, $X_{A2} = 60\ln\dfrac{2h}{a_2}\tan k_2 h$ is the reactive component of the rack input impedance.

The radiation patterns of antennas with a length $L = \lambda/4$, mounted on racks of different heights are presented in Fig. 5.32. The figure shows that a change in the height of the rack leads to a significant change in the directional pattern of the radiator.

Using programs for calculating electrical characteristics of antennas based on the Moments method and calculating the input impedance of the antennas under consideration, it is easy to verify that placing the antenna on a rack significantly changes its input characteristics as compared with placement on a metal plane: resonances are shifted toward low frequencies, and the main lobe of the directional pattern in the vertical plane deviates from the ground at lower frequencies.

The above expressions and calculations are given for a perfectly conductive earth. If the conductivity of the underlying surface is not infinite (for example, the conductivity of the sea surface), then this surface does not create a system of currents equivalent to a mirror image. But when calculating the fields at a large distance from the antenna, we can assume that the surface creates a distorted mirror image equivalent to the current field that in contrast to −1 is

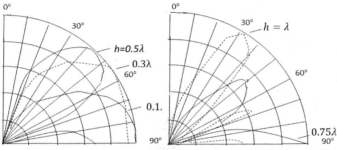

Fig. 5.32. Vertical directional patterns of an antenna with length $L = \lambda/4$ on racks of different heights.

multiplied by some reflection coefficient. Since this coefficient depends on the angle of the beam, the equivalent current of the distorted mirror image depends on the point, at which the beam was reflected, and accordingly depends on the observation point. In this case, the current distribution along the mirror image of the antenna and the rack is determined by the equalities,

$$J(z) = J(h)R\frac{\sin k_1(L-|z|)}{\sin k_1(L-|h|)}, -L \le |z| \le -h;$$

$$J(z) = J(h)R\frac{\cos k_2 z}{\cos k_2 h}, -h \le |z| \le 0. \tag{5.11}$$

Here R is reflection coefficient,

$$R = |R|exp(j\Phi) = \frac{\varepsilon'_r \cos\theta - \sqrt{\varepsilon'_r - \sin^2\theta}}{\varepsilon'_r \cos\theta + \sqrt{\varepsilon'_r - \sin^2\theta}}, \tag{5.12}$$

ε'_r is relative complex permeability of the medium equal to $\varepsilon'_r = \varepsilon'/\varepsilon_0 = \varepsilon_r - j60\lambda\sigma$, ε_0 is the absolute permeability of the air, λ is the wave length, σ is conductivity of the medium.

In Fig. 5.33 directional patterns of antennas with a length $L = \lambda/4$, mounted on racks of different heights are given for three values of the product $\lambda\sigma$: 3 cm, 6 cm and ∞. The relative dielectric constant $\varepsilon_r = 80$. It can be seen from the figures that decreasing the conductivity of the underlying surface leads to a weakening of radiation at small angles to the horizon. Moreover, in the case of a short rack, the directional pattern have one lobe. If the conductivity of the medium is high, then the lobe is pressed to the surface. With a decreasing medium conductivity a dip is formed in the pattern, in the horizontal direction.

In the case of a long rack, the vertical directional pattern consists of several lobes. Decreasing the medium conductivity leads to a sharp decrease in the lower lobe of the directional pattern, i.e., to reduce radiation

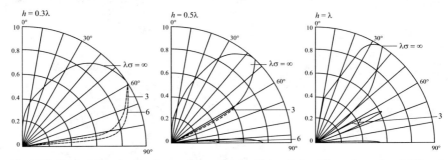

Fig. 5.33. Vertical directional patterns of the antenna with length $L = \lambda/4$, on the racks of different heights depending on surface conductivity.

at angles optimal for radio communications. Thus, the parameters of the underlying surface significantly affect the characteristics of the antennas and should be taken into account when developing and installing them.

The methodology and results of calculating the resistance loss of a metal antenna with a shifted excitation point in the decameter wavelength range depends on the conductivity of the underlying surface are described in [35].

Along with other options, ship's antennas of the HF range are often mounted on a bracket that is installed on one of the ship's decks (Fig. 5.34). Six-meter whip antennas are usually placed on brackets 0.5 m long, and ten-meter whip antennas and other close to them in height are placed on 1 m long brackets.

The characteristics of these antennas mounted on the brackets are substantially differ from the characteristics of antennas located on the infinite perfectly conducting plane. Previously, an influence' study of the ship's hull and other metal structures on these characteristics was carried out, as a rule, experimentally. But measurements of this kind both on the objects and on the models are very time-consuming and do not allow identification of the basic laws of this influence. The analysis method, based on replacing metal surfaces with a system of thin wires, turned out to be more effective [92].

In Fig. 5.35 the wires structure for solving such problems is shown. A whip antenna with a height of $L = 6$ m and an average radius, $a_1 = 0.008$ m is mounted on a bracket of a length $b = 0.5$ m with an equivalent radius $a_2 = 0.1$ m. The height of the bracket above the lower deck is, $h = 8$ m.

The horizontal surface of the deck near the antenna is replaced by structures of 9 radial wires with length R located at the same angles. The vertical surface of the ship is replaced by a system of 7 wires located at the same angles. The mirror image is made similar to the main structure.

The results of calculating the electrical characteristics of an antenna mounted on a bracket depend on the choice of a length R of radial wires. In particular, calculations, performed at a frequency 5 MHz, were shown that directional patterns of a whip antenna of a height $L = 6$ m mounted on the

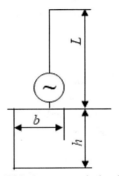

Fig. 5.34. Antenna on the bracket.

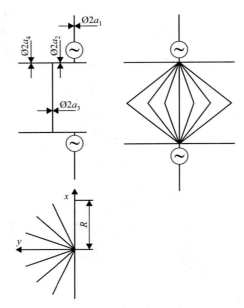

Fig. 5.35. Equivalent wires structure for a whip antenna mounted on a bracket.

bracket substantially change with increasing R, when R is less than R_0, where $R_0 = L + \frac{\lambda}{4} = 21$ m, and in the case of a further increase of R directional patterns ceases to change. It is also expedient to take the same radii of wires in order to exclude the additional calculation errors associated with sharply differing dimensions. Therefore, in the calculations, the lengths of the radial wires were taken equal to R_0, depending on the wavelength and antenna height.

In Figs. 5.36 and 5.37 the directional patterns of the described construction in the horizontal and vertical planes are given at different frequencies. The calculation results showed that the antenna directional patterns on the bracket in the vertical plane are complex, and at high frequencies they are multi-lobe. The corresponding curves for the antenna mounted on the plane (they are indicated by asterisks *) have a simpler shape in the form of a single lobe. Only at a frequency 30 MHz a small side lobe appears.

The calculation of the input impedance of the antenna also shows that installing the antenna on the bracket significantly changes both the active and reactive components of this impedance. In particular, the frequency of the first serial resonance decreases (similarly to the radiator mounted on a rack), and the frequency of the first parallel resonance (in contrast to the radiator mounted on a rack) increases.

Comparison of the described method for calculating the characteristics of the antenna mounted on the bracket with other methods, for example, with

Problems of Design and Placement of Antennas 285

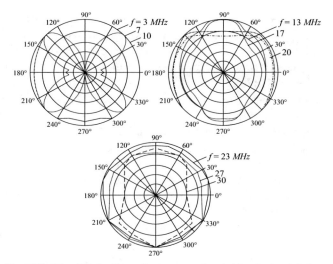

Fig. 5.36. Directional pattern of antenna on a bracket in a horizontal plane.

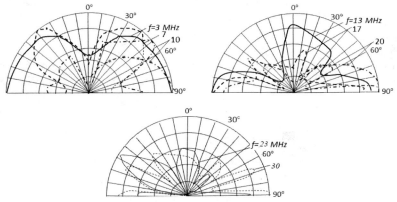

Fig. 5.37. Directional pattern of antenna on a bracket in a vertical plane.

the combined method based on Moments method and geometric diffraction theory [93], shows that the described method allows us to take structural details into account and ensure high accuracy. The combined method also was applied to the radiator mounted on the bracket. This structure was considered as the radiator located on the wedge. But the combine method does not take into account the presence of the bracket and its dimensions.

A significant drawback of this approach was that it turned out to be impossible to calculate the input impedance of the antenna located near the end of the wedge. The presence of the bracket and its dimensions were not taken into account in principle. In the method, described above, the wedge

286 ANTENNAS: *Rigorous Methods of Analysis and Synthesis*

dimensions were taken into account using radial wires of limited length along the assumed directions of the currents. The mutual impedances between short radiators located at a large distance from each other are small, and taking them into account does not refine the result, but leads to gross errors related with rounding of the results. To get rid of gross errors caused by large dimensions of the equivalent vertical metal wedge, it is necessary to limit the length of the wires by a certain magnitude.

The method for solving the problem used in [93] is based on the joint use of two methods, as already mentioned. One of them, the method of moments, provides accurate results. Another method, based on the geometric theory of diffraction, is purely approximate. In connection with this difference, the question of the applicability of this approach to a problem requiring an exact solution, arises. The need for an exact solution follows from the fact that using the approximate method does not allow us to arrive at a useful result. The relevance of the issue raised is associated with the frequent use of the described approach. A negative answer cannot be considered unambiguous; a positive answer requires careful analysis.

In Fig. 5.38 a short-wave antenna installed on a cross-beam (in other words, on a horizontal rod, mounted on a ship's mast) is presented. This installation of a whip antenna was used for designing a ship along with the installation of antennas on racks and brackets. Wire structures of various complexity (Fig. 5.39) were considered in the calculation of such antennas' characteristics. The first of these structures (Fig. 5.39*a*) can be used for a

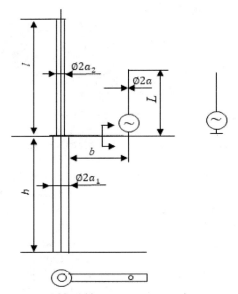

Fig. 5.38. Whip antenna on a cross-beam.

relatively thin mast, and the other two structures are suitable for thicker structures, including superstructures. During a study, various options were considered. In the first structure, the number of wires in different sections were different, but the results weakly depended on these changes. Therefore, while making calculations for the following options, in addition to the main wire, only two wires were used: one each for the upper and lower sections of the mast, respectively.

The study showed that the directional pattern of these antennas in the horizontal plane is primarily determined by the elements of the surrounding structure, located at the same level with the antenna. In the option, presented in Fig. 5.39*a* it is close to the circumference (see Fig. 5.40). In the options, presented in Fig. 5.39*b,c* it resembles the directional pattern of an active radiator located next to passive one. The directional pattern of the antenna in the vertical plane changes significantly when moving it from the ship's deck to the beam-cross and resembles the pattern of the antenna on a rack whose height is close to the height of the mast (Fig. 5.41). In this case, radiation into

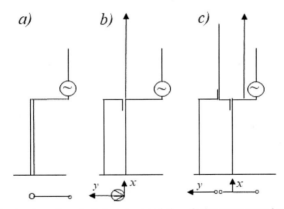

Fig. 5.39. Wire structures for calculating characteristics of antennas mounted on a cross-beam.

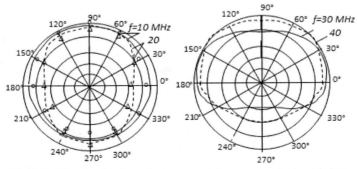

Fig. 5.40. Directional pattern of antenna on a beam-cross in a horizontal plane.

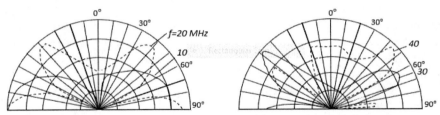

Fig. 5.41. Directional pattern of antenna on a beam-cross in a vertical plane.

the side of the mast is usually greater than to the opposite one. The second lobe of the vertical directional pattern becomes dominant at a frequency of 20 MHz. The currents excited in a horizontal beam-cross create the radiation to the zenith' direction of the zenith'. The input characteristics of this antenna are very different from the input characteristics of an antenna located on a perfectly conducting plane. The *TWR* level decreases, the first series resonance shifts toward low frequencies. On the whole, the characteristics of antennas mounted on the beam-cross are significantly worse than the characteristics of antennas located on the deck.

Experimental verification was carried out on models. The results are given in Figs. 5.40 and 5.41. The coincidence of the calculated data with the experiment confirms the effectiveness of the used analysis method. Details of the calculations were published in [35] and transferred to the designers for use in further work.

The developed programs saved performers from the need to idealize the shape of a superstructure and replacing it with an ellipsoid or an infinitely long cylinder, and also neglecting at that by segments and radii small compared to the wavelength, antenna length and the superstructure height. Analysis of the characteristics of HF antennas located on the rack, bracket and beam-cross allowed the determination of the degree of distortion in the antenna's characteristics depending on the geometric dimensions of these structures. In particular, the inadmissibility of installing antennas on the beam-cross was demonstrated.

The variety of existing conditions on ships for antenna installations makes it advisable to analyze typical installation options, in particular, typical arrangements for the relative positions of antennas near secondary radiators (masts, antennas, chimney, superstructures), which have a decisive influence on their characteristics. As a rule, the characteristics of antennas are especially distorted, if there are two superstructures near them, for example, a chimney and a mast or a ship mast and a free-standing antenna-mast.

Examples of two options of antenna placement and results of calculating their characteristics are given in Figs. 5.42 and 5.43. Figure 5.42*a* is a layout of a whip antenna on the upper bridge of a ship, Fig. 5.43*a* is an arrangement

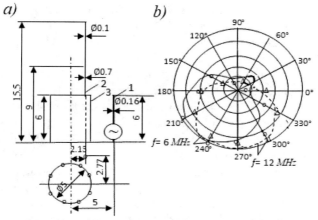

Fig. 5.42. Layout of the whip antenna on the upper bridge of the ship (*a*) and horizontal directional pattern of this antenna (*b*).
1 – whip antenna, 2 – mast, 3 – chimney.

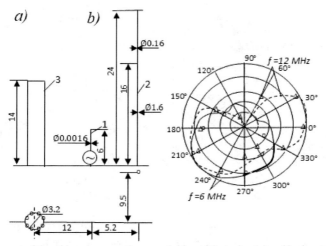

Fig. 5.43. Layout of the whip antenna on the upper bridge of the tanker (*a*) and horizontal directional pattern of this antenna (*b*).
1 – whip antenna, 2 – mast, 3 – chimney.

of a whip antenna on the upper bridge of a tanker. In both cases, the antennas are located near the chimney and mast, the dimensions and location of which are visible from the figures. The dimensions are given in meters, with the numbers marked: 1 – antenna, 2 – mast, 3 – chimney. During the calculation, the chimney was replaced by a wire structure of eight wires located along the generatrix of the circular cylinder and the radii of its butt-end, and the mast was replaced with two parallel wires of different diameters.

290 ANTENNAS: Rigorous Methods of Analysis and Synthesis

The results of calculations and experimental verification of the directional patterns of the whip antennas in the horizontal plane are given in Figs. 5.42b and 5.43b at frequencies 6.5 and 12.5 MHz. It can be seen from the figures that the superstructures significantly change the characteristics of the whip antennas, and experiments confirm the calculations. The Table 5.1 shows the input impedance and the gain of the antennas located in accordance with the figures in comparison with input impedance and the gain of a half-wave asymmetric radiator located in free space (at the same frequencies).

In connection with these results, in order to avoid misunderstandings, it is necessary to recall that, as shown in [27], the use of loads provides freedom of choice of the antenna length, if there are design capabilities. This freedom allows us to obtain the required characteristics in the desired operating range with the help of radiators of different dimensions. The choice of antenna length permits the reduction of the effect of closely spaced metal structures, such as superstructures, on the directional pattern of an antenna or antenna array.

In Fig. 3.16a in Section 3.5 the relative positioning of a metal superstructure and an asymmetric vertical radiator is presented. Two variants of radiators are considered: 1 – an antenna without loads 6 m high and 0.016 m in diameter, 2 – an antenna 12 m high and 0.06 m in diameter, with nine capacitive loads providing optimal electrical characteristics at frequencies from 8 to 22 MHz. Figure 3.16b shows results of calculating directional patterns of both radiators in a horizontal plane at two frequencies in the short-wave range (12.5 and 19 MHz). As can be seen from the figure, the radiation of a standard antenna (curves 1) in the direction of a superstructure decreases sharply. Using antennas with loads reduces the influence of the superstructure.

In Fig. 3.17 similar results are presented for a uniform linear array located near the described superstructure. The same two options are accepted as radiators. The phase shift between the elements of the array is taken equal to zero. The calculation results show that in the upper part of the range the superstructure has a weaker effect on the directional pattern of the antenna array than on the directional pattern of a single antenna because the superstructure does not interfere with the propagation of the electromagnetic wave from the

Table 5.1. Characteristics of whip antennas.

Variant of placement	Z_A, Ohm		G	
	f_A = 6.5 MHz	f_A = 12.5 MHz	f_A = 6.5 MHz	f_A = 12.5 MHz
In free space	6 – j353	39.7 + j20.9	1.9	2.0
In accordance with Fig. 5.42	0.9 – j350	22 + j37.5	5.9	5.9
In accordance with Fig. 5.43	5.7 – j349	40 + j13	5.0	3.7

side radiators. Nevertheless, in this case, the use of antennas with loads due to their greater height also weakens the influence of the superstructure.

From the described examples it follows that modern calculation programs not only reveal regularity of the harmful effect of closely spaced metal structures on the characteristics of ship's antennas, but also allow to propose methods that would substantially weaken this effect or even completely eliminate it.

To conclude this section, we present the results of the characteristics' analysis of one more antenna—a symmetric radiator located in a metal trough near a flat metal surface [94]. The task of calculating the electrical characteristics of a symmetric radiator near such a surface (Fig. 5.44a) is accomplished quite simply. But the task becomes much more complicated if the radiator is located in a metal trough of finite flush length with a metal body (Fig. 5.44b) on the surface of a ship or aircraft. We assume that the trough has the shape of a half cylinder with flat metal ends. The axis of the radiator coincides with the axis of the trough. Another variant of the same problem is the problem of a radiator placed in a dielectric capsule floating along the sea surface.

While solving the problem of a radiator near a perfectly conducting surface in accordance with the method of mirror images this surface can be replaced by a second radiator located behind it at the same distance and having the same geometrical dimensions, but with an oppositely directed exciting emf (Fig. 5.45a). For a radiator near a surface with a trough this method is not very productive.

In this case, an analysis method is useful, in which the metal surface is replaced by a system of thin wires [25, 47]. In such a formulation, as already

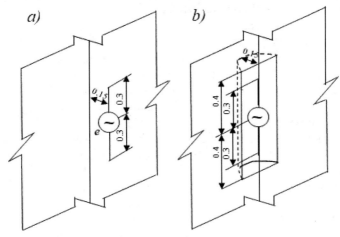

Fig. 5.44. Symmetric radiator near a flat metal surface (*a*) and near this surface with a trough (*b*).

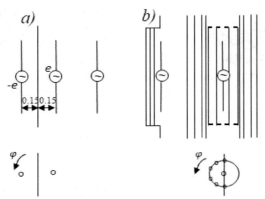

Fig. 5.45. Equivalent wire structure for a radiator near a flat metal surface (*a*) and near this surface with a trough (*b*).

said, the problem is reduced to calculating the current distribution in a wire structure consisting of randomly oriented wire segments. Knowing the currents along the wires, we can calculate all the electrical characteristics of the radiator.

While replacing a metal surface with a wire structure, as shown earlier, the correct choice of the number of wires and their location is of great importance. Increasing the number of wires and their lengths leads to increasing the calculation time (approximately proportional to the square of the number of short radiators into which the wires are divided). An excessive increase in the number of these radiators as already said can worsen the result. An insufficient number of wires and short radiators give a large error. The wires of the structure should be placed along the probable directions of propagation of surface currents. This allows us to reduce the number of wires and minimize the amount of computation.

For example, calculations show that the directional pattern of a whip antenna near a wire structure equivalent to a cylindrical superstructure practically does not change in real variants of the relative position of the antenna and the superstructure if the wire structure is supplemented with horizontal wires. Other antenna characteristics: input impedance, directivity—are also changing slightly. All these characteristics will change with the addition of horizontal wires only, if a distance between the antenna and the superstructure is small. But in practice, these options are of no use, since the antenna loses the properties of a radiator.

With this in mind, to analyze the electrical characteristics of the vertical radiator located along the axis of the cylindrical trough, the wires structure shown in Fig. 5.45*b* was used. As can be seen from the figure, the trough is replaced by a system of wires located along the generatrixes of the cylinder and the radii of its ends. The wires are continued in the vertical direction up and down at a distance equal to a quarter of the wavelength. The lengths of the arcs between adjacent wires forming a cylindrical surface are taken equal to 0.04λ. Outside the trough, the gaps between the wires are also the same. Each half-plane on both sides of the trough is replaced by a structure of three vertical wires spaced from the edge of the trough by 0.08λ, 0.16λ and 0.24λ. These wires are also extended beyond the upper and lower ends of the trough at 0.25λ. Concrete details of the structure are related with the assumption confirmed by calculations that sections of the metal surface closest to the radiator have the greatest influence on the characteristics of the radiator. These areas must be carefully modeled.

The described procedure was applied to calculating the electrical characteristics of a symmetrical vertical radiator with an arm length of 0.3 m, located on the axis of a cylindrical trough with a length of 0.8 m and a radius of 0.15 m. The calculations were performed at frequencies from 150 to 600 MHz, which corresponds to a radiator arm with a length from 0.15λ to 0.6λ. The radius of the radiator wire is 0.005 m. The magnitudes of the radii of the other wires of the structure for simplicity are taken of the same magnitude.

In Fig. 5.46a–c the horizontal directional patterns, calculated by the described method for a radiator located along the axis of the trough, are presented. For comparison, in Fig. 5.46d similar patterns are given for the same radiator located near a flat metal surface at a distance, equal to the depth of the trough (at a distance 0.15 m from it). According to the calculations, the backward radiation (in the direction of the trough) is small, that indicates a

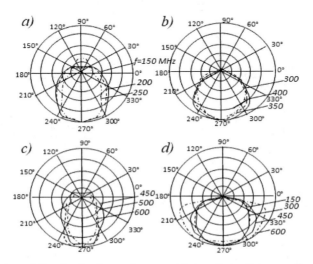

Fig. 5.46. Horizontal directional patterns of a symmetrical vertical radiator near the surface with a trough (*a–c*) and the flat surface (*d*).

294   ANTENNAS: Rigorous Methods of Analysis and Synthesis

good approximation of the metal surface by means of a wire structure. At the same time, the difference between the influence of a flat surface and a surface with a trough is clearly visible from the figure. In the case of the trough the directional pattern has a narrower lobe.

This conclusion is confirmed by the results of calculating the directivity of both radiators given in Table 5.2. The directivity of the radiator near a flat surface varies slightly in the frequency range. The directivity of the radiator placed along the axis of the trough increases with increasing frequency. At frequencies from 300 MHz and higher (with an arm length greater than 0.25λ) its directivity significantly exceeds the directivity of another radiator.

The directional patterns of both radiators in a vertical plane perpendicular to a metal surface are shown in Fig. 5.47. From these patterns, in particular, it follows that the presence of a trough reduces the electric length of the radiator: the second lobe appears at a higher frequency.

A similar conclusion can be made by comparing the input impedances of the radiators (Fig. 5.48). Curves 1 correspond to a radiator located at the surface with a trough, curves 2 – to a radiator near a plane. It can be seen that the second resonance of the radiator in the trough is located at a higher frequency. The resistance of this radiator is significantly reduced except for the area of the parallel resonance.

Table 5.2. Directivity of radiator.

f, MHz	150	200	250	300	350	400	450	500	550	600
near surface with trough	6.3	6.9	5.8	8.9	10.4	11.4	12.4	13.4	14.8	17.4
near plane	7.4	7.3	7.3	7.4	7.5	7.6	7.9	8.1	8.3	8.4

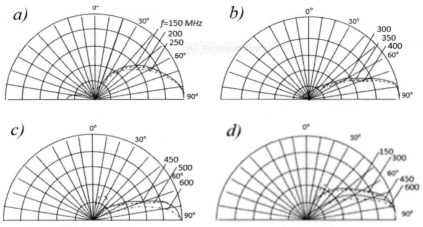

Fig. 5.47. Vertical directional patterns of a symmetrical vertical radiator near the surface with a trough (a, c) and the flat surface (b, d).

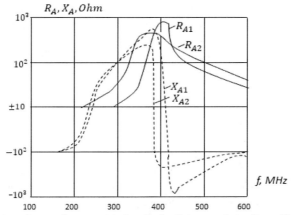

Fig. 5.48. Input impedances of a symmetrical vertical radiator near the surface with a trough (1) and the flat surface (2).

5.5 Multi-tiered antenna with a directional pattern pressed to the ground

As is known, technical development is spasmodic. Antenna technology also develops discontinuously. Two spasmodic processes take place simultaneously. One of them leads to the appearance of new options and new antenna designs. The second process leads to a new understanding of electromagnetic phenomena in radiating and receiving antennas. These processes develop non-synchronously: either one or the other takes a leap in its development. But the movement forward in the development process is not in doubt.

It is need to mention two examples of new technical solutions are: the creation of log-periodic antennas and the development of microstrip antennas. The first one of them took place so swiftly that its result was called inaccurately, and this inaccurate result has survived to the present. As a participant of the events Mushiak wrote, the correct name for the created product is self-complementary log-periodic antenna since without the attribute self-complementarity this antenna would be devoid of a useful result [95]. The second high-profile technical solution was the emergence of microstrip antennas. Despite the avalanche of articles devoted to these antennas, the processes occurring in them are still far from clarity.

It is necessary to understand that the creation of a theory of integral equations for the current in the antenna became the most important new result in studying electromagnetic phenomena. The duality principle made it possible to extend this result to magnetic (slot) radiators, and the oscillating power theorem delivered us from senseless reactive power.

A new step forward in expanding the operating frequency range was the creation of an in-phase current distribution along the antenna wire by incorporating capacitive and capacitive-resistive loads into the wire. This step made it possible, first of all, to ensure a high level of matching in a wide frequency range. The results of applying this technique to the creation of directional patterns, allowed us to send a signal in the direction leading to the maximum communication range, but were less significant. From this point of view, an attempt to create an antenna in which the desired directional pattern does not depend on the ratio of the antenna length to the wavelength is of interest. Such an attempt will be considered in this section of the book.

One of the possible options for implementing this proposal, first described in [96], was considered in [27]. Here a more detailed presentation of the various options of the antenna are given by the main author A.F. Yakovlev.

Vertical linear radiators (for example, whip antennas) are widely used for a short-wave radio communication of mobile objects. A disadvantage of these radiators is that the maximum radiation is directed horizontally only when the electrical length of the antenna is small. With the growth of an electrical length, the horizontal signal decreases. If the antenna height L is greater than 0.7λ, the main lobe of a directional pattern in a vertical plane separates from the ground and the radiation in the horizontal direction drops sharply. Changing of the antenna height, for example, by means of a telescopic construction, allows to improve the antenna directional pattern. But such mechanical tuning consumes a lot of time and complicates the antenna design. Folded structures allow creating an antenna, in which the length of the radiating element is changed without changing the geometric dimensions of the device.

In the simplest version, the such an antenna is designed for operation in two frequency bands (similar to a two-link telescopic antenna, which may have two geometrical heights during operation). The antenna is made as two-tiered and consists of two radiators, which are located one above the other and connected together. The upper radiator is linear, the lower radiator is folded (Fig. 5.49). Excitatory emfs e_1 and e_2 are included in both wires of the folded radiator and may change in amplitude and phase by means of a tuning circuit.

If the antenna operates in the first frequency band, emfs have the same phase, creating currents in one direction in both wires of the folded radiator. These currents become the current of a single unified linear radiator. As a result, the current is created all along the antenna and the height of radiating section is equal to the total height of the antenna.

If the antenna operates in the second frequency band, the current is created only in the wires of the folded radiator. With this aim, exciting emfs are included in anti-phase, and their magnitudes are chosen so that the potential at the point of joining the linear and folded radiators is zero. This eliminates the direct excitation of the linear radiator. The upper section of the antenna,

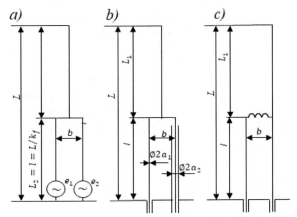

Fig. 5.49. Two-tiered antenna: a – total circuit, b – with coaxial cable, c – with reactive load.

consisting of one wire, can be excited by electromagnetic fields of the currents flowing along the wires of the folded radiator. However, if the length of the upper segment is far from the resonance length (it is not a multiple of $\frac{\lambda}{2}$), the current along this segment is small, i.e., the height of the radiating section of the antenna is equal to the height of the folded radiator.

Changing the height of the radiator allows providing the operation in two frequency bands. If the frequency ratio in each band is equal to k_f, then the total frequency ratio is equal to k_f^2. The height of the fold radiator is chosen to be equal

$$l = L/k_f. \tag{5.13}$$

The circuit of the considered antenna can be generalized for use in the case, when the number of frequency bands is N (instead two bands). For this, the number of antenna tiers should be increased to N. The two upper tiers are similar to the described version. The lower ends of each wire of the folded radiator connect with the upper points of the folded radiator of the next (third) tier. Overall the frequency ratio of the N-tiered antenna is equal to k_f^N. The heights of the lower tier and the remaining tiers are given by expressions,

$$L_n = \begin{cases} L/k_f^{n-1}, n = N, \\ \dfrac{L(k_f - 1)}{k_f^n, n} \neq N, \end{cases} \tag{5.14}$$

where n is the tier number, counting from the top.

Multi-tiered antennas create a new prospect for the development of a wide-range antennas. In this direction, the most significant results were

298   ANTENNAS: Rigorous Methods of Analysis and Synthesis

obtained earlier by means of including concentrated capacitive loads and optimization of these loads (*see* Chapter 3). Calculations show (*see* Fig. 3.12) that the capacitive loads allow extending the range in the direction of high frequencies with a sufficiently high level of matching, ensuring the frequency ratio of the order of 10. But the vertical directional pattern has the required form in a narrower frequency range. Using of the multi-tiered structure and the capacitive loads in the wires of each tier allows us to ensure a high level of matching and the required directional pattern in a wide range.

Let us return to the two-tiered variant of the antenna, more precisely to its excitation in anti-phase mode. The antenna will radiate when it is provided asymmetry in the folded radiator, i.e., it is necessary to obtain different amplitudes of the currents in the left and the right branches of the radiator. With this aim one of the wires must be in the form of a coaxial cable (*see* Fig. 5.49*b*), i.e., the generator must be included not in the lower, but in the upper part of the wire. Another way of creating an asymmetry is inclusion of a reactive load in one of the wires (*see* Fig. 5.49*c*).

In the presence of asymmetry, not only the anti-phase currents, but also in-phase currents will flow in the branches of the folded radiator. In-phase currents are caused by the presence of the ground. Namely these in-phase currents create radiation. However, for the sake of simplicity, we shall conventionally name the mode of antenna operation, anti-phase, when the potential at the point of joining linear and folded radiators is zero.

In order to analyze two-tiered antenna, we shall apply the theory of electrically coupled long lines, described in Section 1.5. This theory allows us to find the currents and the potentials along each wire of the line and emf of generators providing the required operation mode. In this case, the equivalent line (Fig. 5.50) is considered in the general form—with two generators in one

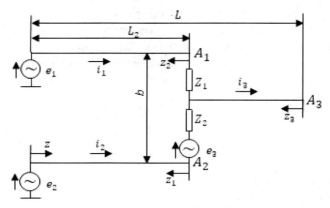

Fig. 5.50. Equivalent asymmetric line.

of the branches and two complex loads. The set of equations (1.50) for the three wires in this case take the form:

$i_1 = I_1 \cos kz_1 + j[U_1/W_{11} - U_2/W_{12}] \sin kz_1$, $u_1 = U_1 \cos kz_1 + j(\rho_{11}I_1 + \rho_{12}I_2) \sin kz_1$,
$i_2 = I_2 \cos kz_1 + j[U_2/W_{22} - U_1/W_{12}] \sin kz_1$, $u_2 = U_2 \cos kz_1 + j(\rho_{12}I_1 + \rho_{22}I_2) \sin kz_1$,
$i_3 = I_3 \cos kz_3 + j(U_3/W_{33}) \sin kz_3$, $u_3 = U_3 \cos kz_3 + j\rho_{33}I_3 \sin kz_3$. (5.15)

The boundary conditions for the currents and the potentials are

$$i_3\big|_{z_3=0} = 0,\ i_1 + i_2\big|_{z_1=0} = i_3\big|_{z_3=L-l},\ u_1\big|_{z_1=l} = e_1,\ u_2\big|_{z_1=l} = e_2,$$
$$u_1 - Z_1 i_1\big|_{z_1=0} = u_2 - Z_2 i_2\big|_{z_1=0} + e_3 = u_3\big|_{z_3=L-l}.$$
(5.16)

Equalities (5.15) and the boundary conditions (5.16) are the set of equations with six unknown magnitudes U_i, I_i ($i = 1, 2, 3$). Substituting (5.15) into (5.16), we obtain:

$I_3 = 0$, $I_2 = -I_1 + j(U_3/W_{33}) \sin k(L-l)$, $U_1 = Z_1 I_1 + U_3 \cos k(L-l)$,
$U_2 = -e_3 - Z_2 I_1 + U_3 \cos k(L-l)[1 + j(Z_2/W_{33}) \tan k(L-l)]$,

$e_1 = I_1 \cos kl\,[Z_1 + j(\rho_{11} - \rho_{12}) \tan kl]$
$\qquad + U_3 \cos kl \cos k(L-l)[1 - (\rho_{12}/W_{33}) \tan kl \tan k(L-l)]$,

$e_2 = -e_3 \cos kl - I_1 \cos kl[Z_2 + j(\rho_{22} - \rho_{12}) \tan kl] + U_3 \cos kl \cos k(L-l)$
$$\left[1 - \frac{\rho_{22} - jZ_2}{W_{33}} \tan kl \tan k(L-l)\right].$$
(5.17)

The remaining two formulas permit to express the magnitudes I_1 and U_3 through emfs of the generators. But the corresponding expressions are cumbersome. They are not used for further analysis and are not presented here. The previous four equalities after substituting into (5.15) allow expressing the current distribution along each wire as a function of magnitudes I_1 and U_3:

$i_1(z) = I_1 \cos k(L-z)[1 + j(Z_1/W_{11} + Z_2/W_{12}) \tan k(l-z)] + j(e_3/W_{12}) \sin k(l-z) +$
$\quad U_3[j(1/W_{11} - 1/W_{12}) \cos k(L-l) + Z_2 \sin k(L-l)/(W_{12}W_{33})] \sin k(L-z)$,
$i_2(z) = I_1 \cos k(L-z)[1 + j(Z_1/W_{12} + Z_2/W_{22}) \tan k(l-z)] - j(e_3/W_{22}) \sin k(l-z) +$
$\quad U_3\left[j(1/W_{22} - 1/W_{12}) \cos k(L-l) - Z_2 \sin \dfrac{k(L-l)}{W_{22}W_{33}} + j \sin k(L-l)\dfrac{\cot k(L-z)}{W_{33}}\right] \sin k(L-z)$,
$i_3(z) = j(U_3/W_{33})$.

Further we shall consider the specific versions of antennas as realizations in particular cases of general equivalent circuits. The circuit of two-tiered antenna with the coaxial cable is shown in Fig. 5.49b. Here a few of the elements of the overall circuit are absent, i.e.,

$$Z_1 = Z_2 = e_2 = 0. \tag{5.18}$$

In the anti-phase mode, in accordance with the boundary condition $u_3|_{z_3=L-l} = 0$ we find that $U_3 = 0$. Then from (5.17) we obtain emf of generators

$$e_1 = j(\rho_{11} - \rho_{12})I_1 \sin kl, e_3 = -j(\rho_{22} - \rho_{12})I_1 \tan kl = -j(\rho_{22} - \rho_{12})\frac{\sec kl}{\rho_{11} - \rho_{12}}. \tag{5.19}$$

It is necessary to emphasize that the relationship between the emf of two generators is an obligatory condition for providing an anti-phase mode in the antenna. The currents along the antenna wires in this mode according to (5.18) are

$$i_1(z) = I_1 \cos k(l-z) + \frac{\rho_{22} - \rho_{12}}{W_{12}} I_1 \tan kl \sin k(l-z)$$

$$i_2(z) = -I_1 \cos k(l-z) - \frac{\rho_{22} - \rho_{12}}{W_{22}} I_1 \tan kl \sin k(l-z), \ i_3(z) = 0. \tag{5.20}$$

The expressions (5.20) confirm that the currents along the first and the second wires contain in-phase and anti-phase components. The total antenna current (sum of currents) varies along the antenna similarly to the current along the linear radiator of the length l –

$$i_1(z) + i_2(z) = -\frac{(\rho_{22} - \rho_{12})(\rho_{11} - \rho_{12})}{\rho_{11}\rho_{22} - \rho_{12}^2} I_1 \tan kl \sin k(l-z). \tag{5.21}$$

Here it is taken into account that (see Section 1.5)

$$\frac{1}{W_{22}} = \frac{\rho_{11}}{\rho_{11}\rho_{22} - \rho_{12}^2}, \ \frac{1}{W_{12}} = \frac{\rho_{12}}{\rho_{11}\rho_{22} - \rho_{12}^2}.$$

The current distribution along the antenna wires in the anti-phase mode is shown in Fig. 5.51a. The currents in the first and second wires are denoted by

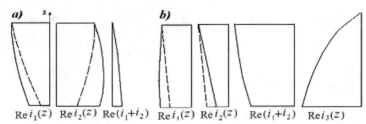

Fig. 5.51. Currents in the two-tiered antenna with a coaxial cable in anti-phase (*a*) and in-phase (*b*) modes.

the numerals 1 and 2, the total currents by the sum of 1 + 2. They have both in-phase and anti-phase components. In-phase components are indicated by the letter *i*, anti-phase components by the letter *a* (in parentheses). Impedance on the output of each generator, which excites the asymmetric long line, is equal to

$$jX_{A1} = \frac{e_1}{i_1(0)} = j(\rho_{11} - \rho_{12})\tan kl \left[1 + \frac{\rho_{22} - \rho_{12}}{W_{12}}\tan^2 kl\right]^{-1},$$

$$jX_{A3} = \frac{e_3}{i_2(l)} = j(\rho_{22} - \rho_{12})\tan kl. \qquad (5.22)$$

These expressions allow us to approximately determine the reactive component of the loading impedance of each generator (similar to the formula for the input impedance of an equivalent to a two-wire long line, which allows approximately determining the reactive component of the input impedance of a linear antenna).

From the viewpoint of radiation, as seen from (5.20) and Fig. 5.51a, the antenna in anti-phase mode consists of two parallel radiators of the height *l* with in-phase currents in the base:

$$i_1^{(i)} = \frac{\rho_{22} - \rho_{12}}{W_{12}} I_1 \tan kl \sin kl, \quad i_2^{(i)} = \frac{\rho_{22} - \rho_{12}}{W_{22}} I_1 \tan kl \sin kl,$$

The radiation resistance of each radiator consists of self-resistance and the mutual resistance multiplied by the ratio of the currents. In particular, for the first radiator, we write:

$$R_{A1} = R_{11} + R_{12} \frac{i_2^{(i)}}{i_1^{(i)}}, \qquad (5.23)$$

where R_{11} is self-resistance, R_{12} is the mutual radiation resistance, and $R_{12} \approx R_{11}$, since the radiator heights are same and the distance between the radiators is small in comparison with the wave length. Thus,

$$R_{A1} = R_{11}(1 - W_{22}/W_{12}). \qquad (5.24)$$

In the anti-phase mode, the electric field strength in the far region and the directional pattern coincide with similar characteristics of the conventional linear radiator of a height *l*. Expressions (5.20) to (5.24) are sufficiently simple and allow us to determine the influence of the geometric dimensions of the antenna on the current magnitude in each wire and the electrical characteristics of the radiator. More precisely, the input impedance of the antenna can be calculated by using an algorithm for the calculation, based on the integral equation for the current, on the Moments method and on the systems of piecewise sinusoidal basis functions.

Let's move on to an analysis of the in-phase mode. For the implementation of this mode one must ensure equality of potentials in both branches of the folded radiator, i.e.,

$$u_1(z) = u_2(z). \tag{5.25}$$

Applying this condition to the set of equations (5.3), we find:

$$e_1 = jI_1 \frac{\rho_{11}\rho_{22} - \rho_{12}^2}{\rho_{22} - \rho_{12}} \sin kl - jI_1 W_{33} \cdot \frac{\rho_{11} + \rho_{22} - 2\rho_{12}}{\rho_{22} - \rho_{12}} \cot k(L-l), e_3 = e_1 \sec l. \tag{5.26}$$

Currents along the wires consist in this case of in-phase components only:

$$i_1(z) + i_2(z) = I_1 W_{33} \frac{\rho_{11} + \rho_{22} - 2\rho_{12}}{\rho_{11}\rho_{22} - \rho_{12}^2} \cot k(L-l) \sin k(l-z)$$

$$+ I_1 \frac{\rho_{11} - \rho_{12}}{\rho_{22} - \rho_{12}} \tan kl \sin k(l-z) + I_1 \left(1 + \frac{\rho_{11} - \rho_{12}}{\rho_{22} - \rho_{12}}\right) \cos k(l-z),$$

$$i_3(z) = I_1 \left(1 + \frac{\rho_{11} - \rho_{12}}{\rho_{22} - \rho_{12}}\right) \frac{\sin k(L-z)}{\sin k(L-l)}. \tag{5.27}$$

The current distribution along the wires is shown in Fig. 5.51b. Impedances on the output of each generator, which excite the asymmetric long line, are,

$$jX_{A1} = \frac{e_1}{i_1(0)} =$$

$$= j \frac{\rho_{11}\rho_{22} - \rho_{12}^2}{\rho_{22} - \rho_{12}} \tan kl \frac{1 - W_{33} \frac{\rho_{11} + \rho_{22} - 2\rho_{12}}{\rho_{11}\rho_{22} - \rho_{12}^2} \cot kl \cot k(L-l)}{1 + \frac{W_{33}}{W_{11}} \frac{\rho_{11} + \rho_{22} - 2\rho_{12}}{\rho_{22} - \rho_{12}} \tan kl \cot k(L-l) - \frac{\rho_{12}}{\rho_{22} - \rho_{12}} \tan^2 kl}.$$

$$jX_{A3} = \frac{e_3}{i_2(l)} = j \frac{\rho_{11}\rho_{22} - \rho_{12}^2}{\rho_{11} - \rho_{12}} \tan kl \left[1 - W_{33} \frac{\rho_{11} + \rho_{22} - 2\rho_{12}}{\rho_{11}\rho_{22} - \rho_{12}^2} \cot kl \cot k(L-l)\right]. \tag{5.28}$$

The resistance of radiation is calculated according to formulas similar to (5.11), but R_{11} is the resistance of the radiator of height L, the current along which is determined by (5.15). Since the derivative of the current has discontinuity on the section border, the calculation should take break of the derivative for current into account. Correspondingly it is necessary to replace the known expression for E_z by an equality of the type,

$$E_z(j) = -j\frac{15}{k}\left\{\frac{2\exp(-jkR_0)}{R_0}\frac{dJ(0)}{dz} - \left[\frac{\exp(-jkR_1)}{R_1} + \frac{\exp(-jkR_{1+})}{R_{1+}}\right]\left[\frac{dJ(l_1+0)}{dz}\right.\right.$$
$$\left.\left. -\frac{dJ(l_1-0)}{dz}\right] + \sum_{m=1}^{M}\left[\frac{\exp(-jkR_{m1})}{R_{m1}} + \frac{\exp(-jkR_{m2})}{R_{m2}}\right]\left[\frac{dJ(l_1+0)}{dz} - \frac{dJ(l_1-0)}{dz}\right]\right\}.$$
(5.29)

where R_{m1} and R_{m2} are the distances from observation point to the sections' borders in the upper and the lower arms of the radiator, M is number of borders, $\frac{dJ(l_m+0)}{dz}$ and $\frac{dJ(l_m-0)}{dz}$ are magnitudes of derivatives for the current from the left and the right of point $z = l_m$.

It should be noted that for the same diameters of the antenna wires the formulas become far simpler.

The circuit of two-tiered antenna with a reactive load is shown in Fig. 5.49c. Here a few of elements of the overall circuit are absent also, i.e.,

$$Z_2 = e_3 = 0, \text{ and } \rho_{11} = \rho_{22}. \quad (5.30)$$

In the anti-phase mode, the boundary condition $u_3|_{z_3=L-l} = 0$ should be executed. This means that the emfs of generators are connected by the relationship

$$e_2 = -e_1 + Z_1 I_1 \cos kl, \quad (5.31)$$

and the currents along the antenna wires in this mode are equal to

$$i_1(z) + i_2(z) = jZ_1 I_1 \frac{\sin k(l-z)}{\rho_{11} + \rho_{12}}, \quad i_3(z) = 0. \quad (5.32)$$

Reactive components of generators' load are

$$jX_{A1} = \frac{Z_1 + j(\rho_{11} - \rho_{12})\tan kl}{1 + j(Z_1/W_{11})\tan kl}, jX_{A2} = \frac{j(\rho_{11} - \rho_{12})\tan kl}{1 + j(Z_1/W_{12})\tan kl}. \quad (5.33)$$

Resistances are calculated in accordance with (5.24), and

$$i_2^{(i)}(0)/i_1^{(i)}(0) = W_{11}/W_{12}.$$

The in-phase mode has salient features. Since the load is connected in the upper section of the folded radiator, it is impossible to produce equality of voltages in both its wires. Let us assume

$$u_1(z) - u_2(z) = U \cos k(l-z). \quad (5.34)$$

For executing this condition, it is necessary that

$$e_2 = e_1 - Z_1 I_1 \cos kl \quad (5.35)$$

The currents along the wires are calculated in accordance with expressions

$$i_1(z) + i_2(z) = 2I_1 \cos k(l-z) + \frac{4I_1 W_{33}}{\rho_{11} + \rho_{12}} \cot k(L-l) \sin k(l-z) + j\frac{Z_1 I_1}{\rho_{11} + \rho_{12}} \sin k(l-z),$$

$$i_3(z) = 2I_1 \frac{\sin k(L-z)}{\sin k(L-l)}. \tag{5.36}$$

Reactive impedances of generators' load are

$$jX_{A1} = \frac{Z_1 + j(\rho_{11} + \rho_{12}) \tan kl - 2jW_{33} \cot k(L-l)}{1 + j(Z_1/W_{11}) \tan kl + \frac{2W_{33}}{\rho_{11} + \rho_{12}} \tan kl \cot k(L-l)}.$$

$$jX_{A2} = \frac{j(\rho_{11} + \rho_{12}) \tan kl - 2jW_{33} \cot k(L-l)}{1 - j\frac{Z_1}{W_{12}} \tan kl + \frac{2W_{33}}{\rho_{11} + \rho_{12}} \tan kl \cot k(L-l)}. \tag{5.37}$$

The current distribution along the wires of the antenna with the reactive load in the in-phase and anti-phase modes is shown in Fig. 5.52.

As an example of the two-tiered antenna with coaxial cable, we consider the antenna with dimensions (in meters): $L = 1.0, l = 0.39, b = 0.037, a_1 = 0.002$, $a_2 = 0.025$. Here, b is distance between the axes of the wires of the folded radiators, a_1 and a_2 are the radii of the wires (*see* Fig. 5.51*b*). Calculation of the antenna characteristics is made by means of the Moments method.

Figure 5.53 shows the calculated curves for the directional patterns in the vertical plane—in the in-phase (*a*) and anti-phase (*b*) modes. A full size model of the antenna was made. The results of the experimental verification are given for frequencies 150 and 300 MHz. As can be seen from the figures, the coincidence of the calculation and the experiment is quite satisfactory. The high level of radiation in the direction perpendicular to the axis of the radiator

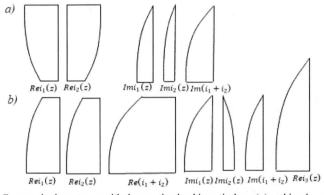

Fig. 5.52. Currents in the antenna with the reactive load in anti-phase (*a*) and in-phase (*b*) modes.

(along the ground) is provided in the double frequency range. However, in the anti-phase mode when the length of the third wire (of the upper segment of the antenna) is a multiple of half the wavelength, i.e., at frequencies 245 and 490 MHz, the main lobe of the directional pattern is located under a great angle to the horizontal. Here the current along the third wire is too large. The dimensions of the antenna must be chosen so that the resonance frequencies lie outside the operating range.

Figure 5.54 shows the current distribution along the antenna wires in the anti-phase mode, including the current distribution along the left wire and the connecting bridge between the wires of the folded radiator, and also the current distribution along the right wire and the upper (third) wire. These distributions of currents are a graphic illustration of the processes in the anti-phase mode of the two-tiered antenna. The results are shown at four frequencies, including

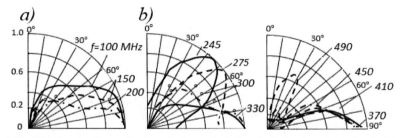

Fig. 5.53. Directional patterns of antenna in the vertical plane in the in-phase (*a*) and anti-phase (*b*) modes.

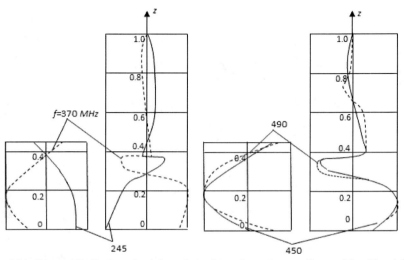

Fig. 5.54. Current distribution along the wires of two-tier antenna with coaxial cable at four frequencies (in anti-phase mode).

frequencies 245 and 490 MHz, where the main lobe is located at a large angle to the horizon, since the current of the third (upper) wire is too great.

Summarizing the results presented in this Section, one should make the following conclusions: The principle of changing electrical height of the radiator without changing its geometric dimensions, realized in the circuit of two-tier antenna, is very promising, quite efficient and requires careful study in order to implement it in real structures.

Instead of a Conclusion

This book is devoted to rigorous methods of calculating antennas characteristics based on solving integral equations for currents along their axes. It describes the main results obtained in recent years. The books of Stratton [1], Balanis [15] and Vainshtein [6] were examples for it, and the integral equations of Leontovich [20] and Hallen [19] were the guiding idea.

Some topics closely related to this book's content have not been considered to avoid repetition and unnecessarily making it voluminous. We give a brief summary of these questions and indicate books in which they are presented in more detail.

1. The method of complex potential and its application to three-dimensional problems, including the calculation of capacitances in various structures consisting of thin wires and cables of delay [10].
2. Losses of linear and multi-wire antennas in the ground and wires [10].
3. Folded and Multi-folded metal and impedance antennas. An antenna with a meandering load, including the calculation of input impedances, currents in wires and voltages on insulators, depending on the number of wires [27].
4. Self-complementary antennas, including three-dimensional ones and antennas with rotation symmetry [27].
5. Mutual compensation of signals, including an antenna for a personal cell phone [10, 27].
6. Antennas for underground radio communications [10].
7. The field of a rectangular loop [27].

References

[1] Stratton, J. A. 1941. Electromagnetic Theory. New York: McGraw-Hill.
[2] Markov, G. T. and Sazonov, D. M. 1975. Antennas. Energy, Moscow. USSR (in Russian).
[3] Miller, M. A. 1954. Application of uniform boundary conditions to the theory of thin antennas. Journal of Technical Physics 8: 1483–1495 (in Russian).
[4] Pistolkors, A. A. 1947. Antennas. Sviyazizdat, Moscow. USSR (in Russian).
[5] Poynting, J. H. 1884. On the transfer of energy in the electromagnetic field. Phil. Trans. Roy. Soc. (London): 343.
[6] Vainshtein, L. A. 1988. Electromagnetic Waves. Sovetskoye Radio, Moscow. USSR (in Russian).
[7] Brown, G. H., Lewis, R. F. and Epstein, J. 1937. Ground systems as a factor of antenna efficiency. Proceedings IRE: 753–787.
[8] Wait, J. R. and Pope, W. A. 1954. The characteristics of a vertical antenna with a radial conductor ground system. Applied Scientific Research 3: 177–195.
[9] Levin, B. M. and Razumov, V. P. 1979. Loss resistance in the ground. Antennas 27: 125–133 (in Russian).
[10] Levin, B. M. 2016. Antenna Engineering. Theory and Problems. CRC Press: London, New York.
[11] Lavrov, G. A. 1975. Mutual Effect of Linear Radiators. Sviyaz, Moscow. USSR (in Russian).
[12] Levin, M. L. 1947. About one new method of finding the thin antenna characteristic reactance. Izvestiya of AN USSR, Ser. Phys. 2: 117–133 (in Russian).
[13] Aizenberg, G. Z. 1962. Short-Wave Antennas. Sviyazizdat, Moscow. USSR (in Russian).
[14] Kontorovich, M. I. 1951. Some remarks in connection with the induced emf method. Radiotechnics 2: 3–9 (in Russian).
[15] Balanis, C. A. 2005. Antenna Theory: Analysis and Design. Wiley & Sons, New York. USA.
[16] Kraus, J. D. 1988. Antennas. McGraw-Hill, Boston. USA.
[17] Levin, B. M. 1992. Once again about induced emf method. Radio Electronics and Communications 2-3: 17–23 (in Russian).
[18] Elliott, R. S. 2003. Antenna Theory and Design. Wiley-IEEE Press, New York. USA.
[19] Hallen, E. 1938. Theoretical investigations into the transmitting and receiving qualities of antennae. Nova Acta Regiae Soc. Sci. Upsaliensis, Ser. IV 4: 1–44.
[20] Leontovich, M. A. and Levin, M. L. 1944. On the theory of oscillations excitation in the linear radiators. Journal of Technical Physics 9: 481–506 (in Russian).
[21] Aharoni, J. 1946. Antennae: An Introduction to Their Theory. The Clarendon Press. Oxford. USA.
[22] Levin, B. M. 2013. The Theory of Thin Antennas and Its Use in Antenna Engineering. Bentham Science Publishers. USA.
[23] Popovic, B. D. 1973. Theory of cylindrical antennas with lumped impedance loadings. The Radio and Electronic Engineer 3: 243–248.
[24] Djordjevic, A. R., Popovic, B. D. and Dragovic, M. B. 1979. A method for rapid analysis of wire antenna structures. Archiv fur Electrotechnic (W. Berlin) 1: 17–23.

[25] Richmond, J. H. 1966. A wire-grid model for scattering by conducting bodies. IEEE Trans. Antennas Propagation 6: 782–786.
[26] Vershkov, M. V., Levin, B. M. and Fraiman, S. S. 1972. Antenna with meandering load. Proceedings CNII MF 151: 73–80 (in Russian).
[27] Levin, B. M. 2019. Wide-range and Multi-frequency Antennas. CRC Press: London, New York.
[28] Leontovich, M. A. 1945. Theory of forced electromagnetic oscillations in thin conductors of arbitrary cross-section and its applications to calculation of some antennas. Proceedings of NII MPSS 1: 1 (in Russian).
[29] Valenti, C. 2002. NEXT and FEXT models for twisted-pair North American loop plant. IEEE Journal Select. Areas Commun. 5: 893–900.
[30] Iossel, Yu. Ya., Kochanov, E. S. and Strunsky, M. G. 1981. Calculation of Electrical Capacitance. Energoisdat, Leningrad. USSR (in Russian).
[31] Cochrane, D. 2001. Passive cancellation of common-mode electromagnetic interference in switching power converters. M.S. Thesis, Virginia Polytechnic Inst. State Univ., Blacksburg.
[32] Pocklington, H. C. 1897. Electrical oscillations in wire. Cambridge Philosophical Society Proc., London: 324–332.
[33] King, R. W. P. 1956. Theory of Linear Antennas. Harvard University Press, Cambridge. USA.
[34] Jones, D. S. 1981. Note of the integral equation for a straight wire antenna. Proc. IEEE – H 2: 114–116.
[35] Levin, B. M. 1998. Monopole and Dipole Antennas for Ship Radio Communication, Abris, St.- Petersburg (in Russian).
[36] Nesterenko, M. V. 2010. Analytical methods in the theory of thin impedance vibrators. Progress in Electromagnetics Research B 21: 299–328.
[37] Levin, B. M. 2015. Antennas with rotational symmetry. Proceedings of Conf. ICATT'15 (April 2015). Kharkov (Ukraine): 30–35.
[38] Levin, B. M. and Markov, V. G. 1997. Method of Complex Potential and Antennas, Ship Electrical Engineering and Communication. St.-Petersburg. Russia (in Russian).
[39] Pistolkors, A. A. 1948. Theory of the circular diffraction antenna. Proceedings of IRE 1: 56–60.
[40] Pheld, Ya. N. 1948. Fundamentals of the Theory of Slot Antennas. Sovetskoye Radio, Moscow. USSR (in Russian).
[41] Glushkovsky, E. A., Levin, B. M. and Rabinovich, E. Ya. 1966. Receiving-transmitting impedance magnetic antenna. Questions of Radioelectronics 29: 35–44 (in Russian).
[42] Glushkovsky, E. A., Levin, B. M. and Rabinovich, E. Ya. 1969. Thin magnetic impedance antennas. Antennas 5: 108–120 (in Russian).
[43] Pistolkors, A. A. 1947. Proceedings of Research Institute of Communication 5(40): 3 (in Russian).
[44] Schefer, G. 1963. Archive der electrischen Ubertragung, 6: 289.
[45] King, R. W. P. 1967. The linear antenna—eighty years of progress. Proc. IEEE 1: 2–16.
[46] Tulyathan, P. and Newman, E. H. 1979. The circumferential variation of the axial component of current in closely spaced thin-wire antennas. IEEE Trans. Antennas Propagation 1: 46–50.
[47] Perini, J. and Buchanan, D. J. 1982. Assessment of MOM techniques for shipboard applications. IEEE Trans. Electromagnetic Compatibility 1: 32–39.
[48] Wu, T. T. and King, R. W. P. 1965. The cylindrical antenna with non-reflecting resistive loading. IEEE Trans. Antennas Propagation 3: 369–373.
[49] Levin, B. M. and Yakovlev, A. D. 1985. Antenna with loads as impedance radiator with impedance changing along its length. Radiotechnics and Electronics Engineering 1: 25–33 (in Russian).
[50] Levin, B. M. 1990. Use of loads for creating a given current distribution along a dipole. Radiotechnics and Electronics Engineering 8: 1581–1589 (in Russian).
[51] Himmelblau, D. 1972. Applied Nonlinear Programming. McGraw-Hill, New York. USA.

[52] Levin, B. M. and Yakovlev, A. F. 1992. About one method of widening on antenna operating range. Radiotechnics and Electronics Engineering 1: 55–64 (in Russian).
[53] Levin, B. M. 2016. Directivity of thin antennas. Proceedings of the 21 Intern. Seminar/Workshop on Direct and Inverse Problems of Electromagnetic and Acoustic Wave Theory DIPED, Tbilisi, Georgia.
[54] Pistolkors, A. A. 1944. General theory of diffraction antennas. Journal of Technical Physics 12: 693–701 (in Russian).
[55] Neyman, L. R. and Kalantarov, P. L. 1959. Theoretical Background of Electro Engineering. Part 3. Moscow – Leningrad. Russia (in Russian).
[56] Chaplin, A. F., Buchazky, M. D. and Mihailov, M. Yu. 1983. Optimization of director-type antennas. Radiotechnics 7: 79–82 (in Russian).
[57] Levin, B. M. 2017. Director antennas with in-phase currents. Proc. of Conf. ICATT'17 (May 2017). Kyiv (Ukraine): 110–113.
[58] Carrel, R. L. 1961. The design of log-periodic dipole antennas. IRE Intern. Convention Record, Part 1: 61–75.
[59] De Vito, G. and Strassa, G. B. 1973. Comments on the design of log-periodic dipole antennas. IEEE Trans. Antennas Propagation 3: 303–309.
[60] Yakovlev, A. F. and Pyatnenkov, A. E. 2007. Wide-Band Directional Antennas Arrays from Dipoles. S.-Petersburg. Russia (in Russian).
[61] Carrel, R. L. 1958. The characteristic impedance of two infinite cones of arbitrary cross-sections. IRE Trans. Antennas and Propagation 2: 197–201.
[62] Buchholz, H. 1957. Elektrische und Magnetische Potentialfelder. Berlin (in German).
[63] Korn, G. and Korn, T. 1961. Mathematical Handbook for Scientists and Engineers, New York, Toronto, London: McGraw-Hill.
[64] Angot Andre. 1957. Complements de Mathematiques. Paris (in French).
[65] Mirolubov, N. N., Kostenko, M. V., Levinstein, M. L. and Tichodeev, N. N. 1963. Methods of Electrostatic Field Calculation. Leningrad: Visshaya Shkola (in Russian).
[66] Levin, B. M. 2014. Reduction of a parabolic problem to a 2D problem and a phantom. Radiotechnics and Electronics Engineering 7: 697–703.
[67] Munson, R. E. 1974. Conformal microstrip antennas and microstrip phased arrays. IRE Trans. Antennas and Propagation: 74–78.
[68] Pues, H. and Van de Capelle, A. 1984. Accurate transmission-line model for the rectangular microstrip antenna. Proc. IEE-H 6: 334–340.
[69] Croq, F. and Pozar, D. M. 1991. Millimeter-wave design of wade-band aperture coupled stacked microstrip antennas. IEEE Trans. Antennas and Propagation 12: 1770.
[70] Menzel, W., Pilz, D. and Al-Tikriti, M. 2002. Millimeter-wave folded reflector antennas with high gain, low loss and low profile. IEEE Antennas Propagation Magazine 3: 24–29.
[71] Tsai, F. -C. E. and Bialkowski, M. E. 2003. Designing a 161-element Ku-band microstrip reflect array of variable size patches using an equivalent unit cell waveguide approach. IEEE Trans. Antennas Propagation 10: 2953–2962.
[72] Levin, B. M. 2006. An antenna directivity calculation on the basis of main patterns. Proc. 18 Intern. Wrozlaw Symp. on Electromagn. Compatibility. Wrozlaw (Poland): 64–67.
[73] Vered, U. 2000. Estimation of intercardinal antenna pattern based on cardinal data. Proc. 21 IEEE Convention of the Electrical and Electronic Engineers in Israel. Tel Aviv (Israel): 37–41.
[74] Widrow, B., Mantey, P. E., Griffiths, L. J. and Goode, B. B. 1967. Adaptive antenna systems. Proceedings IEEE 12: 2143–2159.
[75] Widrow, B. and McCool, J. M. 1976. A comparison of adaptive algorithms based on the methods of steepest descent and random search. IEEE Trans. Antennas Propagation 5: 615–637.
[76] Compton, R. T., Huff, R. J., Swarner, W. G. and Ksienski, A. A. 1976. Adaptive arrays for communication systems. IEEE Trans. Antennas Propagation 5: 599–607.

[77] Gupta, O. P. 2002. Electronically steerable multi-beam antenna for mobile wireless base stations. Circle Reader Service 12: 19–20.
[78] Vershkov, M. V., Levin B. M. et al. 1967. Antenna-mast. Patent of USSR 191651, Bulletin of Inventions, 1967, no. 4.
[79] Vershkov, M. V., Levin B. M. et al. 1973. Antenna-mast. Patent of USSR 328824, Bulletin of Inventions, 1973, no. 45.
[80] Aizenberg, G. Z. and Uriyadko, V. N. 1959. Vertical linear radiator. Patent of USSR 122500, Bulletin of Inventions, no. 18.
[81] Pirogov, A. A. 1972. On the prospects for the use of ballistic antennas. Radiotechnics 1: 83–84.
[82] Stark, A. 1980. Optimum pattern shape of short-wave antennas from radio link computations. Antennas and Propagation. Int. Symp. Digest. Quebec, 1: 306–307.
[83] Low, P. E. 1983. Shipboard Antennas. Dedham, MA. Artech House.
[84] Vershkov, M. V., Levin, B. M. and Feldman, Ya. M. 1973. Antenna. Patent of USSR 341388. Bulletin of Inventions, 1973, no. 45.
[85] Berry, T. G. and Kline, T. F. Self-extending antenna. Patent of USA 3467328.
[86] Dmitrievsky, N. M. and Polinov, Yu. S. 1966. Extandable antenna. Patent of USSR 185970, Bulletin of Inventions, no. 18.
[87] McCorcle, M. Submarine mounted telescopic antenna. Patent of USA 3158865.
[88] Altshuler, E. E. 1961. The travelling-wave linear antenna. IRE Trans. Antennas and Propagation 9: 324–329.
[89] Gole, R. M., Drabkin, R. L. and Mirotvorsky, O. B. 1985. Antenna. Patent of USSR 380229, Bulletin of Inventions, 1985, no. 47.
[90] Halpern, B. and Mittra, R. 1985. Broadband whip antennas for use in HF communications. Proc. IEEE Symposium on Antennas and Propagation. Canada: 763–767.
[91] Levin, B. M. and Fominzev, S. S. 1984. Influence of cylindrical reradiator on parameters of an antenna and antenna array. Radiotechnics and Electronics Engineering 11: 2140–2147 (in Russian).
[92] Levin, B. M. 1995. Asymmetric radiator at the edge of the wedge (of the ship deck). Proc. Intern. Symp. on Electromagn. Compatibility and Electromagn. Ecology (S.-Petersburg): 35–36 (in Russian).
[93] Thiele, G. E. and Newhose, T. H. 1975. A hybrid technique for combining moment methods with the geometrical theory of diffraction. IEEE Trans. Antennas Propagat. 1: 62–69.
[94] Levin, B. M. 1995. Symmetrical dipole on the axis of a trough with finite length. Radiotechnics and Electronics Engineering 5: 689–694 (in Russian).
[95] Mushiake, Y. 1996. Self-Complementary Antennas: Principle of Self-Complementarity for Constant Impedance. Springer, London. UK.
[96] Levin, B. M. and Yakovlev, A. F. 1992. About one method of widening an antenna operation range. Radiotechnics and Electronics Engineering 1: 55–64 (in Russian).

Index

Antenna types
    adaptive 232
    antenna-mast 252
    ballistic 257
    coaxial 274
    conical spiral 274
    curvilinear 107
    discon 185
    folded and multi-folded 49
    for a cellular base station 245
    for personal phone 237
    great folded 268
    inverted L 49
    log-periodic coaxial 181
    loop 116
    microstrip 207
    multi-tiered 295
    on conic surface 100
    on parabolic surface 197
    on pyramid 101
    pneumatic 256
    reflector 169
    resistive 95
    self-complementary 98
    self-complementary log-periodic 175
    self-complementary three-dimensional 196
    ship's 250
    slot 111
    telescopic 262
    transparent 95
    turnstile 274
    V-radiator 137
    volumetric 196
    whip 147, 253, 260
    with bottom excitation 251
    with broadband absorber 266
    with capacitive loads 125
    with concentrated loads 105
    with constant surface impedance 13
    with elastic profile 263
    with discs 266
    with grooved profile 264
    with meandered load 48
    with piecewise surface impedance 13
    with spiral profile 264
    with upper excitation 251
Array 143
    adaptive 155
    linear 218
    reflector 155
    ship's linear 271
    vertical from coaxial radiators 274
    vertical from discon radiators 274

Characteristics
    coefficient of electrostatic induction 18
    decay 97
    directional pattern 23
    directivity 157
    effective length 23
    efficiency 89, 142
    frequency ratio 148
    gain 101
    input impedance 5
    loss in ground 37
    loss in wires 61
    matching level 157
    mutual impedance 47
    pattern factor 142
    permeability 23
    permittivity 23
    potential coefficient 18
    propagation constant 4
    quality 172
    radiation resistance 2
    reflection coefficient 282
    reflectivity 105
    slowing factor 30

314   ANTENNAS: Rigorous Methods of Analysis and Synthesis

transmission coefficient 95
travelling wave ratio 142
weighting coefficient 141
Conditions
   of direct visibility 272
   of Lorentz 4
   of operating 258
   ship's 4, 271
   stormy 251
Criterion
   quasi-Tchebyscheff 142
   root-mean-square 142
Current distribution
   exponential 125
   in-phase 13
   linear 127
   sinusoidal 12

Equation
   of continuity 1
   of electrodynamics 1
   of Kirchhoff 44
   of Laplace 190
   of Maxwell 1
   telegraph 15

Function
   basis 43
   objective 141
   of entire-domain 45
   of sub-domain 45
   penalty 141
   piecewise linear 45
   weight 43

Integral equation for the current
   key 78
   of filament 64
   of Hallen 41
   of Jones 65
   of Leontovich 63
   of magnetic current 115
   of Pocklington 64
   of two radiators 79
   with approximate kernel 64
   with exact kernel 64
      for voltage 111

Line
   of equal potential 190
   of zero potential 202

Long line
   equivalent 12
   impedance 12
   stepped 6

Method
   gradient 143
   of calculating gain, using main directional patterns 224
   of collocation 45
   of compensation 51, 239
   of complementarity 98
   of complex potential 201
   of conjugate gradients 143
   of constants variation 70
   of electrostatic analogy 166
   of Galerkin 276
   of impedance line 124
   of induced emf 38
   of induced emf generalized 42
   of iteration 65, 143
   of long line with loads 133
   of mathematical programming 141
   of mean potentials 18
   of moments 43
   of perturbation 67
   of Poynting vector 20
   of successive approximations 65
   of steepest descent 143

Potential 3
Principle
   of complementarity 98
   of duality 98
   of similarity 70, 172, 208
   of superposition 80

Reducing of
   conic problem to cylindrical 90
   parabolic problem to cylindrical 198
   superstructure impact 154, 275

Surface
   of equal potential 190
   of zero potential 15

Theorem of
   Floquet 222
   oscillating power 34, 111, 295
   reciprocity 39, 215
   Shchelkunov 121
   uniqueness 190, 197, 201